JOHN STEWART BELL AND
TWENTIETH-CENTURY PHYSICS

JOHN STEWART BELL AND TWENTIETH-CENTURY PHYSICS

Vision and Integrity

Andrew Whitaker

Department of Physics, Queen's University Belfast, Northern Ireland

OXFORD
UNIVERSITY PRESS

OXFORD
UNIVERSITY PRESS

Great Clarendon Street, Oxford, OX2 6DP,
United Kingdom

Oxford University Press is a department of the University of Oxford.
It furthers the University's objective of excellence in research, scholarship,
and education by publishing worldwide. Oxford is a registered trade mark of
Oxford University Press in the UK and in certain other countries

Published in the United States of America by Oxford University Press
198 Madison Avenue, New York, NY 10016, United States of America

British Library Cataloguing in Publication Data

Data available

Library of Congress Control Number: 2015956498

ISBN 978–0–19–874299–9

Printed and bound by
CPI Group (UK) Ltd, Croydon, CR0 4YY

Oxford University Press makes no representation, express or implied, that the
drug dosages in this book are correct. Readers must therefore always check
the product information and clinical procedures with the most up-to-date
published product information and data sheets provided by the manufacturers
and the most recent codes of conduct and safety regulations. The authors and
the publishers do not accept responsibility or legal liability for any errors in the
text or for the misuse or misapplication of material in this work. Except where
otherwise stated, drug dosages and recommendations are for the non-pregnant
adult who is not breast-feeding

Links to third party websites are provided by Oxford in good faith and
for information only. Oxford disclaims any responsibility for the materials
contained in any third party website referenced in this work.

to
Mary Bell

Preface

A quarter of a century after his death, and half a century after his most important work, it is becoming clear to most commentators, although still perhaps not to all, that John Stewart Bell was one of the very most important physicists of the twentieth century.

While he performed extremely valuable work on particle accelerators and in nuclear and elementary particle physics, it is clear that his special claim to fame rests with his work on the foundations of quantum theory, and there are two distinct aspects to this.

The first is the important Bell's theorem, or Bell's inequality, a result that startled the scientific community not only by demonstrating a clear distinction between the predictions of quantum theory and the demands of local causality, but also by showing that experimental tests could be performed to decide between the two.

Such experiments have indeed been carried out over the last 45 years with ever-increasing scientific sophistication or, in other words, with a reduction in the number and significance of loopholes, and it is confidently expected that an entirely loophole-free test will be possible in the next year or so. It is all but definite that the verdict will be in favour of quantum theory.

The second aspect, though, is perhaps even more significant. It is that, through his example and his achievements, discussion of fundamental aspects of quantum theory, which had been considered practically the mark of a crank in the years when Niels Bohr and his Copenhagen philosophy held physicists in a pincer-like grip, is now a respected and highly valued field of study, opening up for investigation novel and wide-ranging conjectures about the physical universe.

Study of fundamental concepts has opened up the highly practical topic of quantum information theory, which has already made considerable progress, but is certain to develop in future decades so as to revolutionize information technology.

By the 1960s, quantum mechanics was already the most successful physical theory ever in terms of explaining the behaviour of atoms, molecules, nuclei, solids, and so on, as well as being hugely significant commercially in, for example, the invention of the transistor and the laser. If the achievement of taking such an important field of study and developing it in completely unforeseen directions, adding enormous depth and breadth of understanding and unveiling entire new and fascinating areas of investigation, is to be considered 'great' science, then there are quite reasonable grounds for declaring Bell to be a 'great' scientist.

This book provides a full account of the development of Bell's life and career from its rather humble (if not necessarily 'poor') origins, through the difficulties in obtaining finance for his secondary school and university studies, and onto his enormous achievements at Harwell and then at CERN. However, its main emphasis is on his work, not only on quantum theory but also on what he was actually paid for—the study of accelerators, nuclear physics, and the physics of elementary particles; Bell himself felt that these components of his work were rather neglected. The presentation is non-mathematical, and every effort has been made to provide a thorough but comprehensible account of all areas of his work.

The story is presented decade by decade, so the reader meets quantum theory and its difficulties as Bell himself met them, first at Queen's University Belfast in the 1940s, and then at Harwell in the 1950s. The quantum background is presented quite fully, although in a broadly non-technical fashion and, as successive chapters study the work and ideas of a particular decade, one may appreciate the development of Bell's ideas.

With ideas of such depth, it is not surprising that argument about the precise logical requirements and meaning of the all-important Bell's theorem continues. Indeed, as explained in Chapter 4, differences of opinion appear to have surfaced more obviously in connection with 50th anniversary celebrations. It is not felt that this book is the place for detailed conceptual analysis. Rather, some aspects of the discussion are mentioned, and relevant references given for those wishing to follow them further, but the central approach has been to present the ideas of Bell himself and others, in some cases with rather different views, broadly in their own words. Of course, it might be suggested that Bell's words should be given particular weight!

When Bell entered the accelerator field at the end of the 1940s, the field was new enough that it has been possible here to present a fairly full account of the developments up to that time, and so to set the discovery of strong focussing, in which Bell played a substantial role, in context. Of course, there had been a major widening of the field by the time he returned to the area in the 1980s, but a substantial account of his later work has also been given.

Elementary particle physics is, of course, a much wider subject. A general account of progress in the 1960s and 1970s has been included in order to enable the discussion of some of Bell's most important work in the area, on CPT and gauge theory, to be understood. The equally important discovery of the Adler–Bell–Jackiw (or ABJ) anomaly is also described, although it must be admitted that this is a challenging topic for anybody, however knowledgeable, to appreciate fully!

Bell and his collaborators also dealt with many other problems in nuclear and elementary particle physics. While each topic and the advances made are sketched, perhaps the most interesting point here is that Bell usually chose to work in widely spread areas where novel opportunities and difficulties abounded. Having made, often, a crucial advance, he usually preferred to leave further work in the area, of a broadly clearing-up nature, to others.

As well as describing Bell's work, this book presents fairly substantial accounts of the three institutions connected with his study and work—Queen's University Belfast, Harwell, and CERN. Each had its own special features which helped to form Bell's scientific attitudes and his interests, and it seems appropriate to explain something of their genesis and nature.

As well as of his work, something has been said here of Bell's character, both of possible defects (few enough—perhaps a failure to suffer what he saw as foolishness gladly when really provoked!), but also his manifold strengths—kindness, humour, honesty—both personal and scientific—and courage.

Acknowledgements

There are many people to whom I would offer sincere thanks for help in the writing of this book.

First, I would like to thank Mary Bell for support and interest over a number of years, for her hospitality in Geneva, and for helpful and useful discussions. In Belfast I would like to thank the late Annie Bell for encouragement, and for providing invaluable and useful accounts of John Stewart Bell's first 22 years living at home and of his education, as well as insights into his character and personality; Ruby McConkey, John Bell's sister, for extensive and helpful accounts of the early lives of John Bell and his siblings and for allowing me to use many very interesting pictures of the Bell family; Robert Bell, John's brother, for an extremely interesting and helpful discussion; Dorothy Whiteside, Ruby's daughter, for many useful remarks; and David Bell, John's cousin, for helpful clarifications of some points.

I would like to thank David Bell, John's brother in Canada, for permission to make use of an interesting memoir of John's earlier years.

I would particularly like to thank Reinhold Bertlmann, as well as his wife Renate, for much help and support over a very long period—nearly 20 years! It was a very considerable encouragement, once I had decided I would like to write an account of John Bell's life and work, to discover that Reinhold had also felt, based on my previous writings, that I could be the best person to take on this role. I am grateful for him supporting my travel to Vienna for extensive interviews with him, which provided crucial information and insights on John and Mary Bell's lives and work, and for his hospitality at that time.

I would also like to thank Reinhold Bertlmann and Anton Zeilinger for the opportunity to attend and contribute to the two 'Quantum [Un]Speakables' meetings in Vienna, which were exceptionally useful for me, not only for hearing many authoritative accounts of Bell's work, but also for meeting many of those most important in his career and thus coming to understand many aspects of his ideas and personality.

Renate Bertlmann has also been of tremendous assistance, both in person in Vienna, and for being exceptionally helpful with the provision of illustrations, both those from the first Quantum [Un]Speakables meeting, but also for her pictures of the meetings between the Bells and the Bertlmanns. Gerlinde Fritz of the Austrian Central Library for Physics and Chemistry has also helped with pictures from the meeting.

Many others have been exceptionally helpful in this project: the late Leslie Kerr, for writing an extremely interesting memoir of his friendship with John Bell over a long period, and also the late Raymond Greer, for adding to Leslie's

remarks; Robin Coulter, for providing me with his obituary of Professor Karl
George Emeléus; Brian Gilbody, for giving me a copy of the account, written
by Professor Emeléus, of the work of Dr Robert Harbinson Sloane; Phil Burke,
for allowing me access to the material he gathered from those who worked
with Bell at Harwell in connection with the Royal Society memoir; Kamal
Datta, for giving me a brief account of Bell's time in the United States, the
period during which his most important work was performed; James Gillies
at CERN, for organizing a tour and also providing useful information, and
Torleif and Magda Ericson, Alvaro de Rujula, Tatiana Fabergé, and Marie-
Noëlle Fontaine, also at CERN, for discussions on Bell and his work when he
was there; the late Euan Squires, for conversation and information on Bell and
his ideas over several years; the British Library, for digitizing Bell's PhD thesis
and providing me with a copy; the National Archives, for provision of reports
written by John Bell while he worked in Harwell; Fiona Keates and Rupert
Baker at the Royal Society, for a link to the citation for Bell's Fellowship;
Fran Brearton at Queen's University Belfast, for information on the poetry of
Belfast; Robin Marshall and others at the University of Manchester, for help
with illustrations; Colin Latimer, for providing a publication of the European
Physical Society; the Library at Queen's University Belfast for inter-library
loans and particularly Bronagh McCrudden for access to Special Collections;
and Christopher Llewellyn Smith, Michael Nauenberg and Tony Hey for comments on John Bell and his work.

    During the writing of this book, a very successful exhibition on the life and
work of John Bell was organized at Queen's University Belfast by the curator
of the Naughton Gallery, Shan McAnena. In connection with this, a number
of experts on Bell's work were kind enough to present brief remarks on their
interactions with John Bell, which were later edited for a video presentation.
They did so at the 2014 Quantum [Un]Speakables meeting, and those participating were John Clauser, Reinhold Bertlmann, Anton Zeilinger, Daniel
Greenberger, Alain Aspect, David Mermin, Michael Horne, and Helmut
Rauch. Their remarks were helpful for the writing of this book, and I would
like to thank them all for their time and effort. I would also thank Shan and
Elisa Nocente from the Naughton Gallery for permission to use the gallery's
pictures of Karl Emeléus and William McCrea in this book.

    I would like to thank Reinhold Bertlmann and Travis Norsen for reading
the book in manuscript and making some useful comments and helpful suggestions. Not only Reinhold, as mentioned before, but also Travis also has provided much encouragement.

    Finally, I would like to thank my wife Joan for reading the chapters of the
book as they were written, for making extremely helpful suggestions, and also
for being kind enough to take relevant photographs around Belfast for the first
chapter—as well as for general support and encouragement. Also, I would like

to thank my son Peter, who read sections of the manuscript and made useful comments. My other son John has also provided encouragement.

Of course, there are also many others, far too many to mention here, with whom I have discussed aspects of John Bell and his work over the last 30 years or so. I am grateful to them all.

In connection with the production of this book, at Oxford University Press I would thank Sonke Adlung, Senior Editor, Physical Sciences, and I would also like to thank Ania Wronski, Assistant Commissioning Editor, for constant advice and support, and encouragement through many problems. Also Maegen Reed, Production Editor, has helped with a number of difficulties.

At Newgen, I would thank Saranya Manohar, who has overseen the processes of copyediting, proof preparation and final production of the book.

Contents

1

A Tough Start but a Good One

Belfast and politics

Perhaps surprisingly, Belfast in 1928 was relatively peaceful, which was in marked contrast to earlier in the decade when, between June 1920 and June 1922, more than 450 people had lost their lives in violence in the city [1].

The Government of Ireland Act, which was introduced in Parliament in February 1920, had partitioned the country into Northern Ireland, which consisted of six counties in the north-east, and Southern Ireland, the remaining 26 'southern' counties, actually including the most northerly county of all, Donegal. Each part would have broad control over its internal affairs and its own parliament, although each would also be represented in the national parliament; their relationship to the Crown would be unchanged, and they would have no powers over such matters as war and peace, or external trade [2].

In the north, where two-thirds of the population were Protestant and keen to maintain the union with Britain, the proposition was accepted with some degree of alacrity, but Catholics, in the north as well as in the south, where they were in a great majority, were determined to obtain full independence and a republic for the entire island, and the guerrilla war run by the Irish Republican Army (IRA), which had been in operation since the beginning of 1919, continued, being particularly intense in the north, where it was largely against the Unionists rather than against the British as in the south [1].

Eventually, there was a truce between the British and the IRA in July 1921, leading in December to the Anglo-Irish Treaty, which was signed extremely reluctantly by the Irish side because it confirmed partition and, although it provided for an Irish Free State, gave it only dominion status within the empire. Also, members of the southern parliament would have to take an oath to the Crown [2].

A government in the south was set up under Michael Collins, but strong opposition came from those determined to accept nothing less than a republic. Collins was playing a double game with the British, working with them to establish change and face out the republicans in the south while hoping to keep the two sides in the south together by organizing unrest in the north. But it proved impossible to avert a civil war in the south, and this began in June 1922 [1].

The war made it impossible for southerners to provide any support to the IRA in the north, while in the north itself in April the new Unionist government had introduced the draconian Special Powers Act, under which internment without trial was now taking place. The combined effect of these two factors was to cripple the northern IRA [1].

In the south, the pro-Treaty side had won the civil war by May 1923 and, while regretting partition, they were certainly not prepared to oppose it physically. Rather, nationalists and republicans, north and south, put their trust in a promised Boundary Commission, which they hoped, and Unionists feared, might transfer substantial areas from north to south and thus render Northern Ireland ungovernable; however, by 1925 it became apparent that the Boundary Commission was a damp squib. Ireland was frozen into a pattern none had particularly wanted but which would remain, challenged in word but hardly in deed, until the violence of the late 1960s and early 1970s. (A bombing campaign in England just before the Second World War and an IRA border campaign in the 1950s were largely ineffective) [3].

In Northern Ireland, the Unionist government enhanced its power by abolishing proportional representation for local elections, by some gerrymandering of constituency boundaries, and through some discrimination in jobs and housing, the extent of which may still be debated. The Catholic response was little more than sullen non-cooperation. There were certainly frequent individual sectarian scuffles, and indeed large-scale sectarian riots in 1935, although the riots of 1932 were of Catholics and Protestants protesting side by side against the cutting of employment relief; but, overall, Belfast and the rest of Northern Ireland were probably less troubled than many other countries in the 1930s and 1940s [3]. There was not a single sectarian murder in Belfast between 1923 and 1933, while the ordinary crime rate in Northern Ireland in these years was perhaps the lowest in Europe [4].

All this is to explain why, during the first 22 years of his life, which was spent in Belfast from 1928 to 1950, John Stewart Bell was in more danger from wartime German bombs that from communal disturbance.

Family background

John Stewart Bell (Figure 1.1) was born on Tate's Avenue, Belfast, on 28 July 1928 [5, 6]. He was the first son of Elizabeth Mary Ann, known as Annie, and John, known as Jackie after his mother's pet name for him; to avoid obvious confusion, John Stewart Bell was always known as Stewart at home. Annie and Jackie had been married in 1925 in St Anne's Church of Ireland Cathedral in Belfast. They were quite young at the time—Jackie was 21 and Annie just 18. Their first child, Ruby Jane, was born in December 1925, and after John came two more boys—David Andrew, born in February 1930, and Robert, born in December 1932 (Figures 1.2 and 1.3).

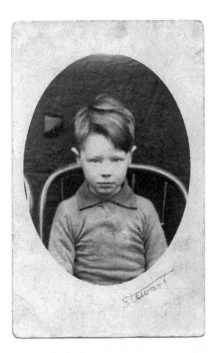

Figure 1.1 A young John Stewart Bell. Courtesy Ruby McConkey.

Both Annie and Jackie were members of large families—eight or nine, and the families had lived in the north of Ireland for many generations. Annie's father came from Enniskillen in County Fermanagh, where his family, the Brownlees, had been well known since the sixteenth century, and her mother from County Tyrone; these were two of the counties that were to become part of Northern Ireland. Before that, at least part of Annie's family had come from Scotland, and the name Stewart was a reminder of that—her mother's family had been called Stewart long ago.

In fact, Annie's family had suffered a drastic fall in status at the beginning of the twentieth century. Her father had owned a prosperous grocery and baker's shop in Enniskillen, and the family was well off at that time. Annie's oldest brother went to Royal Portora School, one of the most famous schools in Ireland, and the family had employed a nursemaid and a housekeeper. But around 1904, disaster struck; the shop was burnt down, there was no insurance, and the family were reduced to penury. Annie's father refused to stay as a 'working man' in a town where he had been a 'gentleman', and that was when the family moved to Belfast to pick up the pieces of their lives. Annie was the first of the children to be born in Belfast.

Figure 1.2 A family outing: Mrs Brownlee (Bell's grandmother), Ruby Bell, and Annie Bell, behind; David, John, and Robert in front. Courtesy Ruby McConkey.

Annie's mother worked as a dressmaker, and somewhat later for a time ran a corner grocery shop on Tate's Avenue. Her father undertook whatever jobs came up. For example, when the Tate's Avenue Bridge was being built in the 1920s, he was in charge of security, and even into his 80s, he undertook security at a local school. Ruby, his granddaughter, remembers that in those days he was still always the gentleman, but he carried a silver-headed walking stick and could deal with anybody! When he died in 1940, he was buried in County Fermanagh, and the town of Enniskillen came to a standstill for the funeral.

Jackie's family came from a rather less privileged part of society. For generations, their main activity had been dealing in horses, which would be bought at various fairs—Comber, Moy, and Lisburn, and sold primarily by word of

Figure 1.3 Tate's Avenue, from the far side of the railway in 1915, before the building of the bridge. The original family home, in which all the children were born, was on the left-hand side, quite close to the railway. Courtesy Public Record Office of Northern Ireland.

mouth in Belfast. Before being sold, the horses would be kept in fields behind the houses in the various roads where the family lived.

Many years later, when Jeremy Bernstein [5] asked John about what occupations members of the previous generations of the Bell family had had, as well as horse dealing, John mentioned carpentry, blacksmithing, manual labour, and farm work. When he was asked whether there were any scientific or academic traditions on either side of his family, he replied that, as far as he knew, there weren't, except that a half-brother of his mother had been a blacksmith and had taught himself something about electricity. However, Ruby remembers Annie's pride that in a previous generation there had been a 'Reverend Brownlee'.

By the early twentieth century, both Annie and Jackie's families were in Belfast, and in fact both Annie and Jackie were born within a few hundred yards of where John himself was to be born 20 years later. They were actually briefly at the same school, St Nicholas' Church School just off the Lisburn Road, but they did not talk to each other, because girls and boys had separate classes and in any case did not bother with each other.

This only lasted until Jackie was 6, when he left school to help with the family finances by buying and selling horses. However, working at this age was against the law and, when he was 8, the school attendance officers caught up with him, and he was returned to school until he was 11 years and 10 months old, in July 1918. His attendance may not have been exemplary, but the law seems to have been enforced rather half-heartedly, and only occasionally did his parents have to pay fines for his non-attendance.

Then, Annie and Jackie fortuitously met again at an Orange bonfire—an event held on the evening before 12 July, when Protestants like to celebrate the Battle of the Boyne and the triumph of William III ('King Billy') at the end of the seventeenth century. Happily, their marriage soon followed.

Up until the Second World War, though, employment opportunities in the less well-off areas of Belfast were modest—in this respect, the inhabitants of Belfast were no different from most working people in the large cities of the United Kingdom. In Annie's word, the depression of these years was 'dreadful'. Belfast was very badly hit. Around the turn of the century, it had been a great industrial city, perhaps leading the world in linen and shipbuilding, but by the 1930s these industries were in desperate trouble [4]. Most ordinary people worked at whatever jobs they could get—labouring, factory work, small-scale trading, and so on.

The Bells were certainly not immune from these pressures, but, as Ruby has said, while Jackie did not much like heavy work, he was prepared to turn his hand to anything. He retained his family task of horse dealing, to which he added dealing in horse-drawn vehicles. Robert has said that, between 1924, when he bought his first horse-drawn vehicle, and when he joined the army in January 1940, he had bought and sold over 350 horses and over 200 vehicles.

In addition, he developed a business as a fruiterer, which was sufficiently prosperous that he was able to employ two of his own brothers. For this purpose, he operated a series of horse-drawn vehicles, including a flat-back lorry, then a butcher's van, and finally a trap. These vehicles were also available for family use, such as for picnics at the Black Mountain on a Sunday after Sunday school. For example, Ruby remembers one 12 July when he took the family out to the country, then returned to Belfast and spent the remainder of the day ferrying people between the furthest tram stop and the 'field' where the Twelfth celebrations took place.

Jackie is stated to be a fruiterer in the records of his own wedding and the birth of David, the couple's second son but, at the births of his children in between and after those dates, he is stated to be a labourer. It seems that, on the latter occasions, the births were registered by an aunt.

Incidentally, at one time Robert was amazed at the large number of neckties that Jackie owned; Jackie explained that a necktie was the standard present from customers to purveyors of fruit and vegetables.

After his death, Annie's view of her husband was that he had 'never had a chance', but that if he *had* had a chance, he could have done well.

Annie herself certainly had far more ambition than Jackie did, perhaps not surprisingly, considering her family's fall from prosperity. John commented that while her family was not actually richer than that of her husband, and indeed not even better educated, they did have more respect for knowledge and getting on in life.

It had originally been intended, following the suggestion of her teacher, that Annie might herself become a teacher. However, this plan was abandoned when her father applied for a job in a local shop. There was no job for him but he was told that 'if he had a wee girl', she could be taken on, and that was the end of any ambitions for Annie to be a teacher.

She worked in several shops, including Woolworth's, but also, following her mother's example, was self-taught as a dressmaker, and at the time of her wedding was listed as a 'stitcher'. In fact, she was a 'sample embroidery stitcher' and, even after her marriage, she maintained very close links with one particular linen manufacturer. When they received a special order from a 'Big House', Annie would be sent for to take charge of the order, much to the annoyance of the full-time workers who did not possess her skills.

Indeed, as a private dressmaker, she was 'stitching night and day' for her customers, and also for her own family, making use of any leftover piece or scrap of material that happened to be available; for example, she was to make John's academic gown out of an old blackout curtain.

Annie was determined to maintain family life as pleasantly as possible within the limited finance. She was undoubtedly intelligent—when her mother ran the corner shop, it was taken for granted that Annie was 'the bright one' who would do the books. She was articulate and resourceful [7]; and she was particularly keen that her children should advance beyond the constraints of Tate's Avenue and the like. In later years, she would happily travel to visit John in Harwell and Geneva, and David in Canada, while her husband preferred to stay at home in Belfast.

Yet, while Annie was certainly the power behind the family, and the success of John and her other children was undoubtedly a result of her hard work, drive, and ambition, John's cousin, David Bell (Figure 1.4), stressed that the ability in the family ran far wider than that. (David Bell's father was one of Jackie's brothers and was around 20 years younger than Jackie.) David believes that both Annie and Jackie were highly able and, indeed, that the same was true, at least

Figure 1.4 Jackie Bell, John Bell's father, assisting his nephew, David Bell, to see over a Tate's Avenue wall, around 1947. Courtesy David Bell.

in broad terms, of their substantial numbers of siblings; and the sad thing is that most of them, as well as the children of the elder ones, were held back by lack of educational opportunities. Once education at secondary and even tertiary levels became available, they flourished and, in fact, while John was at university, three of his cousins were at the prestigious Methodist College.

Just as a typical example, David himself became a Professor of Computer Science at Queen's University Belfast, and his brothers, Jonathan and Paul, became respectively Head Curator at the Ulster Folk and Transport Museum, and a consultant psychiatrist and Medical Director to the South and East Belfast Health and Social Services Trust.

As another example of the success of the Bell dynasty, another of John's cousins, William Bradshaw Bell, always known as Billy Bell [8], son of an elder brother of Jackie, who was born in 1935, became a well-known local politician. According to *Wikipedia*, he is the only person to have been mayor of two different cities in the United Kingdom: Belfast, from 1979 to 1980, and Lisburn, in 2003.

Family life

Tate's Avenue is in South Belfast, which contains most of the more prosperous parts of the city. The road south from the city centre splits into two roads diverging somewhat but both running broadly in the direction of Dublin. To the east, University Road, which becomes Malone Road, is the home of the Ulster Museum, the Botanic Gardens, Queen's University, and much middle-class housing; some of the upmarket streets in this area run down to the more westerly of the two roads, the Lisburn Road, which itself contains housing and a wide variety of commercial activity.

The railway line to Dublin is to the west of the Lisburn Road, and most of the streets that lie to the west of that road, on both sides of the railway, are rather less prosperous than those on the east side. The most substantial road travelling west from the Lisburn Road is Tate's Avenue, which provides one of the few crossing points of the railway. In the 1920s, the area around this road would have been, as indeed it largely remains today, staunchly Protestant and working class. Houses were mostly fairly small, with the provision of facilities in some cases rather moderate.

Yet, living in the area would have been very far from unpleasant, partly because, perhaps as it was a little way from the city centre, the density of housing there was relatively generous, compared to that nearer the city centre. It was also pleasant to be near the amenities of South Belfast, such as Queen's University, whose proximity was handy for John during his time there, and the Lisburn Road shops.

Mostly, though, it was because, as you walk up Tate's Avenue, you are walking towards the delightful vista of the Black Mountain, which is part of the Belfast Hills (Figure 1.5). Whatever the season or the weather, just to have the mountains in view must be uplifting. In his poem 'Spring in Belfast',

Figure 1.5 Tate's Avenue: 'The hill at the top of every street' (Derek Mahon). Courtesy Joan Whitaker.

Derek Mahon [9] wrote: 'We could all be saved by keeping an eye on the hill at the top of every street', and he may well have been thinking of Tate's Avenue! Annie anyway thought that it was lovely. At this time, there were fields at the end of Tate's Avenue and it had practically no traffic. Nearly a century later, it leads to the ring road linking motorways to north and south, and a road of out-of-town shopping, warehouses, and car showrooms, and Tate's Avenue itself is extremely busy; but, at that time, in Annie's words, it was like 'a wee country village', and the children could roam to their hearts' content.

All the children were born at No. 129, on the east of the railway line (Figure 1.3); however, this situation was not ideal as, in the mid-1920s, a bridge had been built over the railway and, as this bridge increased in height towards the railway, the houses near the railway ended up very much in its shadow. Also, there had once been a field beside the house where the children could play; at about this time it was fenced in for a builders' yard, and Annie was not prepared to accept this.

So, for the benefit of the children, in 1936 the Bell family moved along the road to a new house at No. 240 (Figure 1.6), almost at the end of the road, on the far side of the bridge, and well away from it and close to the fields. Robert has written the following about No. 240:

> The house was a new one in a new development. The streets were new and the pavements were new. The block of shops had been recently built. Tate's Avenue Bridge was an impressive modern construction. The whole area was bright, clean, and modern. The green fields started at the end of the block of houses in which we lived, and stretched off into the distance. Those fields were the floodplains of the Blackstaff River. After heavy rain the ponds filled up, and the swans and wild duck arrived with even more than the usual number of seagulls. In the early mornings we heard the call of the corncrake, and in the late evening a curlew. The place at that time was semi-rural and not a bad place to live.

And Annie was delighted that the rent of the new house was three shillings a week less than that for the old one!

The houses in the vicinity of their first home, incidentally, have been knocked down comparatively recently to build a complex of flats.

In later life, Annie would become irritated with commentators who, with the best will in the world and in order to show how far John had risen, emphasized the low financial status—perhaps 'poverty'—of the family. She insisted that they—in particular the children—had everything that was necessary for a good life. She was a great budgeter; Ruby was to say that she made two and sixpence do the work of five shillings, and every Saturday morning she would put out on a window ledge the money for milk, coal, groceries, and so on.

Figure 1.6 The house in which the Bell children were brought up from 1936: 240 Tate's Avenue. Courtesy Joan Whitaker.

She was able to buy the children bicycles—second hand but, as she said, they had just as much fun on them as if they had been new and, in any case, as Robert has pointed out, in wartime there were no new bicycles to be had.

She also tells the story of how, once John had become a vegetarian, at Christmas time when he smelled the turkey cooking he would remark

'I smell a corpse burning.' This story makes the point of his sincerity as a vegetarian, and perhaps also the intolerance of a vegetarian to those who have not
yet taken that leap of conscience, but it does also speak of at least a modest
affluence—even at Christmas, in those days before battery farming, a turkey
was a very good meal.

Indeed, despite financial strictures, life must have been pleasant enough.
Annie, feeling that her own parents might have been too keen on keeping the
children in and not letting them mix, made sure that all the children got out
of the house; they went for cycle rides and took long walks up in the nearby
hills, to the Cavehill further north, or in other directions. Later, when John
was married, his wife Mary also enjoyed these walks. He himself hated actual
sports at the time and, as he said, grew up to be 'a seven-stone weakling',
although he later regretted this, and in Geneva he was to become an enthusiastic skier. Neither was he interested in scouting when he was in Belfast; in
this he was different from his brothers, who were keen scouts—David was to
become a scoutmaster.

All the family had many friends and in those days it was natural for many of
these friendships to be 'cross community', which is to say, between Protestant
and Catholic. As part of her dressmaking business, Annie indeed used to make
First Communion frocks for many of these Catholic friends. Living in such a
staunchly Protestant area, there was obviously scope for getting involved in
political activities, such as the Orange Order, and Annie in particular was a
strong supporter, but John was probably much too interested in books and
ideas to have any concern with what, to him, would have seemed a waste
of time.

The family, though, were brought up soundly in the Protestant faith. They
were Church of Ireland, which is, like the Church of England, part of the
Anglican Communion. John was baptized in St Aidan's Church, which is a few
hundred yards towards the city centre from his home. The choice of church
was to please Annie's mother, as it was near her home, where Annie herself
had been born. However, the Bells' home was in the parish of St Nicholas on
the Lisburn Road, and it was here that the family worshipped, with Sunday
school for the children and then church in later years. John, dutiful as ever,
was awarded a number of prizes for attendance. All the children were confirmed in this church, and indeed Annie herself had been confirmed there
much earlier. However, John was not to remain in the fold for much longer.

Early education

Jackie thought it was natural that the children should leave school at the age
of 14 and get a job. In contrast, Annie was extremely keen on her children

being well educated, so that they might move beyond the type of work of work their parents were limited to. She pointed out the workmen on a wet day and told them that she didn't want her sons to be digging drains for the rest of their lives. She advised them to use their brains and get educated because it's more comfortable—then they could wear their Sunday suits every day! (Years later they were able to tell her that they were always wearing their Sunday suits now, although John, in particular, was usually more casually dressed.)

John needed no encouragement to become absorbed in books and learning. From an early age, he was much in the local Donegal Road public library, absorbing all the information that he could, information that he then proceeded to analyse in depth. He was very generous with the fruits of this analysis, forever dispensing knowledge to all the members of his family, to the extent that they called him 'The Prof'. Even at such a young age, they recognized his brilliance, and by the age of 11 he had read enough to tell his family that he wanted to be a scientist. Annie's natural response was 'What's a scientist?', and he did his best to explain the idea to her.

His interests were not confined to books; in fact, just like his father, he was extremely practical, although he took no interest in his father's actual work, unlike Robert, who enjoyed feeding the horses. John was concerned with experiencing every aspect of life; indeed, when he was knocked over by a car, which was fortunately not travelling too fast, his first action when he was back to his feet was to get paper and pencil and draw the underside of the car.

At school, he was equally outstanding. His first school was Ulsterville Junior School, near Tate's Avenue and the Lisburn Road, and right away he was exceptional. He was always at or near the top of the class—good in all subjects, no better or worse in mathematics than in any other subject.

However, tragedy was quite close to striking. Over the summer holidays, when he was 7, he became exceptionally ill with whooping cough, to the extent that his parents were warned that he might not live. Thankfully, he made a good recovery; but, when school recommenced in September, Annie did not immediately take him back, rather waiting until she thought he was ready.

When she did so, she suggested to the teacher, Miss Nora Rankin, that he should be kept down a class until he was recovered. Miss Rankin replied: 'Oh, but he's one of my best.' For all Annie's confidence in his ability, she was pleased to hear this confirmed by someone she would consider to be more of an authority than herself. John was briefly kept down but moved up at Christmas without a problem and so never got behind.

Ulsterville Junior School was a feeder for Fane Street Elementary School (Figure 1.7), to which John moved at the age of 8. In many ways, Fane Street was and still is an exceptional school. Situated on the less prosperous side of the Lisburn Road, the great majority of its pupils at the time, however positive

Figure 1.7 Fane Street School. Courtesy Joan Whitaker.

their parents may have been, came from backgrounds with little social or educational advantage. Through the years, it has struggled with population movement and with constant changes in educational policy, being successively
elementary, secondary, and then primary in educational level.

Despite this, it has maintained an excellent reputation with employers
and produced quite a number of very successful men and women. Indeed, in
recent years, one of its former pupils has achieved the feat of becoming Head
of the far more upmarket Methodist College, which is definitely on the prosperous side of the Lisburn Road!

Clearly, it was exactly the right school for John, and he thrived. The head-
master, Mr Thompson, called in Annie to say just how good her son was, and
it seemed John had the ability to go on to great things. But here lay the rub!
At 11, pupils could either keep going at Fane Street for a further three years
and then leave to get a job, or they could move on to secondary education,
grammar or technical. But secondary education was not free in Britain at the
time. That was shortly to come to England in 1944, although because of wran-
gling between the various religious providers of education, it did not reach
Northern Ireland until 1947 [4]. But in the 1930s and early 1940s, the only way

for children from backgrounds that were not particular well off to get to secondary school was via scholarships.

In fact, none of John's siblings (Figures 1.8 and 1.9) went to secondary school; all left school at 14. This was definitely not because of any lack of ability. Ruby, the eldest, actually obtained a City of Belfast Scholarship to go to Methodist College but could not take it up because her father in particular thought that it was 'not worth educating a girl'. Annie had thought that she might become a music teacher, and for many years Ruby hurried home once a week to get her music bag for her piano lesson, which cost a shilling. Ruby obtained four certificates for piano, but then, at the age of 14, she began playing with a band and formal study finished; Annie's dream was gone and she was certainly not amused!

Ruby left school to become a shop assistant, but her career developed and she became a fashion buyer and works study expert. After the factory where she was working was bombed in 1969–70, she took a job in telesales with the local Newsletter and stayed there 17 years, rising to become circulation, sales, and PR manager.

Both of John's brothers came to regret leaving school so early. Robert followed John in becoming a laboratory assistant in the physics department at Queen's University, and he enjoyed working with all the equipment. However, to get any qualification, he had to move to one of the largest engineering establishments in Belfast, which he came to detest. Perhaps his commercial nature, inherited from his father and grandfather, came out, as he later became the proprietor of a successful shop selling radios in Belfast; when the bombing made this difficult, he moved to Bangor, where he sold domestic electrical equipment.

David made rather a remarkable climb back to the academic world. He left school to work in a shop, but, when Annie discovered the lowly tasks he was undertaking, she took him away and asked Sam McConkey, who had married Ruby in 1947 and who was a qualified engineer, to take him on as an apprentice. David was happy and successful in this work, but, like Robert, he had to leave to become fully qualified. In his new position, he moved on to study at night school and then to day release, eventually obtaining a degree in electronic engineering by night study.

He spent the years between 1954 and 1956 in Canada with his company, and then in the early 1960s he returned to Canada and became a Lecturer in Lambton College. Discovering that there was a shortage of textbooks, he decided to write his own, and he has since become the author of a considerable number of successful texts. In Canada, David was a staunch trade unionist, and indeed he was chosen to be a candidate for the Social Democratic Party in a coming election but had to withdraw because of the illness of his wife.

Figure 1.8 The Bell children—Robert, David, John, and Ruby—around 1936. Courtesy Ruby McConkey.

Figure 1.9 To the left, Annie Bell, behind, and David and John, in front; to the right, Mrs McGarry, behind, and Sis McGarry, in front (friends). Courtesy Ruby McConkey.

However, to come back to John, it was clear that 1939 would be a fateful year for him. Would he be able to enter a secondary school, preferably a grammar school, where he would be able to develop his obvious ability, or would he be forced out into a world that would be more hostile than that and where progress would be more precarious?

That might be described as rather a parochial view of the world! Clearly, it would also, indeed primarily, be a fateful year for the world, as it lurched into the Second World War. The war would also have specific consequences for the Bell family.

The war and the Tech

During the war, there was never to be conscription in Northern Ireland [10]. The Unionist community might have liked it, as it would have demonstrated and built up the links with Great Britain, but it was felt that the risks of alienating the nationalist community to perhaps open revolt were too great. Nevertheless, many from both communities, and indeed many from Southern Ireland, which remained neutral, did join up, and many others travelled to England to carry out war work. For some, the spur was the desire to support Britain in her hour of need, or to defeat the National Socialist government in Germany. For others, it may have been the lure of a regular job and a steady income.

Jackie was among those who enlisted. He was to fight through the war in the Royal Ulster Rifles, becoming a corporal, although he was seriously wounded in the stomach. It seems that his joining up was something of an accident, since went with two friends to the recruiting office, expecting to be turned down. In fact, although his friends were turned down, one for flat feet and the other for a twisted spine, Jackie was accepted. He had to sell his horses, but he promised Robert that he would recommence the horse dealing after the war. Unfortunately, his injury prevented this from happening.

Ironic as it may seem, after the terrible depression the war brought a measure of prosperity back to Northern Ireland, with a vast number of jobs in war work. During the war, the shipbuilding company Harland and Wolff built almost 170 Admiralty and merchant ships and repaired over 30,000 vessels; they also made more than 13 million aircraft parts, more than 500 tanks, and thousands of guns. The aircraft makers Short and Harland made well over a thousand Stirling bombers and a hundred Sunderland flying boats [4].

In addition, Northern Ireland profited from the presence of an enormous number of American soldiers, who proved to be well-off customers for businesses of all types. Over the full duration of the war, more than 300,000 American troops were stationed at different times in Northern Ireland.

For John, the sad fact was that, for his education to go forward, he was totally dependent on obtaining a source of finance. He passed the qualifying examination for secondary school, presumably with ease but, as already said, tuition at secondary school was not free; it was not even cheap. Within a decade, free secondary education would become available but, at that time, his parents were not able even to consider sending him to any secondary school without financial support, which would probably have to come from a scholarship.

Annie would certainly have loved him to go to the Royal Belfast Academical Institution, known to all as 'Inst' (Figure 1.10). This school occupied a prominent site near the centre of Belfast, and it had an important place in Belfast hearts. It had been founded in 1814 at rather an interesting time in the history of Belfast [11, 12].

In the eighteenth century, there were three main classes of people in Ireland. At the top were members of the Church of Ireland—Anglicans or Episcopalians, who had full right to own land. Decidedly at the bottom were Catholics, who were effectively deprived of rights by the Penal Laws. Rather awkwardly in the middle were Presbyterians, who were at least formally affected by the Penal Laws—for example, their children were illegitimate unless the parents had been married in a Church of Ireland church by a Church of Ireland clergyman—but they were allowed to own land as a

Figure 1.10 Royal Belfast Academical Institution, or 'Inst', c.1960, from John Jamieson, *History of the Royal Belfast Academical Institution 1810–1960* (Mullan, Belfast, 1959). Courtesy Royal Belfast Academical Institution.

privilege. While the other two groups lived in all parts of Ireland, Presbyterians were mainly in the north-east [4].

At the end of the nineteenth century, with revolution in the air in France and America, it was not surprising that there should be a rebellion, that of 1798, in Ireland, but what may be surprising is that the leaders of the so-called United Irishmen included Presbyterians as well as Catholics. Indeed, in the north, those taking part were virtually all Presbyterians. The rebellion was put down violently, the main result across the country being the Act of Union, under which the Irish Parliament was closed down and the governments and parliaments of Great Britain and Ireland were united. The Union was to last until 1922 [4].

In Belfast, the rebellion heralded a period of liberalism, and it is interesting that some of those who must have come quite close to being hanged in 1798 rapidly became pillars of the establishment. Several liberal institutions were founded which have lasted to the present day, including the well-known Linenhall Library.

The most important was Inst, then, and until 1831, without its royal title, as the powers that be were perhaps justifiably suspicious of the radical republican motives of its founders. As well as a school department (for boys only), it doubled as a college, on the model of a Scottish university, and at the time had medical and theological faculties. James Thomson, father of William Thomson, later Lord Kelvin, who taught at both levels and was Professor of Mathematics in the Collegiate Department, was the dominant personality until his move to the University of Glasgow in 1832 [11].

The school had been founded by public subscription among Belfast's citizens, specifically to provide a liberal and non-sectarian education, and also a practical one, dedicated 'to diffuse useful knowledge, particularly among the middling orders of society, as one of the necessities rather than of the luxuries of life; not to have a good education only the portion of the rich and the noble, but as a patrimony of the whole people'. Its building is perhaps the finest example of late Georgian architecture in Ireland [12].

By 1939 the college and medical areas had long gone to Queen's College (later Queen's University) and the theological work to the Presbyterian Church's own college, but the school was strong academically and had a good interest in physical science—two of the houses were named after Kelvin and Larmor (Sir Joseph Larmor, a former Inst boy and Queen's student, later Lucasian Professor of Mathematics at Cambridge).

It will be clear that the school would, and probably should, have been ideal for the (comparatively) poor but brilliant John, who was certainly desirous of good middle-class employment. But it was not to be. He took the examination for the scholarship but was unsuccessful, and this was repeated when he took scholarship examinations at Methodist College and all the other grammar schools in Belfast. John was later to wonder whether this gruelling experience put his brothers off going through the same process.

One may wonder why such an obviously gifted boy was so uniformly unsuccessful in being awarded a scholarship. The disappointment certainly could not be explained by any feeling that he was admittedly very good at mathematics but perhaps not so good at the other subjects—as already stated, he was an all-rounder academically, good at everything.

The explanation must lie in the fact that the grammar schools had preparatory departments, which parents paid for their children under the age of eleven to attend. If children were reasonably bright, when they were 11 their parents would enter them for the scholarship examination, so as to avoid having to continue paying fees for secondary education; clearly, these children would have 'an inside track' for the process—useful hints and much-needed confidence. Also the private tutor system was well established; parents could and would pay out money for tuition directed solely at the exam, in the hope of their child being awarded a scholarship, thus saving money in the long run. The boy from Tate's Avenue and Fane Street School, however talented, would probably stand little chance. Again, one must think with sadness of the loss of talent.

Yet, at the last moment, money did turn up—not for any of Belfast's prestigious grammar schools, but for Belfast Technical High School. The source is not quite clear, but, from John's application to Queen's University in 1945, it seems that he had a scholarship of around seven guineas for at least three years, a guinea being one pound and one shilling. Today, the scholarship might be worth around £350. For Annie in particular, sending her son to 'the Tech' rather than, say, Inst, must have seemed a great let-down. As things turned out, though, it gave John all that he needed.

Strangely enough, the Technical High School [13] has had a symbiotic relationship with Inst. In 1899 the Belfast Corporation felt that it was essential that the city should have a fine, up-to-date technical college (Figure 1.11), largely for part-time evening instruction of apprentices and workers in the shipbuilding, textile, and machinery industries. Unfortunately, though, there was no suitable site in or near the city centre.

At roughly the same time, the governors of Inst found that their debt, which had been large for many years, was spiralling upwards. They were told that the land in front of the school was worth between £100,000 and £200,000, perhaps between 8 million and 16 million pounds today, and they were eventually persuaded in 1902 by the Lord Mayor to lease in perpetuity a strip of the front lawn of size 240 ft by 205 ft for £1,350 a year to be used for the new technical college.

The college—originally the Belfast Municipal Technical Institute—was completed in 1907. It was built from Portland limestone and is in a fine baroque style, but its uncompromising massive frontage, blocking the view of the unassuming but stylish Inst building, has been described as 'an act of extraordinary

Figure 1.11 A view of the College of Technology at the time of its foundation *c.*1907, from *Municipal Technical Institute Belfast: Description and Illustrations of the Buildings and its Equipment* (Belfast Municipal College of Technology, 1907). Courtesy Belfast Metropolitan College.

architectural vandalism' and 'the largest and most ornate cuckoo's egg ever laid in a songbird's nest'.

This would have been of little concern, of course, to John (Figure 1.12), who must have been far more concerned about the course of study he would follow, how he would get on with his teachers, and how well he himself would fare in an environment which might be rather more challenging than those he had been used to.

In fact, he was to follow a course leading to matriculation (Figure 1.13), which would allow him to enter Queen's University in the city, still with the same proviso—if he were able to generate the necessary money. Many of the subjects he would study would be the same and probably have much the same

Figure 1.12 A youthful John Bell. Courtesy Ruby McConkey.

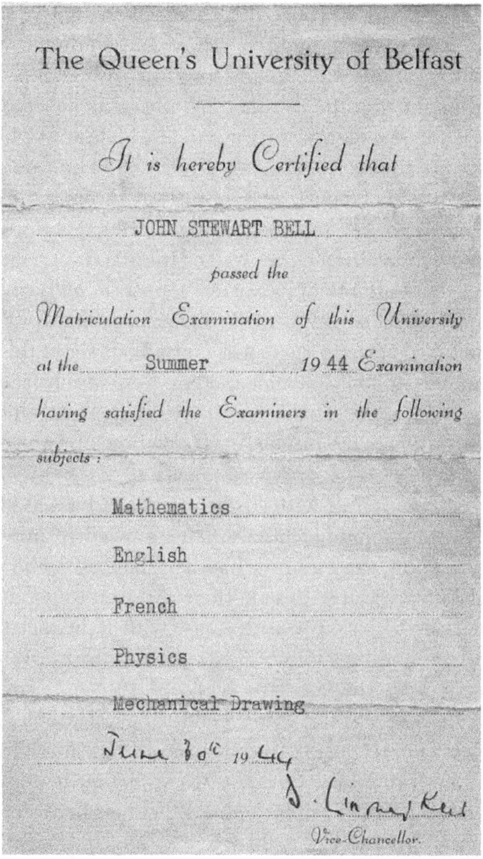

The Queen's University of Belfast

It is hereby Certified that

JOHN STEWART BELL

passed the

Matriculation Examination of this University

at the Summer 19 44 *Examination*

having satisfied the Examiners in the following

subjects :

Mathematics

English

French

Physics

Mechanical Drawing

June 30ᵉ 19 44

J. Linsay Keir

Vice-Chancellor.

Figure 1.13 John Bell's matriculation certificate for Queen's University Belfast. Courtesy Ruby McConkey.

content as he would have had to follow at a grammar school—English, mathematics, French, physics, chemistry, and so on; but, whereas a grammar school boy would have studied Latin, and perhaps Greek, the curriculum at the Tech included instead bricklaying (though theoretical only), carpentry (which he enjoyed), and bookkeeping (which he didn't!).

As to his teachers, in fact he was able to assert to Bernstein that he had found all of them, not only those at the Technical School but also those at Ulsterville Avenue and Fane Street to have been not only excellent in imparting and explaining information but also helpful to the individual student. He felt that he had been very lucky throughout his education.

And, if he had been at all concerned about his ability to do well, he need not have been. Annie was to find this out when he brought home the first list of marks of the various students and she saw that his name was at the top. Initially, she thought that the list must have been alphabetical and that his name was at the top just because he was a 'B'. 'No, Mummy', he said; 'It can't be like that because there's a fellow called Adams and he's about seventh!' In fact, John was to be top of his classes right the way through the course.

Ruby and John travelled into Belfast together at this time, Ruby to her work and John to school. In fact, they walked the first tram stage to save a little money in order that John might have a few coppers in his pocket.

Being an unaggressive soul, it is unfortunate but perhaps not too surprising that John was bullied at some stage. Ruby found out about this and told him that, while she herself could not hit the boys responsible, she would come with him while *he* hit them. This is exactly what happened, and consequently there was no repetition of the trouble; in fact later on John and one of the boys became great friends.

John enjoyed having fun with his brothers but he was also very close to Ruby. He was conscious of how much he owed to her and sad that, in terms of education, she had not 'had her chance'. So he was delighted when she and Sam were able to send their children to good schools; both children were later to go on to university, with Isabel becoming a nurse and incidentally obtaining a pilot's licence, and Dorothy going on to work in building technology [14].

Meanwhile, the war was not going well for Belfast people. The shipyards presented an obvious target for German bombs but, for some reason, the government did not think the Germans would bother to come so far. They thought that it would be far more natural for their bombers to drop their loads on targets in England than to fly right over England and attack Northern Ireland. It was a fatal miscalculation [4, 10].

By February 1941, John Andrews, by then Prime Minister of Northern Ireland for only three months, realized that the situation was dire. He reported that he was extremely concerned about the lack of anti-aircraft defences, and also the state of mind of the public, who were both physically and psychologically totally unprepared for bombing.

In September 1940, Belfast had been provided with a light balloon barrage, and by the spring of the following year it was protected by 16 heavy and 6 light anti-aircraft guns. Even that scant effort, though, was better than other aspects of preparations of the defence of the city. On 20 July 1940, a Royal Air Force squadron equipped with Hurricane fighters was transferred from Edinburgh, Scotland, to Aldergrove, which is a few miles from Belfast. Unfortunately, though, they could operate fully only in daytime. It seems that official advice was that no major German raid would be at night. In addition, there were to be no searchlights in the city until 10 April 1941 (too late for at least the first attacks), and no provision for a smoke screen.

On the ground, too, provision for bombing was derisory, even after Coventry and other British cities had been pulverized. (In Coventry alone, 50,000 houses had been destroyed and over 550 people killed, the worst attack being in November 1940.) People in Belfast were told that they could use the city's two underground toilets in case of attack but, in fact, shelters could protect only one-quarter of the population. There had been a plan to evacuate 70,000 children from the city in July 1940, but fewer than 9,000 had taken advantage of this opportunity and, of these, half had returned to Belfast before the bombing began in the spring of 1941. There were plans to deal with only 200 bodies and only 10,000 made homeless by bombing.

As early as November 1940, a single unobserved German plane had flown over the city, bringing back high-definition photographs of suitable targets and reporting back that the defences of the city were totally insufficient. The first bombing mission was on the night of 7 April 1941. A relatively small squadron of German bombers attacked industrial targets—the shipyard and the docks. Again, the report back of the airmen was that Belfast's defences were 'inferior in quality, scanty and insufficient'.

On the night of 15 April came Belfast's night of horror (Figure 1.14). One hundred and eighty German bombers flew from northern France, over Britain and then the Irish Sea, to Belfast. Over a period of 5 hours, wave after wave of bombers swept in, and overall more than 200 metric tons of bombs and 800 firebomb canisters were dropped on the city. The telephone exchange, which provided all communication with Britain, and the anti-aircraft operations room were destroyed, and in this raid, whether on purpose or by mischance, the bombs fell mainly on civilian housing, principally in the north of the city. Many of the smallest houses of the poorest people in the city were destroyed.

On that night, the death toll was more than 900, with around 600 seriously wounded. Up to that point in the war, no city apart from London had so high a death toll in a single night, and indeed through the rest of the war only Liverpool (possibly) was to do so. The air raid warden system operated reasonably efficiently, but rescue service, hospitals, and mortuaries were overwhelmed.

So, unsurprisingly, was the population. All day on the 16th, thousands upon thousands of Belfast citizens 'trekked', leaving the city by car, train, or bus, sometimes the matter of the destination being a secondary consideration to getting out of Belfast for the coming night. Many were billeted in towns and villages in the country, and others crossed the border. By the end of the month, more than 200,000 people had left the city, and many of those who had to stay for their daytime work took to 'ditching' at night—walking out

Figure 1.14 The results of bombing in the High Street, on 16 April 1941. Courtesy Belfast Telegraph.

to suburbs, spending the night in fields, parks, or ditches, where they felt safe, and returning to Belfast in the morning.

Among those on the move were Annie and her children. The children had not taken part in the evacuation of 1940 but, from the beginning of the war, Annie had been keen that the children had friends in to play rather than them being out in the blackout. She hated the idea of them being away if there were to be an air raid and thought they were much safer inside than elsewhere; so, she let them bring any friends in, and most nights there would be two or three extra children in the house.

Then came the raids. For the first, on 7 April, Annie was on her own with the children. They had no shelter and so just sat together all night. They were able to look out of the window and see the planes dropping flares. Most of the bombs fell in the north of the city, but a stray bomb fell in Blythe Street, which was about half a mile from the Bells' house, killing ten members of the same family and demolishing a terrace of houses.

Jackie had been in London for the Blitz there, but he was at home on leave for the raid of 15 April. On the following day, it was like a holiday—nobody worked; nobody did anything; everybody was dazed. But Jackie took action; he decided that the family should stay with friends in Enniskillen near the border with Southern Ireland for a while.

There were further air raids on Belfast on the nights of 4 and 5 May 1941. These raids were to be known collectively as the 'fire raid'. Two hundred aircraft dropped around 100,000 incendiary bombs followed by high explosives, the targets being chiefly industrial—the docks, the shipyards, and the aircraft factory—as well as the city centre. Because of the trekkers and ditchers, there were fewer fatalities than for the previous raid—under 200—but great damage was done to industry. Harland and Wolff suffered greater damage this night than any other British shipyard on a single night of the war.

By the end of the night, there was not a series of fires but one enormous fire. The fire brigade was completely unable to cope and appealed for help. Rather surprisingly, fire engines were sent up from Southern Ireland on the authority of Eamon de Valera, the prime minister at that time, who put neutrality at risk in obedience to his concept of a united Ireland. In fact, because of the fracture of water mains, there was little that could be done and many fires had to be left to burn themselves out.

After that, there were to be no more major raids on Belfast. The Germans felt that they had devastated the industry of Belfast to such an extent that there was no hope or risk of recovery, although in fact the regeneration of industry, as distinct from housing, was reasonably quick and thorough. Gradually, the ditchers regained the confidence to spend the nights in the city, and the trekkers, including the Bell family, came back to Belfast. Jackie had long since returned to his duties and, for the rest of the war, the family would neither see him nor have any idea of his whereabouts, which were kept strictly secret. John returned to his studies and his other intellectual interests.

As has already been said, he was always very keen to expound his knowledge to anyone who would listen. An amusing example was when the other members of the family were out at the cinema, and a man called to collect the money for the football pools, which, like most people in the United Kingdom, the family did every week. John gave the unfortunate man a lengthy lecture on statistics and probability, pointing out the extremely small possibility of winning.

For a number of years, John and his two brothers were something of a gang with, of course, John as the intellectual leader (and Ruby the unpaid nursemaid who would be in trouble if the boys themselves got into trouble!) This situation may have lasted until John was about 13, when the various interests gradually diverged.

David has said that John was better at card games than anyone else he knew and that he also taught the younger boys how to play chess. David was keen to beat his older brother, and studied numerous library books about chess, eventually working out an original routine which involved sacrificing his queen to produce a checkmate position. John was almost fooled—but not quite, and David had to admit that he would never beat him! John also constructed an attractive chess set, with the pieces stained black and red.

The boys developed a great interest in photography. John first made a pin-hole camera, with the aid of a mustard tin painted black inside and with a pin-hole in the lid, a piece of photographic paper, and a bathroom/darkroom with a dim red light. Later, he introduced his brothers to the procedure for developing films and making prints. They progressed from Brownie box cameras to cameras with better lenses, and they learned how to tint and hand-colour photographs. John constructed an enlarger so they were able to produce conveniently sized prints, and he was always keen to learn more—he was a regular purchaser of the *Amateur Photography* magazine.

He also introduced his younger brothers to a variety of hobbies—stamp collecting (which entailed learning about the nations of the world), fretwork, constructing boats in bottles, and conjuring tricks. He also taught Robert to ride a bicycle, taking the opportunity to explain the principle of the gyroscope.

One of their great interests was radio, building radio sets from parts. The great place in Belfast for this activity was the Smithfield Market, a wonderful place of covered cobblestoned streets dating from Victorian times, where new and second-hand books, clothing, records, and so on, could be purchased. Sadly, it was burnt down in the 1970s, and the modern Smithfield is a pale imitation of past glories.

At one shop in Smithfield, old radios were taken apart and the parts were put up for sale. Here the boys purchased valves, chokes, transformers, condensers, resistors, wire, headphones, and Morse code keys and used them to construct receivers that could be listened to through headphones. John provided instruction on circuit diagrams, the use of the soldering iron, and how radios actually worked. With more valves, they moved on to sets that did not require headphones, and the coal hole under the stairs became the 'radio room', with coal being relegated to the backyard.

Finally, John purchased a microphone, but he used this to shock his family. Secretly, he connected it to the family radio; subsequently, the news reader 'broadcast' that Mrs Bell of Tate's Avenue had won a million pounds in a magazine contest. It took the family a little time to realize that it was only John playing a joke.

There were occasionally other episodes when John may have tried the patience of his elders. When the other boys were young, he taught them not to ask for something; wait and then cry, he told them, and then you will be sure to get it!

Later, in his mid-teens, he must have been bored, because he decided to engrave his initials outside the front door. The carving remains to this day (Figure 1.15).

Figure 1.15 John Stewart Bell's mark ('SB', for Stewart Bell; as his father's name was also John, at home John was always called by his middle name, Stewart) at 240 Tate's Avenue, made Friday, 11 April 1941. Courtesy Joan Whitaker.

A roving mind was to lead him, although not by the straightest of routes, to an interest in physics. He had been brought up in the church but, perhaps inevitably, came to wonder whether what he was taught there was true. Annie was very religious, and he used to be sent to two Sunday schools: Plymouth Brethren, and Church of Ireland. Sadly, the two Sunday schools did not agree with each other, the Plymouth Brethren in particular saying that the Church of Ireland was quite wrong!

John was around 11 or 12 at this time and, to decide which was correct, he decided to read the Bible solidly for a week, at the end of which he announced, much to Annie's consternation, that he did not believe in God! Annie called in the local minister, Reverend John Mercer, to 'sort him out'. He arrived fully confident that he could carry out his mission—and left some time later angry and deflated. He and Annie had to content themselves with the understanding that John would 'grow out of it'. Actually, he never returned to belief, but through the rest of his life he enjoyed conversations and friendships with religious people of many faiths. For example, while he was studying at Queen's University Belfast, he had discussions with a Catholic friend, Denis McConalogue, and he attended some meetings of the Student Christian Movement, although just for argument.

Loss of religious belief may have taken him to a study of philosophy. Probably the most well-known philosopher at the time was C. E. M. (Cyril) Joad [15], who appeared on the Brains Trust, a strangely very popular radio programme, in which a panel of worthies answered questions on a range of intellectual topics. Joad's response to any question inevitably began 'It all depends on what you mean by ...' He was something of a charlatan, but he wrote many readable books, and he did encourage ordinary people to take philosophy seriously.

John's study of philosophy began with Joad and the Brains Trust but, in an effort to find out 'what life is all about', he moved on to more professional accounts of the basic ideas, and eventually to taking out from the library thick books on Greek philosophy. But in the end he became disillusioned with philosophy, as it did not seem to lead anywhere. The task of 'good' philosophers seemed to be merely to refute 'bad' philosophers, and that, he thought, was all there was to the subject. There were no answers.

Physics seemed to be the next best thing. Admittedly, the questions were less interesting than those answered by the philosophers but, unlike philosophy, physics did ask what the world is like, and it did progress and come to conclusions. Each generation built on the work of the previous one.

Slowly, it seemed that physics, and in particular the ideas and theories of physics, were becoming his main intellectual interest. He read popular books on physics—Einstein on relativity [16], and books by such authors as Bertrand Russell, Lancelot Hogben, and J. G. (James) Crowther.

At the Tech, he enjoyed, and found reasonably easy, his physics courses, which contained basic mechanics—Newton's laws of motion; simple statics and dynamics; an introduction to electricity and magnetism; introductory atomic physics; and so on. His mathematics would have got as far as basic calculus—differentiation and integration.

Already, he was thinking deeply and making clear moral decisions about his life. He did not smoke or drink, and, as already mentioned, he became a vegetarian. He also did his very best to put his family off eggs by telling them about the behaviour of hens when their waste was put down. Annie, of course, was far from sure that a vegetarian diet would keep him healthy. Occasionally, she would coax him to eat an egg and, whenever she was making vegetable soup, she would put a little meat in. It is difficult to be a vegetarian when you are living in someone else's house and they do the cooking!

In his decision to become a vegetarian, John was very much influenced by George Bernard Shaw (so it is rather amusing that Bernstein noted his long, neat, red hair and his pointed beard and characterized his appearance as somewhat Shavian). Shaw and Joad had a fair amount in common. In fact, at the beginning of the First World War, they and Bertrand Russell had formed an alliance totally rejecting the war.

To this trio might be added H. G. Wells, completing a group of people for whom John had great respect and whom he wished, at some level, to emulate. Of course, by no means were they a homogeneous group intellectually, Russell being far superior to any of the others, and Joad somewhat inferior. Yet, all were deep and earnest thinkers, and all expected the results of their thoughts to lead to moral conclusions and to definite action in their own lives and in politics. Russell and Wells were extremely concerned with science, although Joad and Shaw less so, and Wells in particular had a vision of science transforming life, particularly perhaps that of the humble, for the better. It is possible to trace many of these characteristics in John's life.

A year of transition

In 1945 Jackie returned from active service to Belfast with his stomach wound. He was awarded a small military pension and he was in and out of hospital for some time. Indeed, he never returned to full health. For a short period, he worked in the stores at the shipyard, where he earned rather better money than he had been used to but, after a couple of months, his health let him down and he had to leave that job.

Fortunately, being a returned soldier, he was able to gain employment with the British Legion, starting as a Park Ranger but swiftly being promoted to being in charge of the British Legion car parks throughout Belfast. Also, in the summer, he was responsible for sending squads of men to Portrush on the North Coast, to service the golf tournaments that took place there.

Even before then, John had to make hard decisions or, to put things more accurately, he had had to wait and see what possibilities the financial circumstances might allow for him. In summer 1944, he passed his matriculation, which was set and awarded by Queen's University, his subjects of choice being English, French, mathematics, mechanical drawing, and physics.

There was no doubt as to what he would have liked to do over the next few years: his aim was to study for a degree in physics at the local university, Queen's University Belfast. However, there were two problems.

Figure 1.16 Portrait of Karl George Emeléus, painted by Tom Carr in 1965; collection of Queen's University Belfast. Courtesy Queen's University Belfast.

The first one was merely a matter of timing. He was only 16 and the minimum age for entering Queen's was 17. The second was more significant—the question of whether he should be getting a permanent job and, if he did get to Queen's, how he might support himself and avoid putting an unfair financial burden on his parents. Of course, his father being on military service may, from this point of view, have been a blessing.

On the actual facts of the next few months, there is no uncertainty. He applied for various jobs in the city—office boy in a small factory, a junior job at the BBC—but he did not get appointed to any of them. At one interview, he was told he was overqualified; at others his body language may have given the message that he did not really want the job.

Then, presumably coming up to the beginning of the university term in September, he was taken on as a laboratory assistant in the very place where he wanted to study the following year—the physics department at Queen's. Annie went with him to the university and met Professor Emeléus, the Head of Department (Figure 1.16), and saw where he would be working. Then, in the *subsequent* summer, he did indeed register to start his studies.

However, there are different stories about how all this was arranged and how the financial problems were solved. At Queen's itself, there is a long-standing belief, or perhaps myth, that his position as a laboratory assistant was intended to be permanent but that the powers that be recognized his great ability, arranged the necessary finance—and the rest is history! More than 50 years later, Annie told rather a different story. For her, the laboratory assistant job was always going to be only for the year he had to wait until he would be 17, and there was never any doubt that he was going to move from the status of a technician to that of a student. Most of the finance, she said, was covered by his earnings during this year, although she was able to help a little.

More recently, John's brother Robert has been able to add further information, resulting from a visit that Dr Sloane, second-in-command to Professor Emeléus, paid to his shop in Bangor 30 years later. Dr Sloane told Robert that it was not the case that there was a job waiting to be filled. Rather, it may be that Annie and John themselves had asked if any work might be available, that Professor Eeméus and Dr Sloane had then asked them to visit the department, and that the professor and Dr Sloane had been so impressed by John that they had *created* a job for him specially for the year.

It is undoubtedly true that Eeméus and Sloane recognized John's great ability and helped him every way they could during his year as a laboratory assistant. He was the attendant for the first-year laboratory, where the equipment and duties were presumably fairly basic—putting out equipment, tidying it away, and so on. Eeméus lent him the well-known introductory text by Millikan, Roller, and Watson called *Mechanics, Molecular Physics, Heat and Sound* [17], and also a book by the physicist, J. J. Thomson, Nobel Prize winner, sometime discoverer of the electron, and Cavendish (Cambridge) Professor [18]. John learned much from the book by Millikan and colleagues, although he found Thomson's book extremely difficult.

Perhaps even more importantly, Eeméus allowed John to attend the first-year lectures in physics and take the exam at the end of the year. This was presumably part of his plan after the original interview. With a little more tweaking, it was to enable John to save a year on his physics course, and this in turn was to be extremely beneficial for him. Certainly, in many ways, this preliminary year at Queen's would turn out to have been exceptionally fruitful.

However, at the time, it did not solve the financial problems in becoming a full-time student. John's equivalent for the third-year laboratory was another young man, Reggie Scott, who, in completely opposite fashion to John, was to stay as a technician at Queen's, always in the physics department except for a brief spell in one of the medical departments, right up to the 1990s, when he retired as Chief Technician. He was still in post in 1988 when John returned to Queen's to accept an honorary degree.

Even after retirement, Reggie remembered [19] that, back in the 1940s, the senior laboratory assistant at the time, a man called Percy Jenkins, had a sister who was a prominent member of the Belfast Co-Operative Society, and

Professor Eméleus was able to arrange for her to obtain a grant for John from the education committee of the society. The grant was for £20, perhaps worth £1,000 today, and was probably the key to making his education possible at all, or at least to giving him the financial independence he would have liked. The grant appeared sometime between July 1945, when John applied to Queen's, and October 1945, when he commenced his studies.

Whatever the precise details, it is absolutely clear that Professor Eméleus and Dr Sloane deserve enormous credit for fostering the young John Bell. In his visit to Robert's shop, Dr Sloane expressed great pleasure in John's successful career up to that point. However, he did also mention other slightly less pleasant matters, which we shall return to shortly.

Queen's: The background

Queen's University [20] had been originally founded in 1845 as Queen's College Belfast, one of the three Queen's Colleges founded in that year, together with those at Galway and Cork, to balance Trinity College, Dublin, which was the only university in Ireland at that time; it had been founded as early as 1591 and it had a very strong Anglican ethos. Nominally, the three Queen's Colleges were not tied to any denomination and, indeed, were criticized as 'the godless colleges', as they were not provided with chapels, but it was accepted that the two colleges in the south of the country would have a mainly Catholic student body, while that at Belfast would be mainly Presbyterian.

Many thought that James Thomson, father of the future Lord Kelvin, would have been an excellent choice for the post of President of the new college at Belfast, but it was preferred to have a Presbyterian minister in that position; so an inoffensive clergyman, Dr Pooley Henry, was appointed. It may have been a big opportunity missed. Thomson was actually offered the post of Vice President at a salary of £500 as compared to £800 for the Presidency, an offer which he rather angrily turned down [11, 20].

Fortunately, an equally good candidate was appointed to the position of Vice President. This was Thomas Andrews [21], who was medically qualified, Professor of Chemistry, and soon to be famous for the so-called Andrews' experiment, which demonstrated the difference between a gas and a vapour, introduced the idea of the critical point, and thus clarified which gases could be liquefied by pressure alone and which could not.

The original idea of the Queen's Colleges was that the colleges themselves would have provided the teaching, with a central Queen's University in Dublin carrying out the examinations. However, this arrangement became extremely unpopular, and several other arrangements that were tried during the nineteenth century were also not remotely satisfactory.

Figure 1.17 Front of Queen's University Belfast, around 1920. Courtesy Queen's University Belfast.

Then in 1908, with the Irish Universities Act, the government put a completely new scheme into effect, with two new universities. The first was in the south of the Ireland; the Queen's Colleges at Cork and Galway were linked with University College Dublin, which had been itself based on the former Catholic University, to form the National University of Ireland. In the north, Queen's College Belfast was given full university status as Queen's University Belfast. This arrangement has worked well ever since and is in existence today, although subsequently many more institutions have been founded [20] (Figure 1.17).

In Belfast, Queen's was very soon taken to the hearts of the local people. The poet Philip Larkin [22] worked in the library at Queen's from 1950 to 1955, after a brief spell at the University of Leicester and before his long period as (Chief) Librarian of the University of Hull. His Belfast days were the happiest and perhaps the most successful creatively of his life. He enjoyed both the city and the university. At Leicester he had felt that the university was regarded 'if at all, as an accidental impertinence', while in Belfast the university had both local character and integration: 'Your doctor, your dentist, your minister, your solicitor would all be Queen's men, and would probably know each other. Queen's stood for something in the city and in the province.... It was accepted for what it was.'

By the early 1940s (Figure 1.18), the number of students in attendance was around 1500, but of these roughly half were studying either medicine or dentistry—Queen's was sometimes described as a medical college with a university attached [23]. Engineering numbers were around 300, although this faculty was based at the Tech until the 1950s. (Both medical and engineering students had to study physics in their first year, so the numbers taking the subject in that year were pushed up enormously.) There were around 250 in arts and half as many in science, with the remainder in law, agriculture, and theology. On the teaching side, there were around 30 professors and around 120 others—readers, lecturers, clinical lecturers, assistants, and demonstrators [24].

Academic standards, perhaps a little mixed in the Queen's College days, rose steadily with university status. In fact, in the years before the war, physics and mathematics were particularly strong, with a great deal of relatively young talent. The Professor of Mathematics at Queen's from 1936 when he was 31, was William McCrea [25] (Figure 1.19). In 1943 he was given leave to work on operational research with Patrick Blackett, the founder of the subject, at the Admiralty. At the end of the war, he went to Royal Holloway College in London, and then in 1966 to the then new University of Sussex; by that time, his interests had moved to astrophysics and cosmology, and he made the Sussex Astronomy Centre, which he led, a world centre of excellence. He was elected a Fellow of the Royal Society in 1952, and became Sir William in 1994.

From 1933, when he was 25, the Australian Harrie Massey [26] was Lecturer in Charge of Mathematical Physics at Queen's. He was a man of enormous energy, initially working mostly on the theory of atomic and ionic collisions but later also being a pioneer of space science. In 1938 he moved to University College London, where he was to become one of the most influential scientists in the country, being elected a Fellow of the Royal Society in 1940 and being knighted in 1960. For several decades, there was a very close academic relationship between the physics department at University College and that at Queen's.

In fact, in the late 1930s, there was an extraordinary flowering of talent in mathematical physics at Queen's, even though, unfortunately, much of it dispersed to carry out war work. Those involved are shown in Figure 1.20, along with Eméleus and Sloane, who were to survive into the Bell era. Samuel Francis (Frank) Boys [27] was yet another who was to be elected a Fellow of the Royal Society, although only just before his death in 1972. He was appointed to Queen's as Assistant Lecturer in Mathematical Physics in 1938 but left for war work on explosives research at the Admiralty. After the war, he was a lecturer at Cambridge from 1949, pioneering the use of electronic computers and the use of Gaussian orbitals in quantum chemistry.

Figure 1.18 When Bell was at Queen's, things were less tranquil than as shown in Figure 1.17; the quadrangle and lawns were used for staff to grow vegetables during the Second World War, and there were several major events supporting the troops. Courtesy Queen's University Belfast.

Figure 1.19 William McCrea. Courtesy Royal Society.

David Bates, James Hamilton, and Richard Buckingham were all protégés of Massey. Bates [28] went with Massey to London, and, after war work, he returned to Queen's in 1951 to set up the renowned School of Theoretical Atomic and Molecular Physics; he was another who both became Fellow of the Royal Society and was knighted.

James Hamilton [29] divided his time in the early 1940s between lecturing at Queen's and as a fellow working with Erwin Schrödinger and Walter Heitler in the Dublin Institute for Advanced Studies. Later, he became one of the leading elementary particle physicists in the world; after spells at the University of Manchester, Cambridge University, and University College London, he was at NORDITA—the Nordic (Scandinavian) Institute for Theoretical Physics—from 1964 to 1986. NORDITA had been founded by Niels Bohr and was an international organization; Hamilton was Director of NORDITA from 1984 to 1985.

Richard Buckingham [30] had an important formula named after him—the Buckingham potential—as early as 1938, when he was 27. He was to become Professor and Director of the Institute of Computing Science in the University of London and then in 1963 the first chair of the Technical Committee for Education of the International Federation for Information Processing.

So it is clear that, although things were naturally quiet in the war period and then the immediate post-war period, when John was at Queen's, the

Figure 1.20 Teaching and research staff of the Department of Physics, Queen's University of Belfast, 1938–9: front row, from left to right: S. F. Boys, J. Wylie, H. S. W. Massey, K. G. Eméléus, R. A. Buckingham, and R. H. Sloane; back row, from left to right: T. McFadden, D. R. Bates, E. B. Cathcart, J. W. Wilson, K. H. Harvey, L. S. Leech, and J. Hamilton. Courtesy Queen's University Belfast.

university had a very strong tradition and, indeed, with the return of Bates in particular, a very strong future in physics and mathematical physics.

Student days

John officially entered Queen's as a student on 4 October 1945 [31]. He would be expected to study for four years, and in the first year he was required to take four subjects. Because he had passed the first-year physics examination while working as a technician, he was able to enter the second-year course, lectures, and practical and, since he had just taken the higher mathematics matriculation examination, he was able to enter the second-year course in that subject as well. However, for chemistry and mathematical physics, he had to register for the first-year courses. In the case of chemistry, this would not matter, as he would be able to drop the subject, but having to take first-year mathematical physics could potentially have limited his final achievement, although fortunately the problem was to be sidestepped.

Figure 1.21 John Stewart Bell; registration at Queen's, October 1945. Courtesy Queen's University Belfast.

Yet, the picture (Figure 1.21) taken on the occasion of his registration brings one up short. Anybody used to the urbane image presented to the outside world later in his life may be surprised to find perhaps insecurity, perhaps even a hint of truculence. At first sight, this attitude may be surprising. After all, he could have had no genuine academic insecurity—as first-year technician, he would have had an excellent opportunity to judge the abilities of the various students and to have realized that he was academically quite able to cope—to say the least.

However, socially, it may have seemed a big step. His background was not well off and, up to that point, all his education had been at institutions, which, whatever their definite merits, would have seemed, in the eyes of the world, not in the top class. Now, he would be mixing with students who, in the main, would have come from far more privileged backgrounds and some of the top schools in Northern Ireland. In some ways, it was the leap of his life. The fact that he would have known the students when he was a laboratory assistant, perhaps socially rather a lowly position, could even have made things worse.

Happily it seems that any such fears were quite unnecessary. Leslie Kerr, who was in the year below him at Queen's and was to return to Queen's as a lecturer in 1957, has provided an extremely useful memoir [32] of John at this time. Leslie suspected that Eméleus had arranged for financial support for John, although perhaps this was merely the Co-operative money already mentioned. He noted that, when in 1984 the department raised money for a portrait of Eméleus and a medal to be presented annually to the best physics student, John gave the largest amount—£300, perhaps worth nearer £1,000 today.

In the 1940s, it took Leslie only a very short time to recognize John's intellectual quality; he was impressed by the simplicity and brilliance of his insight, and the ease in which he was able to reach a solution of the most complicated problem. Leslie was one year behind John at Queen's, but he became a member of a group of four or five students who would meet in the student's union for lunch and might discuss physics and philosophy.

John, of course, was dedicated to physics and thinking about physics, and his brother Robert says that the students he had little time for were the medics and, particularly, the rugby players, who, at least in his view, had no interest at all in their work.

Leslie and John went to 'hops' in the union; also, as part of a group, they walked up the Cavehill and then down via a popular place for dancing at the time, the Floral Hall. Leslie particularly remembered that, on one occasion while on this walk, they came across a series of quite large depressions in the ground and that, in at least one of these depressions, there were several large boulders, mostly at the bottom but some higher up. John announced that the distances of the boulders from the bottom of the depression must be in a Boltzmann distribution (the type of distribution of atoms among their energy levels) and was highly aggrieved when Leslie argued that there was no reason that they should be in such a distribution, even if there were an ensemble of depressions.

They also discussed philosophy. John was at first impressed by logical positivism, which was very much in vogue at the time. Its most famous exponent was A. J. Ayer in his book *Language, Truth and Logic* [33], and it was felt to have solved, or rather to have made irrelevant, all the great philosophical problems of the past. However, John soon came to believe that it must lead to solipsism, and Leslie found John's refutation of logical positivism beautifully simple and convincing. At about this time, John was reading Bertrand Russell's *History of Western Philosophy* [34].

The group also discussed quantum theory. John had already assembled many of his later arguments concerning wave–particle duality, two-slit interference, and the inappropriateness of the concept of measurement having a central role in quantum theory, although he had not yet come across the Einstein–Podolsky–Rosen paper (for all of which, see later in this book). Leslie remarked on John's irritation when the discussion returned to matters on which he had already expounded, as he felt, clearly enough.

Leslie met John regularly until he left England at the end of the 1950s and, after John's death, Mary Bell, John's widow, placed Leslie in a prime position at the closing dinner of the symposium on quantum theory held at CERN [35]. Leslie regarded John as easily the most brilliant physicist he had ever known, of a completely different order even to other highly gifted scientists. The only person who ever reminded him of John, both in apparent personality and in intellect, was Richard Feynman.

It was while he was at Queen's that John first discussed the political situation with Catholic students. Studying at the Tech, he may have had more opportunity to meet with the 'other side' than the very great majority of students in Northern Ireland, who go to school only with their co-religionists; even trainee teachers of different persuasions are not felt to be mature enough be taught together! At the Tech, presumably because it had slipped through the net, all denominations were welcome, but perhaps John was probably not particularly aware of this at the time, and indeed it may be that Catholic students were dissuaded from attending the Tech by their Church.

At Queen's, though, he did have discussions with Catholic students, and he came to realize that there *was* discrimination and that these fellow students *did* have grievances. He certainly thought that these grievances should be removed but, probably like most well-meaning Protestants, judged the level of discrimination to be rather less than that believed to be the case by most equally well-meaning Catholics. Both groups hoped, though, and probably believed, that the situation would gradually improve, and that, as living standards improved, the tribal aspect of Northern Ireland society would diminish. Sadly, of course, this was not to happen.

Let us now turn to John's teachers and his studies while at Queen's. While at this period there were usually one or two temporary assistant lecturers or demonstrators, usually graduates of Queen's, often working for PhDs and who did much of the laboratory work at lower levels, there were two permanent strong personalities in the Department of Experimental Physics: Professor Karl George Emeléus [36], and Dr Robert Harbinson Sloane [37]. Both men were usually known by their second names—George and Harbinson.

Emeléus was of Finnish descent, but he was born in 1901 in London and had been brought up in Sussex. He studied physics at Cambridge University, where he worked with Ernest Rutherford and James Chadwick, taking some of the first pictures of particle tracks and helping to develop the then youthful Geiger counter. He spent a brief period working on plasma physics with Edward Appleton at King's College London, and then, in 1927 and on Appleton's recommendation, he was appointed to a Lectureship in Experimental Physics at Queen's. He was then second-in-command to Professor William Blair Morton, who had been in post since 1897, and, on Morton's retirement in 1933, Emeléus was appointed to be Professor of Experimental Physics.

He was an excellent research worker, publishing over 250 papers in his field of conduction of electricity through ionized gases [plasmas] at low pressures. In 1929 he wrote the definitive monograph on the topic, *The Conduction of Electricity through Gases* [38], which was reprinted in 1936 and 1951; in fact, he was to continue publishing two or three papers a year on the topic right up to his death in 1989.

He was said to have been disappointed that, unlike his brother, Harry Julius Emeléus [39], who was Professor of Inorganic Chemistry as Cambridge University from 1945, he was not elected Fellow of the Royal Society. But the truth is that, to a very great extent, he dedicated his life to undergraduate teaching. He delivered an extremely high proportion of the lectures in the department and he was known for his ability to present the most difficult topic in a helpful and indeed captivating way. As Robin Coulter [36] said in his obituary of Emeléus, the local medical practitioners, all of whom had attended his first-year lectures, preferred to discuss Emeléus' lecturing style to dealing with their patient's aches and pains.

He was a conscientious administrator in the department, the university, and beyond but also, although a rather formal person, he was concerned and helpful with the problems and requirements of his students. We have already seen how he had gone out of his way to help John during his spell as a technician, and this was typical of his approach to John and to others through their courses. As John said, although Emeléus was very polished, he was always ready to answer questions.

Sloane was equally dedicated to the university, the department, and his students. Born in 1911, and a protégé of Emeléus, he carried out research in the same general area—conduction of electricity in gases at low pressures, although Sloane was particularly skilled in constructing the highly sophisticated equipment that was required. In John's day, Sloane's teaching was mostly restricted to laboratory work at the higher levels and also to a course for the final-year students (of which, more shortly). He also did most of the general and financial administration of the department and, when a new building was planned and built in the 1950s and early 1960s, he was to exert an enormous amount of effort persuading the architects of the needs of physicists and checking every detail of the work as it was performed.

Sloane was an excellent physicist, becoming Reader, equivalent to Associate Professor elsewhere; there were only a handful of such positions at Queen's. He also became a Member of the Royal Irish Academy, again a high award for one who was not a professor. Probably, being only around ten years younger than Emeléus, he was a little frustrated in career terms, for he could not become a professor while Emeléus was in post, but would be too old to take over when Emeléus retired.

While John was at Queen's studying physics and mathematics, an extremely interesting figure spent a short period teaching mathematics. This was John

Herivel, a native of Belfast and former student of Methodist College; he had been one of the most important of those working at Bletchley Park [40], and his 'Herivel tip' (or 'Herivelismus'), which took advantage of an imagined laziness of the Enigma operator at the beginning of the day, had allowed those at Bletchley to decode many messages in a period when no other methods were available [41].

At the end of the war, Herivel spent one year teaching mathematics in a local school but was completely unable to maintain discipline; so he took up a lectureship at Queen's, soon moving to the history and philosophy of science, where he was extremely successful [42].

John's studies proceeded exceptionally well. As a result of his excellent performance in his first exams, he was awarded both a Second Foundation Scholarship, which was worth £60 per annum and which he could hold for two or three years, and a Belfast Education Board Scholarship, which was worth £20 per annum and which he was to hold for two years. His total income might be around £4,000 in today's terms. If not actually well off, at least he presumably needed to have fewer qualms about living off his parents.

For his second year at Queen's, in 1946–7, he registered for third-year courses in maths and experimental physics. Also, although he had not taken second-year mathematical physics, he was allowed to proceed to third year in that subject as well. Presumably, the Dean of Science, Professor Paul Ubbelohde [43], himself a very well-known chemist, Fellow of the Royal Society, and later Professor of Thermodynamics at Imperial College London for over 20 years, felt, as Miss Rankin had at Ulsterville Primary School 12 years earlier, that substantial ability would enable a problem such as missing studies to be circumvented. In fact, the only examination John had to sit that year was in mathematical physics, and passing it put him on a pass degree level.

The experimental physics led into the advanced course, which included substantial amounts of practical work, and also many more lectures from Emeléus. Whereas the first two years of the course had taken further topics that the students would have met at school, such as mechanics, heat, light, electricity, and magnetism, the students were now meeting new, theoretically based subjects such as thermodynamics, statistical mechanics, and quantum theory.

It must certainly have been quantum theory that attracted most interest, and perhaps surprise and concern from John. Incidentally, with a little money in his pocket by this stage, he had taken to frequenting the second-hand bookshops in Smithfield. As well as being able to purchase his course books second-hand, he bought a number of popular books on science, particularly quantum theory; but it was to be the forthcoming academic year before some of his frustrations over the theory were to come to the surface.

In this subsequent year, his third at Queen's, John would take a further course in mathematical physics, essential in order that he could go further with the subject in the year after that. However, his main task this year was to complete his studies in experimental physics to honours degree standard, which required, as well as lectures, two days laboratory each week.

However, it was the lectures that led to some dismay and even conflict. Dr Sloane gave a main course, which John described to Bernstein in somewhat disparaging terms. Quantum mechanics is, of course, intimately connected with atomic physics and atomic spectra and, indeed, it may be said that it was developed, to a very large extent, to explain otherwise inexplicable experimental information in these areas.

Yet, of course, once one moves beyond hydrogen in the periodic table, atomic energy levels and hence atomic spectra can be calculated only by lengthy approximate calculations or computations that are quite unsuitable for an undergraduate course; so an easy method of generating a general understanding of atoms has been developed. Essentially it is based on the Bohr atom which was the semi-classical way of handling atomic phenomena and which predated rigorous quantum theory; in this method, each angular momentum—orbital and spin—for each electron is treated as a rather classical vector, and the vectors are added together according to strict rules, to produce the possible states of the particular atom.

As a way of obtaining a good general understanding of atoms, the vector model is certainly useful, although John would have been repelled by its *ad hoc* nature. But his main grouse was that the method was worked through in such detail—as he said, right the way through the periodic table! This approach, he thought, was because of the long-standing interest in the behaviour of atoms, for example in collisions, at Queen's.

Kerr [32] subsequently pointed out that John was a little unfair in his remarks to Bernstein. First, he rather ignored the fact that, by the time of Sloane's course, Eméléus had already taught a course on formal quantum theory. Also, the second half of Sloane's course dealt with radio frequency atomic spectroscopy and accelerators, both topics that had only come into any form of prominence since the war; so, Sloane could fairly be praised for being very much up to date. Leslie would admit, though, that John was not the only one to be bored by so much concentration on atomic spectra!

Early struggles with quantum theory

In a sense, though, John's criticism of Sloane's lectures was only one aspect of a general unhappiness he felt about the way quantum theory was presented. He would have accepted that Eméléus had, as always, given a masterful account of how to work with quantum theory, along the lines of 'here is the Schrödinger equation, put in the hydrogen potential, obtain the energy levels, use the differences of these to produce the spectrum, and so on'. But John was gravely dissatisfied with such an approach. Mathematical manipulations were all very well, but did quantum theory not require a philosophy? What is our actual perception of the physical world? What actually lies behind and justifies the equations?

One particular aspect of this general malaise worried him especially. What everybody knows about quantum theory is what is usually called the 'uncertainty principle'. In most popular accounts, it is described as limiting how accurately we may *measure* or *know* each of position and momentum. The more accurately we know the position, the less accurately we may know the momentum, and vice versa. If we know either one of them exactly, we can know nothing about the other.

John was knowledgeable enough to appreciate that this statement of the principle is, according to any conventional understanding of quantum theory, completely unsatisfactory. It suggests that both the position and the momentum of the particle under consideration do actually exist with total accuracy and that our limitation is in knowing them both or being able to measure them both. However, conventional understanding tells us that, in fact, the momentum and the position do not even *exist* to greater combined precision than allowed by the principle. This is why the principle is much better termed the 'indeterminacy principle', although calling it the 'Heisenberg principle' avoids the requirement for complicated explanations where they are not necessary!

As has been said, John appreciated all that. But his quandary was as follows—in an actual case, what determines how accurately each of momentum and position exist, even before we might make any measurement ourselves? It was certainly not the kind of question Emeléus would have answered in his lectures. As for books, the first on these matters John would have studied was Max Born's *Atomic Physics* [44], a pleasant book with a lot of interesting information but which certainly does not answer tricky questions (Figure 1.22).

Then, in John's final year on the physics course, the important book by Leonard Schiff, *Quantum Mechanics* [45], was published. In his preface, Schiff acknowledges the help of Robert Oppenheimer, who had pioneered the teaching and practice of quantum theory in the United States, and the book was to become probably the standard treatment of the subject throughout the world. Towards the beginning of the book, there are four pages for 'Uncertainty and Complementarity' and 'Discussion of Measurement'. Yet, John was not able to gain very much from these brief pages, for reasons we shall come back to shortly.

John also reported glancing at Paul Dirac's famous book, *The Principles of Quantum Mechanics* [46], which is certainly a masterly exposition of the mathematical and logical structure of quantum theory, but it was of no use to John at this time in his quest for more down-to-earth information.

Figure 1.22 Max Born. Courtesy of Max Planck Institute at Berlin-Dahlem.

In the end, John discussed the matter with Sloane. It must be remembered that Sloane and Emeléus were able physicists, well read across many areas of physics. Yet, of course, they were by no means specialists in quantum theory. Emeléus would have had to pick up the theory while in post at Queen's, and Sloane would initially have been taught it probably by Emeléus while the theory was in its infancy. Certainly, Emeléus would have been quite competent in lecturing on the general theory, and Sloane, for example, could have answered the questions of the usual student with ease.

But John, it goes without saying, was no usual student. To his searching question, Sloane attempted an answer that, in fact, would probably have been the expedient of many lecturers at the time: you can take the indeterminacy of either momentum or position to be what you choose but you are then stuck with the indeterminacy of the other provided by the Heisenberg relation.

John found it totally unbelievable that the state of the system could be determined subjectively by the wish of the person analysing it. Angry with himself for struggling to understand, angry with Sloane for coming out with what he felt was patent nonsense, he effectively accused Sloane of dishonesty; intellectual dishonesty might be a better way of putting it. Sloane, a man of great dignity and moral stature, and one who had been very kind and helpful to John, was scandalized—'You're going too far!' he cried.

In retrospect, Sloane's inability to answer John's question—his inability, as John saw it, even to approach the issue in an intellectually respectable way—must have been a great good fortune to John [47]. Here was an extremely competent and conscientious physicist, who must have claimed to understand and agree with the governing quantum philosophy of Niels Bohr, and yet John felt his grip on the topic was weak in the extreme. Perhaps Sloane had failed to understand fully arguments that many others were quite in command of; but, perhaps, as another famous Dane would have said, the emperor's clothes were not that substantial after all. John would definitely have been encouraged to continue with his own thoughts.

Sloane definitely remained extremely angry. In his meeting with Robert in the Bangor shop that was mentioned before, while full of praise for John's subsequent work, he still 'exploded' with anger over the altercation 30 years before!

Students in other places may also have been dubious of the orthodox approach but were not permitted the intellectual space to challenge accepted ideas. An interesting example is Euan Squires (Figure 1.23). In the 1950s, he was a research student at Manchester University, where Léon Rosenfeld (Figure 1.24) was Professor of Theoretical Physics.

Rosenfeld [48] had worked extremely closely with Bohr (Figure 1.25) in the 1930s and was justifiably regarded as Bohr's main cheerleader. Indeed, with the founding of NORDITA in 1957, Bohr encouraged Rosenfeld to take up a chair there, a suggestion which was taken as an indication that Rosenfeld was Bohr's choice of a successor, at least as far as quantum interpretation was concerned. After Bohr's death, Rosenfeld was often referred to as 'Bohr's representative on earth', and sometimes, perhaps even more riskily, as 'more Catholic than the Pope'.

At Christmas dinners, Squires [49] reported that the students would sing the following song, to the tune of 'The boar's head in hand bear I':

> *At Bohr's feet I lay me down*
> *For I have no theories of my own,*
> *His principles perplex my mind*
> *But he is so very kind.*
> *Correspondence is my cry; I don't know why; I don't know why.*

Figure 1.23 Euan Squires. Courtesy Emilio Segré Visual Archives.

(It should be said that 'correspondence' here is a technical term at the centre of Bohr's ideas; it refers to the relationship between classical and quantum ideas.)

They didn't know why and, as Squires went on to remark, they were afraid to ask. And what is quite clear is that, if they *had* asked, Rosenfeld, or behind him Bohr, would have been able to give an answer of great erudition. It would probably not have helped them actually to understand in any way the Bohr/ Rosenfeld point of view but would certainly have made it quite clear that the fault was definitely their own, rather than anything lacking in the point of view itself. Fortunately for John, Sloane was not able to close down the questioning apprentice so easily.

The paths of John Bell and Euan Squires were to intersect a few times. They were both at Harwell in the later 1950s and wrote joint papers, which will be discussed in Chapter 2. After brief spells at Cambridge, Berkeley, and Edinburgh, Squires became Professor of Applied Mathematics at Durham University in 1964 at the very young age of 31. Despite the name of his chair, he was always at heart a physicist and built up a strong group of particle physicists at Durham.

Then, in 1985, his career took a different direction. He wrote a charming popular account of quantum theory [50] and, in the course of doing so, became aware that many of the more philosophical issues concerning quantum theory

Figure 1.24 Léon Rosenfeld, from N. Blaedel, *Harmony and Unity* (Science Tech Publishers, Madison, WI). Courtesy Science Tech Publishers.

were unresolved and extremely interesting. In a sense, it might be said that he recovered the questioning spirit he had shown at Manchester 30 years before.

He studied these issues for the rest of his life, starting with a book [49] which was very different from his previous one and which examined the conceptual problems at the heart of quantum theory rather than just the issues it handled so well. Later, he wrote a third book suggesting that some of these problems might be usefully addressed by explicitly introducing human consciousness into the measurement process [51].

Squires had a pleasant personality and, from the mid-1990s, was the leading spirit behind a series of annual UK conferences addressing the issues initially championed by John, bringing in a number of people who went on to work actively in the area. Sadly, though, he was to die young in 1996.

Early views on quantum theory

Sloane had not been able to answer John's question, and it took John quite a struggle to realize what the answer was. There are in fact two approaches, often thought of as 'orthodox', to the conceptual problems of quantum theory. The

Figure 1.25 Niels Bohr. Courtesy MRC Laboratory of Medical Biology, Cambridge.

first, constructed in 1928, within a couple of years of Werner Heisenberg and Erwin Schrödinger producing in 1925–6 the actual mathematical structure of quantum theory, was the brainchild principally of Niels Bohr.

This approach uses what Bohr calls the 'framework' of 'complementarity' and is often called the 'Copenhagen interpretation' of quantum theory since Bohr was based in Copenhagen, although it is not given this name by its strongest advocates, for whom this set of ideas is not just one among several interpretations but an essential part of quantum theory itself [52, 53]. The ideas are subtle, elusive, and perhaps even obscure, so it is not surprising that, for many decades after 1928, most physicists, while claiming to agree with Bohr's position, in fact would have made little if any attempt actually to understand it.

We shall come back to Bohr's arguments later in this section and, in fact, throughout the book, but here we make a general comment. Bohr was definitely a great physicist [54]. His 1913 work on the 'Bohr atom' was an enormously important stepping stone to the full quantum theory produced in 1925–6, even though, as a stepping stone, its details had to be largely discarded when the full theory emerged. In the 1920s, he was central in understanding the atomic basis of the periodic table, and he was the undisputed leader of those coming to grips with the new quantum theory. Later, he was able to elucidate the liquid drop model of the nucleus, and the mechanism of nuclear fission.

John would have acknowledged all these facts, but to Bernstein he remarked that he liked to speak of two Bohrs—one, a pragmatic fellow, the other, a 'very arrogant, pontificating man' who was convinced that he had solved the problems of quantum theory and, in so doing, had illuminated not just atomic physics but epistemology, philosophy, and indeed humanity in general.

In fact, this second Bohr, John felt, was rather patronizing towards the ancient philosophers of the Far East, claiming to have solved the problems that had defeated them. In particular, John could not agree that Bohr's approach to the interpretation of quantum theory was at all successful. Thus, in this book, for all Bohr's undoubted merits, not very much good will be said about him!

However, for the moment, we turn to the other 'orthodox' approach to quantum ideas current in the late 1940s. It is usually referred to as the approach of John von Neumann (Figure 1.26), which is a little misleading (and unfair to von Neumann if you feel the approach is unhelpful), as it was quite common in the early days of quantum theory. In a famous book [55] of 1932, written in German and not translated into English until 1955 (a fact that was to be extremely frustrating to John, as we shall see in Chapter 2), von Neumann spelled out the ideas and actually pointed out the difficulties they inevitably caused.

While the arguments of Bohr were subtle and mainly verbal, those of von Neumann were much more direct and mathematical than Bohr's and, consequently, much easier to understand and criticize. Opinions differ on whether von Neumann has merely made more explicit what was elusive in the work of Bohr or produced a set of ideas of his own. They would also differ on whether von Neumann's directness merely vulgarized Bohr's elusiveness or whether it showed up a certain evasiveness, previously hidden by lofty language, on Bohr's part.

Here, we sketch the ideas in simple terms. We first introduce the idea of a wave function, a mathematical function that, at least according to orthodox views, tells us all that we *can* know about the system, although not necessarily all that we might like to know.

For example, suppose that we wish to perform a measurement of the energy of a particular system. In the general case, the wave function of the system prior to the measurement cannot tell us what result we may obtain in the measurement. It can only give us the probabilities of obtaining each of a number of different results. In fact, we can break down this initial wave

Figure 1.26 John von Neumann. Courtesy of Marina Whitman, from I. Hargitai, The Martians of Science, OUP 2006.

function into the sum of a number of different functions, *each* of which *will* lead to a *specific* result in the measurement. The amount of any particular one of these special functions in the wave function at a particular time will tell us the probability of getting the corresponding result in a measurement made at that time.

So far, this is just quantum theory, but we now come to the new ingredient particular to this interpretation. At a measurement we must, of course, obtain a particular result. The new point is that, at the measurement, the wave function of the system *collapses* to the particular function that would have been *certain* to lead to this result. The implication, of course, is that a subsequent immediate measurement of the same quantity must lead to the *same* result, and this, of course, is the point of the exercise. The idea of this *collapse of wave function* or, as it is often called, the *projection postulate*, is the central point of the von Neumann scheme.

As has been said, von Neumann did point out the difficulties of the scheme. In the absence of measurement, the wave function changes in a smooth and regular way, following, in fact, the fundamental equation of quantum theory—the Schrödinger equation; von Neumann called this *Type 2 behaviour.* Yet, at a measurement, it behaves in a completely different manner; it collapses in a totally abrupt way, which von Neumann called *Type 1 behaviour*. There seemed to be no explanation of this dramatic change of behaviour.

We should mention two other conceptual problems of quantum theory. The first is exemplified if we ask what the value of the energy was before we perform the measurement. All we know is that the wave function tells us there was, in a sense, a potential of obtaining a number of answers; we must also remember that, according to an orthodox approach, we can obtain no information that is not in the wave function, so we have to admit that, prior to a measurement, the energy just does not have a value. Physicists usually call this phenomenon a loss of *realism*. Classical or pre-quantum physics takes for granted that any physical quantity has a definite value at all times, but it seems that this is not the case in quantum theory.

The second casualty of orthodox quantum theory is *determinism*. Determinism tells us that, if we know the state of a system, or indeed that of the full universe, at a particular time, we can calculate its state at any later time. It should actually be said that the fact that we may calculate it is not really the point; the fact is that the future is determined, whether anybody knows it or not. In the absence of measurement, the behaviour of the quantum system under the Schrödinger equation is totally deterministic but, at a measurement, the collapse postulate tells us that the wave function changes abruptly to a state determined only by which measurement value has been obtained. Clearly, the collapse process must not be deterministic.

Now, let us turn back to John Bell. The von Neumann scheme would actually have given him the answer to his initial problem easily. Clearly, what determines the state of the system, that is, for a free particle, the accuracy with which each of position and momentum are known, is the wave function that the system has collapsed to after the previous measurement. John's difficulty in realizing this fact was that the von Neumann approach was not discussed in most books—and certainly not in the book by Schiff [45], which concentrated on Bohr's approach. It is not clear whether John eventually came across the von Neumann approach in some book or paper, or whether he reached his answer through his own deep thoughts.

But the fact was that, even when he realized what this approach to quantum theory told him, he found it grossly unsatisfactory. John disliked the formal role played by the concept of 'measurement' in quantum theory.

He recognized of course that, in science in general, the action of measurement was central. To find out about a particular scientific quantity we must measure it. But that was exactly John's argument. When we measure a quantity in the general scientific context, we are endeavouring to obtain a value that exists before the measurement. The subject of science, he felt, should not be restricted to the results of measurements; it should study what exists even in the absence of measurement.

This is just another way of stating that John was, in the terms discussed earlier, a realist; he believed in a real world with its own properties existing entirely independently of whether we might measure them. He announced himself to be very much a 'follower of Einstein' [56].

It is well known that Einstein (Figure 1.27) had great reservations about quantum theory. In fact, contrary to what is often believed, he accepted the *formalism* of the theory totally, and his reservations concerned the interpretation [57]. Again, it is often taken for granted that Einstein's main objection concerned the lack of determinism in the theory—his saying 'God does not play dice' is famous. In fact, though, while that was his principal objection initially, in his considered response over time lack of realism was more important than lack of determinism, although the latter was still significant.

Thus John was indeed a follower of Einstein in their joint determination to retain realism, although they differed on determinism—John worried about that hardly at all. Another demand, made by both Einstein and John, was for an 'observer-free realm', that is, that physics should exist without observers, without the necessity for measurements [57].

We have seen that von Neumann himself pointed out the awkwardness of the fact that the behaviour of a system during a measurement (Type 1

Figure 1.27 Albert Einstein in 1921. Courtesy of Hebrew University of Jerusalem.

behaviour) was completely different from that when there was no measurement (Type 2 behaviour). But John took the argument further, to where it could be said to be inconsistent rather than just awkward. He pointed out that the term 'measurement' was not actually a *primary term*. Although a particular series of operations might have a specific purpose in the mind of the experimenter—to find out a fact about the universe—in practice, the 'measurement' could only consist of a number of Type 2 processes; so, how could the behaviour of the system be totally different between Type 1 and Type 2?

Another question about the von Neumann argument is as follows. The term 'measurement' is usually used to mean an experiment carried out by an observer. If so, this definition could be taken to imply that no measurements could be made, and in a sense physics could not start, until human beings developed. And maybe, since the term usually suggests an observation with some skill, perhaps a good scientific training is required—does the experimenter require a PhD to get a 'licence' to collapse wave functions, or will a BSc suffice? While these questions may seem somewhat jocular, it is actually difficult to avoid them in a von Neumann scheme.

However, John found no more satisfaction in Bohr's approach to these matters. Schiff [45] provided rather a brief account of Bohr's approach, concentrating on measurement rather than the state of affairs between measurements. While John at least appreciated the straightforward nature of von Neumann's views, he found Bohr's rather annoyingly vague and, indeed, felt that, for Bohr, lack of precision seemed to be a virtue.

Just as for von Neumann, measurement was central for Bohr. As we have seen, one of the set-piece conceptual problems of quantum theory was why one could not allocate values simultaneously to the position and the momentum of a system. Bohr explained this in terms of *complementarity*. This approach forbade discussion of the momentum of a particle except in the presence of an apparatus set up to measure it. Similarly, position could not be discussed except in the presence of a measurement of this quantity.

But, as Bohr pointed out, it is impossible to set up an apparatus to measure both quantities simultaneously; so, the two quantities cannot be discussed simultaneously, and the problem is solved! John was not attracted to this argument, which he felt was evasive rather than perceptive. Neither did he appreciate the fact that Bohr's description of measurement was almost entirely verbal.

However, there was one aspect of Bohr's argument that he very much agreed with. For Bohr, the system itself was definitely quantum mechanical. It would be described using a wave function. But he insisted that the measurement itself and the measuring system must be treated classically. By this, he meant that the description of the basic experimental results must be in terms of classical concepts, such as positions of pointers on dials, or marks on photographic plates.

John very much approved of this argument but very much disapproved of Bohr's discussion of the division between the measuring system, to be treated classically, and the measured system, to be treated by quantum theory. This division is often known as the *Heisenberg cut*, and the rather peculiar feature of it, from John's point of view, was that it was not possible to define its location unambiguously. He often called it the *shifty split*.

From Bohr's point of view, this impossibility was due to the difference between the way in which energy and momentum would be exchanged between observing and observed systems on the classical and the quantum cases. Classically, the exchange could be performed and monitored continuously because infinitesimal amounts of energy and momentum could be exchanged. However, in the quantum case, because of the non-zero nature of Planck's constant, the exchange of energy and momentum is discontinuous and irregular. Bohr therefore deduces that there is no possibility of an unambiguous statement of the position of the Heisenberg cut.

(Planck's constant, by the way, is the constant that was discovered by Max Planck (Figure 1.28) in 1900 at the very beginning of the development of quantum theory. Classically, its value would have been zero; the fact that Planck discovered it had to be taken as non-zero was a very great shock!)

Bohr's conclusion implies that, despite the fact that, for him, the observing system must be discussed classically while the observed system must be discussed quantum mechanically, there is no strict division between them—they are really a single system. Bohr tried to make this seem reasonable by thinking of a person investigating some area of ground with a walking stick. If the stick is held tightly, it is natural to think of it as part of the observer. However, if it is held loosely, it is more natural to think of the observer ending at the hand, and the stick as being part of the observed system.

It is fair to say that John was totally unimpressed by what he called these 'parables' of Bohr. He wanted not verbal dexterity but hard facts and mathematics—how large does a measuring object have to be to act classically?

In any case, his thoughts about the situation were entirely different from those of Bohr. If the measuring apparatus has to be classical (on which point they agreed), and if there is no satisfactory way of deciding where the Heisenberg cut lies, then it seemed clear to John that the observed system must be broadly of the same nature as the observing system.

Of course, this does not mean the observed system will obey classical laws. Indeed, we know that it obeys quantum laws. The usual way of thinking about this point is that quantum laws apply in a very straightforward way to small systems, such as electrons or atoms; when we apply them to large systems, the same quantum laws should be equivalent to the classical laws we are familiar with. This is called the *classical limit* or the *macroscopic limit* of quantum theory, with the word 'macroscopic' applying to large systems, just as 'microscopic'

Figure 1.28 Max Planck in 1906. Courtesy of Max Planck Institute at Berlin-Dahlem.

applies to small systems. There has been considerable success in establishing this limiting argument, although there are still difficulties with it [57].

What John's argument *did* mean is that quantum and classical laws should be of the same general nature, or in other words, for quantum systems, just as for classical systems, all properties of the various particles should have definite values at all times. This is just the idea of realism as we already defined it.

At first, this idea may seem to contradict what we have already learned about quantum systems. We know that the state of a quantum system is represented by a wave function and that this wave function tells us a certain amount about the properties of the system but cannot give us precise values of all these properties. We have also learned that no information exists about the system other than that given by the wave function, but the crucial point is that this extra point is *not* a part of quantum theory itself; it is only a part of the orthodox Copenhagen interpretation of the theory.

If we do not restrict ourselves to this interpretation, we are quite entitled to hypothesize the existence of extra information about the system, over and above that which may be obtained from the wave function. This information will be in the form of what are known as *hidden variables*.

For example, the wave function of an electron may give us the probability of its being found in any particular position if a measurement is made at a particular time. However, the hidden variable might tell us the actual value of the position before any measurement is made. It will be clear that this approach solves the problem of the lack of realism—the electron has a position at all times. It also solves the problem of the lack of determinism, since the value of position exists before the measurement; it does not emerge at random from several possibilities in a non-deterministic way at the measurement.

It will be freely admitted that we are here looking at a particularly simple situation, where the use of a hidden variable solves problems in rather a trivial way. In complicated systems, any use of hidden variables is, at the very least, rather more difficult than in a simple system. Indeed, much of the significance of John's work on quantum theory is in elucidating when hidden variables might be used and the problems that they themselves might introduce.

But one may still ask—why was the very possibility of hidden variables ruled out by Bohr and the 'founding fathers' of quantum theory? (John often used the term 'founding fathers' to refer to Bohr, Heisenberg (Figure 1.29), and the third member of the group that established the Copenhagen interpretation, Wolfgang Pauli (Figure 1.30). It was a term used to imply just a minute amount of menace on the part of the fathers themselves, rather as the term 'godfather' is used today.)

From the history of science, there is an episode suggesting the foolishness of such a stance. In the nineteenth century, the properties of heat, work, and energy were investigated by two rather distinct routes. The first [58], developed by Rudolf Clausius, William Thomson (Lord Kelvin), and MacQuorn Rankine, started from very general principles, which were generalizations of experience known as the laws of thermodynamics. The chief ones were that heat always flows from a hot object to a cool one, and that it is impossible to transform all the heat from a source of fuel into useful work—some must be deposited as heat at a reduced temperature. From these precepts, the whole theory of classical thermodynamics was developed.

The second route, used in particular by James Clerk Maxwell [59] and Ludwig Boltzmann [60], uses specific atomic and molecular models for the systems involved, and essentially produces the same results. This theory is known as statistical mechanics.

Figure 1.29 Werner Heisenberg. Courtesy of Max Planck Institute at Berlin-Dahlem.

The two theories actually complemented each other nicely. Classical thermodynamics was independent of any, perhaps transitory, atomic view-point, while statistical mechanics gave an explanation at the atomic level. Unfortunately, though, positivists in particular Ernst Mach, who wished to banish from science any ideas not directly available to the senses, were highly critical of atomic ideas and, after Maxwell's death in 1879, this criticism was felt mostly by Boltzmann [60].

Boltzmann was to commit suicide in 1906 [60]. It is certainly not the case that the criticism of his physics was directly responsible for his taking his own life; he was ill, he was suffering from heart and eyesight problems as well as asthma, and he had manic-depressive tendencies and family problems. Nevertheless, it

Figure 1.30 Wolfgang Pauli. Courtesy of CERN Pauli Archive.

is still possible to feel that he was severely affected by the criticism. Ironically, in a few years after 1906, the whole field of atomic physics opened up experimentally and it was clear that Boltzmann had been broadly correct.

There seemed to be a good analogy between the atomic hypothesis in the nineteenth century and that of hidden variables in the twentieth.

It is interesting to surmise why the Copenhagen philosophy was so opposed to any idea of hidden variables, and in that context it is useful to go back to the early days of radioactivity [61]. Radioactivity was discovered in 1898 and, for the first five years of its study, the main question was whether it was caused by some external agency or whether it was an internal process in the atom itself.

By 1903 the external possibility had been ruled out, and research turned to relating the process of radioactivity to atomic models and, in particular, to the then prominent Thomson 'plum pudding' or 'current bun' model, in which electrons were embedded in the atom's positive charge, which was thought to be distributed throughout the atom. However, all this work proved abortive and, from about 1910, it was largely abandoned. It was assumed that no atomic model that explained radioactivity was either possible or required and that the radioactive emissions occurred spontaneously and probabilistically. It could be said that, between 1903 and 1910, there was a search for some kind of hidden variable that would dictate when a radioactive emission would occur; however, in 1910, this search was given up.

Helge Kragh referred to this failed search in the title of his paper, 'The origin of radioactivity: From solvable problem to unsolved non-problem'. In other words, a quest for a hidden variable explanation became a 'non-problem'. Kragh noted that it had been confirmed as a non-problem 'because quantum mechanics says so', but this phrase really meant 'because the Copenhagen interpretation says so'.

When Niels Bohr produced his great atomic model in 1913, he introduced atomic transitions between different energy levels, and it was natural, perhaps automatic, to take over the approach from radioactivity—no hidden variables; in turn, this motivated the Copenhagen interpretation of modern quantum theory.

Another argument concerns the so-called Forman thesis [62]: an American historian of science, Paul Forman, has argued that the physicists who produced quantum theory and the Copenhagen interpretation were reacting to contemporary intellectual trends in the German Weimar Republic in the early 1920s. There was a hostile atmosphere towards determinism, realism, and materialism, and Forman contends that this environment contributed to the emergence of the Copenhagen interpretation of quantum theory in particular. Of course, it diametrically opposed any consideration of hidden variables.

While the Forman thesis itself has been extremely widely discussed by historians of science, it is quite clear that, well before the 1940s, the Copenhagen interpretation was effectively sacrosanct. It would seem to be impossible for anybody—John Bell or anybody else—to challenge it without being regarded as a crank and disregarded by the scientific community.

Last year at Queen's, and Peter Paul Ewald

At the end of his third year as a student, John graduated on 9 July 1948 with first-class honours in experimental physics. Actually, of the seven students who graduated in experimental or mathematical physics, six students were awarded firsts.

Because he had saved a year on this course, and this in turn was because Professor Eméleus had allowed him to take the first-year examinations while he was working as a technician, he was entitled to stay at Queen's for another year, and he also kept the Foundation Scholarship for this year although not the award from the Belfast Education Board, which was the smaller of the two. Since he had already passed so many courses in mathematical physics, he was able to complete an honours degree in that subject in this additional year. In fact, he merely had to take one further course and undertake a project.

During this year, he had the great fortune to work closely with the head of the Mathematical Physics department, Peter Paul Ewald [63, 64] (Figure 1.31),

Figure 1.31 Peter Paul Ewald. Courtesy Royal Society.

who had been an international leader of physics research but who, as John said to Bernstein, had been 'washed up on the shores of Ireland' as a result of the political catastrophes of the 1930s. John added that one would not have expected to have a man of Ewald's calibre in Belfast. Ewald had been involved in the very foundations of the technique of X-ray crystallography; indeed, as we shall see, it may be said that he was involved even *before* the foundations had been laid.

Ewald was born in Berlin in 1888 and spent his first university year in Cambridge from 1905 to 1906. (His family was extremely Anglophile and, when he applied for the position in Belfast, one of his referees, William Henry Bragg, mentioned that he was the best speaker of English and the best suited to English conditions of any of the German refugees seeking a position at the time.)

Ewald studied from 1906 to 1907 in Göttingen, where he enjoyed the course of the great mathematician David Hilbert, but he then moved to Munich, where he equally enjoyed the hydrodynamics course of the very famous physicist Arnold Sommerfeld. From then on, he regarded himself as a theoretical physicist and, in 1910, he asked Sommerfeld for a thesis problem.

Sommerfeld produced a substantial list, out of which Ewald picked one with the title 'Double refraction of light produced by the anisotropic lattice arrangement of atoms in a crystal', as it chimed in with a long-term interest he had in using light to determine the ultimate atomic structure of matter. At this time, the idea that crystals are regular geometric arrangements of atoms was only a hypothesis.

His thesis would turn out to be the first step in the dynamic theory of the behaviour of electromagnetic radiation in crystals. The wavelength of the light he was considering was, of course, far greater than the interatomic distance in the crystal; but, in late 1911 or early 1912, Ewald had an interesting conversation with Max von Laue, who was working at Munich at the time. It was more interesting, it should be said, to von Laue than to Ewald, who had hoped to get advice on the thesis but did not get any. Rather, von Laue seemed astounded that the crystal would behave as a regular lattice, and he interrogated Ewald on the dimension of the lattice and the behaviour expected if electromagnetic radiation of a very short wavelength were to be used.

By Easter 1912, von Laue, together with Paul Knipping and Walter Friedrich, had used X-rays, which are a type of electromagnetic radiation that has the same nature as light but with wavelengths comparable with the interatomic spacing in crystals, to demonstrate X-ray diffraction by crystals or, in the long term, the study of crystal structure by X-ray crystallography, which grew to be an enormous field of great importance over the next century. Von Laue was to be awarded the Nobel Prize for Physics in 1914 [65], but he clearly owed much to Ewald.

Ewald then devoted the whole of his scientific life to both theoretical and experimental work in this area of physics. In fact, within a few hours of hearing of von Laue's work, he had produced much of the mathematical framework of X-ray crystallography—the reciprocal lattice, basic in any discussion of crystals; the 'Ewald sphere' (or sphere of reflection, as Ewald called it); and the 'Ewald method' for summing electrostatic energies in crystals [66].

Ewald appreciated very much the structure analyses of William Henry Bragg and his son William Lawrence Bragg, for which they were jointly awarded the 1915 Nobel Prize in Physics, and he was friendly with them for the rest of their lives; but he felt that the analysis of von Laue and the Braggs was simplistic, since theirs was a 'kinetic theory' in which they merely traced rays through the crystal, allowing scattering at each atom. Ewald constructed a much more satisfactory 'dynamical theory' in which the interactions between the fields at the various atoms were included and, soon after the First World War, deviations from Bragg's law in favour of Ewald's calculations were discovered.

Ewald had short spells as an assistant to Hilbert and then to Sommerfeld; then, from 1921, he was a professor at the Technical University of Stuttgart, where he rapidly rose through the ranks.

His best-known assistant was Hans Bethe [67], who was to be awarded the Nobel Prize in Physics in 1967 for his work on solar energy. Bethe's graduate work had been under Sommerfeld; it had commenced in 1926, only a few years after the discovery of the wave nature of the electron, and of electron diffraction by crystals, and Sommerfeld suggested that Bethe should effectively repeat for electron diffraction what Ewald had carried out for X-ray diffraction a decade before. Bethe did so and, in this case, it was absolutely essential to use Ewald's dynamic theory rather than the kinetic theory. This is because the interactions are much stronger for electron diffraction than for X-ray diffraction, essentially because, in electron diffraction, electrons are not only the scattered particle but also a part of the lattice.

When Bethe went to Stuttgart in 1929, he was treated as one of the family, and Ewald's daughter Rose, 11 years his junior, must have been especially attracted to him. Bethe later returned to Munich and was eventually forced out by the Nazis, but when he obtained a visa to emigrate to the United States, he visited the Ewalds, and Rose, then 17, asked him to take her with him. He assumed she was joking and must have been amazed when, two years later, she turned up at a lecture he was giving at Duke University. They were to be married in 1939. Rose became housing officer for the Manhattan Project; Hans and Rose became a particularly well-known couple in the American physics community and, in 1949, Rose was to become involved in Hans's crucial decision on whether to build the hydrogen bomb.

In the 1920s, much of Ewald's own time was spent on building up the subject of X-ray crystallography. In 1923 he wrote the seminal work *Crystals and X-Rays*, which was updated in volumes of the *Handbuch der Physik* in 1927 and 1933, and he was a highly active editor of the new journal *Zeitschrift für Kristallographie* from 1924 until 1939, when the journal lapsed. Associated with the *Zeitschrift* was the *Strukturbericht*, or *Structure Reports*, for which Ewald was one of two editors and which ran to 800 pages in 1931. He was also the main instigator of the decision to prepare *International Tables for the Determination of Crystal Structure*, which was published in 1935.

In 1932 he was voted Rector of the Technical University and should have been in post for one year. However, it was during this period that Hitler took over in Germany, and all Jewish professors were dismissed. Ewald was a quarter Jewish, which was counted by the regime as Jewish, but his service in the First World War, when he had been unfit to join the army because of a heart condition but had served on the Eastern front as an X-ray technician, allowed him to stay in post. However, since his wife was a full Jew and because he himself had liberal opinions and definitely did not wish to fight the British, it was inevitable that there would in time be a parting of the ways with his university.

In April 1933 he resigned as Rector but, even though the student body was Nazi, his popularity with the students was such that they expressed their regret, saying that they appreciated his fairness and popularity, even though they could not share his views. Over the next few years, he was allowed to teach, although not to examine. He travelled to the United States and taught summer schools there, principally in hope of being offered a reasonable position, but none with sufficient salary was forthcoming. He was probably too old to walk into a junior job and, of course, there was massive competition for senior positions from others in the same position as himself.

Eventually, in December 1936, he was told in Stuttgart that his 'objectivity was no longer valid or acceptable' and that he had to stop teaching. Still, everything was surprisingly cordial and he was awarded a pension, which would come in extremely handy over the next few years.

Lawrence Bragg was able to obtain a temporary research grant for him in Cambridge—£400 a year, or perhaps £20,000 today, and he moved there in October 1937, with his family, including his mother Clara, who was a famous painter, following on in April 1938.

However, obtaining a permanent job was still essential, and the one that he was offered and which he accepted was at Queen's. He was replacing Massey, and the job was advertised as Lecturer in Mathematical Physics. At his interview, he was told, though, that, since he was such a distinguished scientist, he would definitely be promoted to Professor as soon as possible, but unfortunately there soon came out an edict that there would be no promotions for the duration of the war. (The salary of a lecturer was £300, while that of a professor would have been something like double that.) He was appointed from April 1939 and his family followed in August. He was eventually to be promoted in 1945.

In his final year at Queen's, John interacted with Ewald quite closely, both in the final-year course and, probably more particularly, on the final-year project. The project was an extensive study of the quantum mechanics of long-chain molecules, and Ewald was highly impressed. Kerr [32] remembered that there was talk of this work being published, either under John's name only, or as a joint publication under both names, but it seems that nothing came of this.

John found that Ewald would talk to him about anything and was very informal in his manner. In this way, Ewald was very different from Emeléus and Sloane, who were conscientious and helpful but very dignified and perhaps a little aloof; in fact, Ewald even told Bell that one of his assistants was mad [5]. In Ewald's obituary [65], it is said that his first assistant was James Hamilton [29], an excellent scientist, as already discussed, and certainly not mad. His last assistant was José Moyal [68], who later did useful work on the foundations of quantum theory and also carried out important work in Canberra—again, certainly not mad. In between, though, there was an unnamed assistant who refused to climb any stairs—this could perhaps be the mad one.

Ewald would have enjoyed talking to John, because his conversations on physics in Belfast must have been limited. Unlike Massey, he had no scientific interest in common with Emeléus, there was no one with a common interest in crystallography, and quite a lot of his teaching was to large classes of engineers. The latter did prove a boon when, around 1940, there was serious talk of his being interned; teaching physicists or mathematicians would not have prevented this happening, but teaching engineers was obviously essential for the war effort, and he was spared. His youngest son, Arnold, was interned, however, and was sent to Australia, where he was to make his life after the war.

One pleasant feature of the Ewald's time in Ireland was that they were able to interact with Erwin and Anny Schrödinger [69]. The Schrödingers had spent a brief spell in Stuttgart in the 1920s and the two couples had become very friendly, so when Schrödinger came to Dublin in 1939 as founding Director of the School of Theoretical Physics at the Dublin Institute for Advanced Studies, it was natural that the couples should meet frequently. However, there was also occasional awkwardness. Throughout his marriage, and perhaps particularly in Dublin, Schrödinger had many affairs and, perhaps in compensation, Anny became rather too fond of Ewald, who had to show considerable tact to handle the situation pleasantly.

At the end of his final year at Queen's, John was to receive another first-class honours degree; it would also be Ewald's last year in Belfast. After the war, he had received several good job offers and, although he and his family were happy in Belfast, he recognized that, if he stayed, he would have to retire in 1953. His mother had not wanted to leave Belfast, but she died in 1948; incidentally, there is a plaque on the house the Ewalds occupied in Belfast, but it is for Clara, rather than for Peter Paul. He then felt free to move and did so in the following year to become Chairman of the Physics Department at the Brooklyn Polytechnic University (now New York Polytechnic University).

Ewald was then able to continue his pre-war activities. As early as 1944, he had been able to travel to London to suggest the formation of an International Union of Crystallography and, after the war, he became a central figure in the new organization; he also took up major roles in new journals in that scientific area. He had become a British citizen in 1945 and so was eligible to become Fellow of the Royal Society in 1958, on which occasion his proposers were a list of the great and good of crystallography [70]. He continued at the Polytechnic until 1959, working on at the academic love of his life, X-ray crystallography, until his death in 1985 at the age of 97.

Like Ewald, John had to make decisions about his future in 1949. Deep in his heart, he would have loved to have gone on thinking about quantum theory, possibly removing some of what he saw as its defects by introducing hidden variables. But he then read Max Born's popular science book *Natural Philosophy of Cause and Chance* [71], which was published in 1949. This book is a readable and descriptive account of some of the issues around quantum theory. It discusses the difficulties in constructing hidden variable theories that could duplicate all the predictions of quantum theory; however, it then says that there is no need to discuss the difficulties in a general way, because von Neumann had proved that it was *impossible* to construct a hidden variable account of quantum theory.

Von Neumann's argument was in the 1932 book [55] mentioned in the section 'Early views on quantum theory'. By that time, it had still not been translated from German, a language John did not know, so it must have been frustrating for him not to be able to examine the argument. In any case, it seems that he saw no reason to challenge it. Von Neumann was exceptionally accomplished, but a more experienced person might have recognized that he was a mathematician, not a physicist. It was extremely unlikely that there was a mathematical mistake in his argument, but maybe not quite so unlikely that the physical assumptions might be challenged. The latter would turn out to be the case, but not for more than a decade.

John recognized that 'it was a big risk that I would get hung up on these questions once I learned about them', so he decided to 'walk away' from consideration of the foundations of quantum theory. This decision must have meant more than just not trying to build a career around it, as it must have been obvious that doing so would have been completely impossible. Even had he obtained a PhD and an academic post, and thus the freedom to carry out whatever research he wished, he would have been aware that the Copenhagen interpretation held such a stranglehold on beliefs about quantum theory that any attempt to challenge it would have immediately labelled him as a crank. 'Walking away' must have meant, at least for the time being, not even worrying about the ideas in his spare time, but concentrating fully on tasks that were obviously more fruitful than challenging the Copenhagen interpretation. He must have revised this opinion, at least to the extent of thinking about the issues, in the next two or three years.

He had now, of course, to make a large decision. What he would have loved to have done would have been to have studied for a PhD. One possibility would have been to do so at University College London, which, as has been said, had strong links with physics at Queen's. Ewald's suggestion, though, was that John should study at Birmingham University under Rudolf Peierls, and he offered to put John in touch with Peierls. At the time, Birmingham under Peierls was probably the leading department in England for theoretical physics.

Yet, John had rather a guilty conscience about financial relationships with his family. It is true that his scholarship income of £80 in the middle two years of his course would presumably have enabled him to make a reasonable contribution at home. It is also true that, had he started a PhD in England, he would certainly have been awarded a grant and would have been self-supporting. Probably though, he wished to be in a position of genuinely making a contribution at home, for which he would need a job, and thus the facility, in a typically Irish pattern, to send a little money home. He was always very generous to his parents; much later he was able to buy them outright the Tate's Avenue house.

Without a PhD, he wondered whether there would ever be a possibility of making his way in scientific research. He was comforted by Ewald's reply [32] that it did not matter where he started 'as long as he had elbows'.

There remained the question of actually getting a job. There were times that his mother Annie had actually been somewhat concerned about his appearance. Obsessed with his studies, he may have been careless about how he presented himself and, at one time, he had long hair and a beard. Occasionally, children shouted at him in the street, a situation which Annie would definitely not have liked! However, he must have spruced himself up for whatever interviews came along as, when his wife-to-be met him in his new job, he was clean-shaven.

In job-hunting, he would have received extremely strong support from both Eméleus and Ewald, and he obtained a very good job as a Scientific Officer in the British Scientific Civil Service at the Atomic Energy Research Establishment at Harwell, not too far from Oxford in England. He was never to live in Ireland again.

Ruby was extremely concerned when he left for England. He had never lived away from home—'Who would look after him?' But she felt that he never looked back—he would be doing what he loved.

Note added in proof: Bell's Theorem Crescent and John Bell House

'The Tech', more recently Belfast Metropolitan College, moved to the so-called Titanic Quarter in September 2011. The road round the College has been named 'Bell's Theorem Crescent'. (In Belfast it is not permissible to name roads after people, in case anyone may disapprove of the people concerned. It is permissible, though, to name them after the people's theorems, those people of course, who have produced theorems.)

www.bbc.co.uk/news/uk-northern-ireland-31536765

The original Tech buildings have received a £16 million facelift to become upmarket student accommodation, which will be known as John Bell House.

http://www.futurebelfast.com/1a-college-square-east.html

2

The 1950s

Progress on All Fronts

Harwell and Klaus Fuchs

In many ways, John Bell was entering the world of physics at a most favourable time. While the First World War had been the chemists' war—centred on poison gas and explosives—the Second World War had most definitely been that of the physicists [1]. The atomic bomb had ended the war and, even before that, radar had helped to win it for the allies, or at least, from the British point of view, saved it from being lost before the entry of the Americans and the Russians. But physics had also played an important part in the war in many other ways, contributing to developments such as the jet engine, the proximity fuse, degaussing of ships, rockets, electronics, computing, and meteorology and it had also played a large part in the development of operational research.

At the end of the war, physicists, nearly all of whom had been involved in war work, were mostly keen to get back to their laboratories to devote the techniques developed for war purposes to pure science [2]. One example among many of the results is the technique of magnetic resonance. Following from the developments in the use of microwaves and radio waves during the war, the former in particular being central in the application of radar, the invention of electron spin resonance and nuclear magnetic resonance, or ESR and NMR, came shortly after the war ended.

These techniques enabled very important and useful information to be obtained about atoms and nuclei, respectively, and were of tremendous use, not only in physics but also in chemistry and many other branches of science. Felix Bloch and Edward Purcell were awarded the Nobel Prize in 1952 for the invention of NMR, and, in the fullness of time, this technique has become today's magnetic resonance imaging (MRI), a crucial tool in modern medicine [3].

Just as most scientists were keen to return to research, wartime successes had created a generally favourable view of science among governments [4–6], and financial support and employment opportunities in the discipline would be healthy for some years. Public perception was also generally good, with the prospects of cheap atomic energy being widely welcomed, although in a decade or so this perception would change to a certain extent, as the Cold War took its grip and the anti-nuclear movement grew.

While the path back to the laboratory was obvious for many physicists, for some the 'laboratory' would not be a small-scale university or industrial grouping but might soon become a large-scale dedicated centre for research, as 'big science' took off [7]. In particular for those, American or British, who had been involved in Los Alamos in making the atomic bomb, while many were probably keen enough to put this work behind them, others wished or were persuaded to continue in the development of atomic energy, either for peaceful purposes or for the continued development of weapons.

One who definitely had to be persuaded was William Penney [8] (Figure 2.1), already recognized as a brilliant mathematician before the war, who had become probably the most valued British scientist at Los Alamos. He was the only Briton in the small group deciding on targets in Japan for the bomb; subsequently, he viewed the dropping of the second bomb on Nagasaki from the air and went on to observe and measure the destruction from the ground. General Leslie Groves, who was in overall charge of the American bomb project, wanted Penney to stay at Los Alamos for good, but Penney wished to return to academic life—he had been offered the highly prestigious position of Sedleian Professor of Mathematics at Oxford. Yet, as we shall see, he was persuaded to turn it down for national work.

As has been said, there were two aspects to developments in atomic energy: power for peaceful purposes and construction of weapons. In Britain

Figure 2.1 From left to right: William Penney, Otto Frisch, Rudolf Peierls, and John Cockcroft. This picture was taken after the war. Courtesy Daily Telegraph.

the decision whether to become involved in either or both of these activities was down to the new prime minister [8, 9]. Following the end of the war in Europe in May 1945, the coalition between Winston Churchill's Conservative party and the Labour party of Clement Atlee had broken up, and in the resulting election of July, Labour had achieved a substantial majority of seats, perhaps surprisingly if one thought of Churchill's achievement and popularity in the war, but maybe less so if one reminded oneself of the voters' hopes for a more equal society in the future.

Atlee was very familiar with the work on atomic weapons that had been carried out in Britain in the early years of the war, and the contribution British scientists had made to the atomic bomb project, although it should be immediately admitted that many of the 'British' scientists were refugees from Adolf Hitler's regime and were, of course, particularly keen that Germany should not win the war.

The first crucial step in the path to the bomb had been taken in the so-called Frisch–Peierls memorandum of early 1940 [10, 11]. Frisch was Otto Frisch, who, around Christmas 1938 and with his aunt Lisa Meitner, had made the dramatic discovery that the experimental results of his aunt and Otto Hahn only made sense if nuclear fission had occurred; when bombarded with neutrons, uranium nuclei must have split into two small nuclei, with an emission of energy. The *idea* of constructing an atomic bomb using a *chain reaction* was clear, although it was far from obvious whether it would actually be possible and, if it was, how difficult it might be.

By this time, both Frisch and Meitner were refugees. Frisch, from Germany, was working in Copenhagen, and Meitner, originally from Austria, in Sweden.

Rudolf Peierls (Figures 2.1, 2.2) was also a refugee from Germany [12, 13]; by 1936, with a string of important discoveries already behind him and still only 28 years old, he had become Professor of Mathematical Physics at Birmingham University. As we shall later see, he was to play a significant part in Bell's career, supporting him vigorously and having the greatest respect for his ability but still disagreeing strongly with many of his ideas!

By 1940 Frisch was working in Birmingham with Peierls, and the memorandum produced by the two men concerned the amount of uranium required to make an atomic bomb—the so-called critical mass. Until this point, it was thought to be of the order of tons, much too heavy to be practicable—thankfully, many may have thought.

Frisch and Peierls considered the possibility of obtaining and using a single rare isotope of uranium-235, rather than naturally occurring uranium which consisted largely of uranium-238, and found, to their great surprise, that the amount required was actually as little as a few pounds—for better or worse, a bomb might be possible. It would certainly be extremely difficult to separate the required uranium-235 from the uranium-238—a hundred thousand

Figure 2.2 Rudolf Peierls (left) and Francis Simon. After the war, both men became professors at Oxford and were knighted. Courtesy Daily Telegraph.

diffusion tubes might be needed—but it was by no means impossible and might not even be particularly expensive [10].

At the time, Frisch and Peierls were still enemy aliens, but they passed on their memorandum to the influential physicist, Mark Oliphant, Professor of Physics at Birmingham; as a result, a small committee, the so-called MAUD committee was set up to discuss what action was required [9]. It included Nobel Prize winners James Chadwick and George Thomson, as well as John Cockcroft, who would be awarded the prize after the war for his work on the first particle accelerator, as well as Oliphant himself.

Franz (later Francis) Simon [14, 15], yet another refugee from Germany and at the time Reader in Physics at Oxford University, was asked to carry out theoretical and experimental studies of the required diffusion plant and by December 1940 he had produced a report that Richard Rhodes [10] describes as 'nearly as crucial to the future of uranium bomb development as the original Frisch-Peierls memorandum'.

The highly detailed reports of the MAUD committee were influential in the American decision to move ahead with the bomb project and, in October 1941, Vannevar Bush, Director of the Office for Scientific Research and Development, met with President Franklin Roosevelt and Vice President Henry Wallace, with the sole topic of the talks being the MAUD report [10].

Events took a considerable time coming to fruition, and it was not until November 1943 that an American project was initiated at Los Alamos, and a group of British scientists including Chadwick, Frisch, Peierls, Penney, Egon Bretscher, and Klaus Fuchs became part of the team of physicists.

Others were to be sent to the Anglo-Canadian side of the project, at first at Montreal and later at the nearby Chalk River reactor. Cockcroft [16], who had already played a large part in the production of equipment for radar as Chief Superintendent of the Air Defence Research and Development Establishment, became Director at Chalk River.

Fuchs would play a significant role in Bell's career; more importantly than that, his actions would be of importance for the whole world. Yet another German refugee, he had been born in 1911 and, as a student, he became an ardent Communist, although, at least once he was past his student days, he was able to avoid making this obvious. In 1933 he fled to Britain, working first in Bristol University and then in Edinburgh with Max Born, one of the pioneers of quantum theory. Fuchs was briefly interned in Canada; but then, with the support of Born, he was released and moved to Birmingham to work with Peierls, investigating further the Frisch–Peierls blueprint for the bomb. Thence he moved to Los Alamos [10].

Here he had a wide-ranging brief, becoming familiar with many aspects of the manufacturing process of the bomb, including implosion, multipoint detonation, the necessary conventional explosives, and the study of different bomb configurations. The only possible sign of something untoward came when he was asked by Bretscher to smuggle a classified document on hydrogen bomb theory out of the laboratory for the benefit of the British. Being warned that security had been tightened at the main gate, he replied: 'I'm used to carrying secret papers'! [10].

Overall, the British made a major contribution to the creation of the atom bomb, a contribution that was, at least at first, well recognized by the Americans. Cooperation with the Americans on atomic matters after the war in any case seemed to be assured by the Quebec Agreement [10] negotiated by

Churchill in August 1943. This agreement stated that the two nations would not use the bomb against each other, they would not use it against third parties without each other's consent, and they would not communicate any information about the bomb to third parties without each other's consent.

In the hectic rush to make the bomb, practically nobody had the imagination to look forward to the international situation after the war. Niels Bohr [17] was the exception. He, together with his son, Aage, who was himself later to become later an important physicist and win the Nobel Prize, did spend time at Los Alamos, but he was less occupied with technical matters and more with the future of nuclear weapons. If the Americans and British attempted to maintain exclusive possession of the atom bomb, he felt that they would certainly be disappointed. First the Russians, but after that many other nations would build their own atom bombs, and weapons more powerful than the atom bomb would surely follow—the hydrogen bomb had already been discussed. There would be an arms race followed highly possibly by a catastrophe.

Bohr felt that some advance disclosure, leading to an agreement between states to ban the weapons, was a far better option. However, Churchill in particular found the suggestion of informing any other nation of the work on the bomb ridiculous and appalling, and suggested that Bohr was close to committing treason. It is true that Bohr had communicated with the Russian physicist Peter Kapitza, but Bohr was always totally scrupulous in reporting such encounters to the authorities [10].

Confident of a good relationship with the Americans, Atlee had to plan the future of British atomic weapons and atomic energy. There was certainly no doubt in the minds of the committee that Atlee summoned in December 1945 that Britain should pursue the goal of peaceful atomic energy. They decided that a reactor or 'pile' should be constructed at Sellafield in one of the northerly counties of England, Cumberland. Soon Sellafield would be renamed Windscale [8, 9]. Hopes were high, in retrospect probably much too high, that the construction of this reactor would be the first step leading to the provision of cheap energy and the solution to Britain's fuel shortage, which became particularly important during the extremely bad winter of 1946–7, when much of British industry had to be shut down for several weeks because of lack of fuel.

It was imperative that British research was not left behind and, as early as November 1943, at the time when Oliphant and Chadwick were persuading Cockcroft to take command at Chalk River, the three agreed that after the war a British research establishment investigating atomic energy, to be known as the Atomic Energy Research Establishment (AERE), would be essential. Atlee only formally announced that the government was setting up such a centre in October 1945, but Cockcroft had already been appointed in July and he had been given 'the utmost measure of freedom' in its control. This included the freedom to decide on its location, and he insisted that it should be near a major university with a strong nuclear physics facility, it should have good transport links to London but still be fairly remote, and it should have a good supply of cooling water. Cockcroft was already looking ahead to recruiting staff of high quality, in competition with universities or other research centres, and so he insisted that it should have pleasant surroundings.

It was assumed that a former RAF airfield would be the natural choice, as it would have good roads and large hangars to house the atomic piles. Somewhere in the neighbourhood of Cambridge would have seemed an obvious choice, since nuclear physics was so strong at Cambridge University. However, with the Cold War growing in intensity, the RAF did not want to give up any of its airfields in the east of the country, and the choice fell on RAF Harwell [18].

Harwell was 16 miles south of Oxford, on the border of Berkshire and Oxfordshire. The station had been built in 1937, with a rather small amount of pleasant housing, and had played a full part in wartime operations, but the RAF left it at the very end of 1945 and it was handed over to the recently formed AERE. It fulfilled most of Cockcroft's criteria, although Oxford physics was far inferior to that of Cambridge. The countryside was indeed pleasant, although locals rather thought that 'The Atomic', as it was often called, very much spoilt the view!

Around 50 buildings were left by the RAF, together with four large hangars, but a vast amount of new building was needed. The essentials were carried out extremely quickly, and the first scientists and engineers were able to start work in April 1946. However, construction continued for at least ten years, so the scientists had to put up with all the inconvenience of working on a building site, which sometimes in winter was something of a quagmire. Many of the building requirements were severe, with the so-called hot lab requiring walls of 10 ft thickness and loaded with lead and steel [18].

Provision of housing was also a major task. The number of employees was around 1,000 by 1947, and increased to 3,000 by 1953 and to 6,000 by the end of the 1950s [18]. Some of the basic grade workers could be recruited locally, but most came from further away and needed accommodation. Cockcroft and the heads of division were allocated the RAF housing, and the various RAF messes were taken over by Harwell for single employees. Class distinctions were maintained; the 'A' or officers' mess was allocated to scientists, the 'B' or NCOs' mess to scientific support staff and postgraduate scientists, and the 'C' or airmen's mess to industrial support staff.

For the large number of incoming families, two large estates consisting of several hundred prefabricated bungalows, or 'prefabs', were erected. These were a speciality throughout Britain in the post-war years, designed to meet the severe housing shortage caused, at least in part, by bomb damage during the war. Prefabs were single story structures manufactured in parts. They were usually found to be quite pleasant, if often too hot in summer and too cold in winter. When they were put up, the government promised that they were a

Figure 2.3 Harwell, around 1955. from K. E. B. Jay, *Atomic Energy Research at Harwell* (Butterworth, London, 1955). Courtesy UKAEA.

temporary expedient and would be demolished in ten years. This promise was probably less to assure the tenants that they would not be left in second-rate housing than to assure the building industry that their jobs were not being taken away permanently.

However, in ten years' time, the housing shortage was still acute, and it really did not seem that the best way to improve the situation was to knock down a very large number of prefabs, especially when it would be readily admitted that they were still much more comfortable than a great deal of the older housing stock. Not surprisingly, the prefabs lasted for many decades, those at Harwell not being demolished until the 1990s [18].

Cockcroft was absolutely clear what kind of establishment he wanted. The specific requirement was to produce the necessary data and understanding to support the atomic energy project. Research was carried out on prototype atomic reactors, and much effort was put into chemical processing techniques for the extraction and purification of uranium and other nuclear materials, as well as for separating the constituents of irradiated fuel. The specialist divisions included Chemistry, Nuclear Physics, Reactor Physics, Theoretical Physics, Isotopes, and Engineering. Metallurgy and materials were also highly important. Theoretical work on materials was included as part of the Theoretical Physics Division [18] (Figures 2.3 and 2.4).

However, Cockcroft saw Harwell in much grander terms than these. He thought that it should be a great research institute for all branches of nuclear

Figure 2.4 The director of Harwell, and the heads of the divisions at Harwell, in 1949; from left to right, Klaus Fuchs (Head of the Theoretical Physics Division), Herbert Skinner (Head of the General Physics Division), B. Chalmers (Head of the Metallurgy Division), H. Tongue (Head of the Engineering Division), Egon Bretscher (Head of the Nuclear Physics Division), R. Spence (Head of the Chemistry Division), and John Cockcroft (Director of Harwell). Courtesy UKAEA.

science, rather like a postgraduate university, working to the highest standards, possessing excellent facilities and technical backing and able to attract scientists and engineers of the highest quality. Among the early directors of divisions were two men who had returned from Los Alamos: Bretscher, who was in charge of Chemistry, and Fuchs, in charge of Theoretical Physics. Another excellent appointment was Herbert Skinner [19], who was in charge of General Physics [18] as well as being deputy to Cockcroft.

Cockcroft was particularly keen to recognize and encourage young talent. After his death, John Adams (Figure 2.5), who we shall meet shortly and whose career was given a great boost at Harwell, wrote that 'for me and for many of my generation, [Cockcroft] was the great patron who provided the conditions in which we grew in stature as scientists and engineers' [20]. Bell was one who would certainly come into that category.

Both Windscale and Harwell, incidentally, were placed under the Ministry of Supply, rather a catch-all organization which had provided a very wide range of 'goods' during the war.

Figure 2.5 John Adams. Courtesy Royal Society.

While Atlee was prepared to be open about the peaceful use of atomic energy, he had also to consider whether Britain should build its own bomb. Initially, he was very attracted to Bohr's idea of international control but, by the end of 1945, he was convinced that any such control, which it might be hoped would be by some arm of the new United Nations Organization, would, to say the very least, be a very long time coming. In the meantime, there was fairly general agreement that a British bomb must be built. The official line was often that it was for self-defence against other nations who would certainly come into possession of their own bombs.

Probably much more significant was the belief, of Labour as much as Conservative politicians, that Britain should seek to maintain its status as one of the Big Three alongside the Unites States and Russia, and an essential component of that was possession of the atom bomb. (Russia, it was assumed, would relatively soon be able to make its own.) Ernest Bevin, the Foreign Secretary, is reported to have said at an important meeting of October 1946 that he wanted no future holder of the post to be talked down to, as *he* had been, by the American Secretary of State. 'We've got to have this thing over here, whatever it costs. We've got to have a bloody Union Jack flying on top of it' [8].

The decision to go ahead with making the bomb was not made formally until January 1947 but, for almost 18 months before that, there was general understanding in government that it would happen. One aim of Windscale was certainly, as has been said, the production of energy for peaceful purposes, but another one, probably the primary one, was to provide plutonium for bombs.

As early as November 1945, Penney had been persuaded by C. P. Snow, novelist and Civil Service Commissioner, to turn down Oxford for the seemingly mundane position of Chief Superintendent of Armament Research (CSAR), with the promise that Britain would soon be committed to making an atom bomb and that Penney would be vital to the success of the project. He was to become CSAR in January 1946 but was not instructed to begin work constructing a bomb until April 1947. Even after this date, research on the bomb was to be kept as secret as it could be in a democracy [8].

In the meantime, the hopes of the British that they might share technical information with the Americans lay in tatters. It appeared that the Quebec Agreement applied on the American side only to the now-dead President Roosevelt.

President Harry Truman vacillated. He was under great pressure from many Americans to keep the bomb for themselves. There were good reasons for this, unfair as it seemed to the British. Americans felt that they had paid for the bomb, so they should keep it, together with its full commercial possibilities, and it was felt that a bilateral agreement with Britain would make any eventual international control of nuclear weapons more difficult. Also General Groves, who had the overall responsibility for construction of the bomb, thought that Britain was vulnerable for attack from Russia, and he also thought that Britain was liable to harbour spies. Indeed, in March 1946, Allan Nunn May, a British scientist who had worked on the Chalk River reactor, was arrested for passing information to Russia.

So, when the McMahon Bill prohibiting the sharing of atomic information with foreign powers was passed in July 1946, Truman signed it into law. (Ironically, Brien McMahon, the author of the bill, admitted much later that he had not known the full story of Britain's contribution to building the bomb and that, if he had, he would have acted differently.) It was clear that Penney and his British associates were on their own [8].

It had been suggested that the design and production of the bomb might be carried out at Harwell, but this was ruled out for security reasons and also because many of those employed at Harwell had no wish to take part in work on the atom bomb. The government, in any case, wanted to present Harwell as an institution totally devoted to peaceful work. Instead, the main work was to be carried out at the Armament Research establishments at Woolwich and at Fort Halstead in Kent until, in 1951, it was moved to its own site at Aldermaston, a former RAF base in Berkshire [8].

Indeed, Cockcroft was extremely keen to keep Harwell out of any aspect of weapons work. For example, at a meeting of July 1948, it was proposed that the initiator for the bomb would be built at Harwell. This proposal was turned down, but it was conceded that Harwell would help with the research. Indeed, since Harwell, Windscale, and the Armament Research Division, which was actually producing the bomb, were all under the Ministry of Supply, it was almost inevitable that there would be cross-fertilization of ideas at the very least.

Much later, Bell was to tell Jeremy Bernstein of his concerns at the time about the role of Harwell [21]. Harwell was not supposed to be doing work on nuclear weapons, he said, although he came to believe that such work was going on.

As an example of what definitely seems to be weapons work carried out at Harwell, the first substantial sample of plutonium arrived from Canada at the end of 1951. Because the 'hot labs' at Aldermaston were not ready, the study of this material was done by Armament Research Division metallurgists at Harwell. Metallurgy in itself may be applied to peace or war, but it could scarcely be denied that the work was directly related to construction of the bomb [8].

Certainly, Penney was helped by scientists, including those now at Harwell, who had worked at Los Alamos. He himself, of course, remembered much general information, and when, despite the bad state of British-American relations, he and seven other British scientists were invited to take part in the first post-war American bomb tests in Bikini, he doubtless took the opportunity to glean as much information as possible. Again, when American bomb scientists visited England, it was his duty to wine and dine them, and he would certainly have tapped them for any useful hints.

This was not always easy. Luis Alvarez, an American colleague of Penney's and a friend of his from the Los Alamos days, reports having dinner with Penney, who told him of his intense frustration that a good number of man-years of his best metallurgists had been spent trying to find what crucible material had been used at Los Alamos to melt plutonium, still without success. Alvarez could have told him outright but was not able to do so [8].

In these circumstances, Fuchs was a godsend. With his security clearance from Harwell and his wide experience at Los Alamos, he appeared able to remember information about such vital matters as the assembly of the bomb, the mechanism of implosion, and how to calculate the yield of the bomb. He was certainly regarded as a star at Harwell, and he made frequent trips to Fort Halstead to give lectures on the construction of the bomb [8].

It was at this stage, in mid-1949, that Bell applied for a job at Harwell. Competition must have been fierce, and he was without a PhD and came from a university which, while perfectly respectable, was not among the best known. Yet, at his interview, it seems that the question was not whether he should be offered a job—that was apparently taken for granted—but which of the members of the interviewing panel would get Bell's services!

Fuchs wanted him for his own group, while Bill Walkinshaw, who was in charge of accelerator theory, was also hugely impressed and wanted him for *his* group. Fuchs was Chair of the Committee and pulled rank—Bell was to work on reactors in the Theoretical Physics Division [22].

In many ways, the set-up at Harwell was absolutely ideal for him. He had a permanent job—there are few more secure positions than in the British Civil Service. The salary was modest by many standards but highly respectable compared to the great majority of those from his background. Probably more important to him was that it gave him the opportunity to get into research with excellent colleagues, even without a PhD. It was a research environment of excellence, where the facilities were good, a great deal was possible, and indeed much was expected.

Yet, for a few months, unfortunately, he was neither fulfilled nor happy. Walkinshaw has said that he had expected this. The accelerator group at this time was not actually based in Harwell itself but in an outstation at Malvern, roughly 80 miles to the north-west of Harwell; but, whenever Walkinshaw had to visit Harwell, he made a point of looking Bell up and found him intensely miserable [22]. Perhaps he was lost among the prima donnas of the reactor world. Incidentally, in his memoir of Bell, Kerr [23] recalled, much later, that Bell offered him some notes on neutron diffraction, presumably made at this time, in case Kerr might find them useful in his work on plasma afterglows.

Then came an event which turned out to be particularly good news for Bell, although very bad for Penney and even worse for Fuchs. Since September 1949, Fuchs himself had been under intense suspicion of spying for the Soviets and, in January 1950, he finally confessed that he had been spying since 1941. During the war, he had thought that, since the USSR was an ally, there was nothing wrong in giving them information about the bomb. What seemed worse was that he had actually been in America after the war for the early work on the hydrogen bomb, and he had passed on this information as well, although this information almost certainly dealt with designs of bomb that were later rejected [8].

He was charged with espionage because Russia was an ally at the time. Otherwise, he could have been charged with treason and executed. The maximum sentence for espionage was 14 years, and this was his sentence; however, he was released after nine years and then went to East Germany, where he had a successful career, dying in 1988 [8].

He was luckier than two other members of his spy ring in America, Julius and Ethel Rosenberg, who were executed in 1953. The judge sentencing them to death blamed them for the 50,000 American casualties in the Korean War. Russia had obtained the atomic bomb in September 1949; had it not been for that, and the Rosenbergs' assistance in achieving it, he felt that Russia would never have dared to take on the United States.

As Brian Cathcart says in his book *Test of Greatness: Britain's Struggles for the Atom Bomb*, the Fuchs case is rich in irony. Fuchs is often called 'the man who stole the atom bomb', but in fact he stole it twice: the first time for Russia, and the second, maybe using exactly the same documents, for Britain. And then the British locked him up [8].

Penney also was furious. For practically two years since the beginning of 1948, relations between Britain and America on atomic matters had been slowly thawing, the result of the growth of the Soviet empire and the desirability of a strong relationship across the Atlantic. Intense negotiations had been taking place from September 1949, but these were blown out of the water when the Fuchs case came to light [8].

For Bell, though, the news was good. Incidentally, he had confided to Kerr that the only strange thing he had noticed in the previous months was that Fuchs had had to go to London regularly, presumably for interviews with the security services [23]. With Fuchs out of the way, Walkinshaw was able to put in another bid for his services in the accelerator group, this time successfully. Although Bell was only to stay in this group a few years, it was definitely his stepping stone to a research career at the highest level.

Accelerators

The age of the accelerator [24] is usually said to have begun when Ernest Rutherford, the undisputed world leader in nuclear physics, and the head of the Cavendish Laboratory in Cambridge, in his Presidential Address to the Royal Society in 1927, appealed for the development of sources of atoms and electrons with energies much greater than those of naturally occurring radioactive particles—alpha particles and beta particles. Since the energy of an alpha particle from a radioactive atom is around 5 MeV (million electron volts), this seemed a very difficult task.

However, in a visit to the Cavendish the following year, George Gamow, the Russian theoretical physicist, reported a recent result of quantum theory: because particles penetrating or being expelled from nuclei were represented by wave functions, they could 'tunnel' out or in and could do so with far less energy than if they had to climb over the potential barrier, as had been previously assumed. Also, from the detailed analysis of tunnelling, a proton would have the same penetration power as an alpha particle with 16 times as much energy [24]. It seemed that an instrument accelerating *protons* to somewhat less than 1 MeV might satisfy Rutherford's aspirations.

The first such accelerator was built at the Cavendish by John Cockcroft and Ernest Walton in 1932 [20, 25]. Their device used an AC circuit with a clever combination of rectifiers and capacitors; capacitors were charged in parallel at low potential and discharged in series through a load resistor, thus generating the high potential. Protons of around 700 keV were produced, and the current was several times higher than any that might be obtained from radioactive sources. The protons were used to bombard a target of lithium and succeeded in causing nuclear disintegration, so Cockcroft and Walton became known as the scientists who had 'split the atom'.

The Cockcroft–Walton type of accelerator would be developed further both before and in the years immediately after the war, but it turned out that such machines were larger and so less useful than other types of machine developed later that produced the same range of particle energy [24]. The maximum energy they could produce was around 4 MeV, and today they are mainly used for preliminary acceleration before particles are injected into machines producing much higher energies.

Another interesting type of accelerator was the well-known Van der Graaf generator, first demonstrated by Robert Van der Graaf in 1929. In this technique, electric charge is sprayed onto a moving insulated belt and conveyed to a large sphere. Many young people and their elders have enjoyed the experience of touching the sphere and feeling their hair standing on end, and these machines are used today to give energies up to 10 MeV. However, the breakdown of voltage means that the technique cannot be extended to high energies [26].

The two main types of accelerator that were developed in the 1930s and had the potential to reach considerably higher energies were the cyclotron and the linear accelerator [24]. In the cyclotron, the accelerating particle follows a circular path around a strong magnetic field. The path passes through two semicircular boxes called dees, between which there is an electric field oscillating at high frequency, and the field performs the actual acceleration (Figure 2.6).

The key to the scheme is that the frequency of oscillation of the rotating particles depends only on the change and mass of the particles, as well as the strength of the magnetic field, and so it is a constant—provided, of course, that the mass of the particles is itself constant. It is independent of the radius of the orbit and of the speed of the particles because, while the frequency is actually proportional to the speed divided by the radius, as the particles are accelerated both radius and speed increase proportionally, so the frequency remains the same. The frequency of the alternating field is, of course, arranged to be equal to this frequency of oscillation.

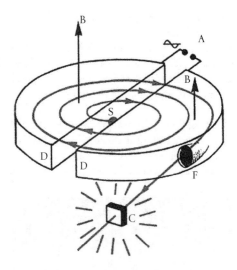

Figure 2.6 Acceleration in a cyclotron: an ion is emitted at S and accelerated at each passage between the D shapes; the magnetic field draws it towards the exit, and it leaves through F to irradiate the target C. From Robert Gouiran, *Particles and Accelerators* (Weidenfeld and Nicholson, London, 1967).

In the second type of accelerator, the linear accelerator (or linac), the particles pass through a series of hollow cylindrical electrodes in a straight line. Alternate tubes are connected electrically (ABABABAB ...), and the two sets of tubes are connected to a high-frequency oscillator. The accelerating particle is only acted on by the electric field when it is in the vicinity of a gap between tubes, and the length of particular tube, the speed of the particle when it reaches that tube, and the oscillator period are related so that, every time the particle reaches a gap, it is indeed accelerated.

The schemes are good but it is fair to say that, by the Second World War, both methods were at an impasse. This was obviously true for the linear accelerator, for which the lack of high-frequency oscillators of sufficiently high power meant that only modest energies were achieved which were insufficient to cause nuclear disintegration, and the method was essentially shelved [27].

In contrast, cyclotrons had been extraordinarily successful [28]. Under the inspirational leadership of Ernest Lawrence (Figure 2.7), equally skilled at designing accelerators and obtaining the funds to pay for them, with Stanley Livingstone also extremely important and the main actual builder, the size of the machines increased dramatically from a magnet of pole diameter of 4 inches in 1930, to one of 11 inches in 1931—giving a machine that was capable of causing nuclear disintegration (only just after Cockcroft and Walton)—to

37 inches in 1937, and 60 inches in 1939, the latter machine producing protons with an energy of around 20 MeV.

Lawrence was deservedly awarded the Nobel Prize for Physics in 1939, and his project for 1940 was a cyclotron with magnet pole piece 184 inches in diameter; it would use 3,700,000 kg of steel and was intended to accelerate deuterons (nuclei of deuterium, consisting of one proton and one neutron) to energies of 100 MeV [28].

Yet, for all this success, there were also problems. Future Nobel Prize winner Hans Bethe warned Lawrence that the effects of relativity effectively increased the mass of particles, as their speed became comparable to that of light, and this spoilt the cyclotron argument explained above, which relied on the mass

Figure 2.7 Ernest Lawrence, with an early 27-inch cyclotron. From Robert Gouiran, *Particles and Accelerators* (Weidenfeld and Nicholson, London, 1967).

being constant. Bethe claimed that the 37-inch machine was optimal; the time and vast amounts of money spent on increasing the energy any further would actually be wasted [28, 29]! Lawrence laughed off this warning.

During the war, it may be said that Lawrence's reputation was saved somewhat by luck. The 184-inch machine was built and it was possible to convert it into a 'calutron', effectively a mass spectrometer, to perform the isotope separation required to construct an atomic weapon. Most of the uranium-235 and plutonium-239 required for the two bombs was obtained by this method [28]. Plutonium had actually been discovered using the cyclotron, although its existence had already been suspected for a few years. It was first detected experimentally by a team led by Glenn Seaborg and Edwin McMillan in 1941 at the Berkeley Radiation Laboratory when they bombarded uranium-238 with deuterons.

Nevertheless, at the end of the war, it was clear that there were definite hurdles to be surmounted if either cyclotrons or linear accelerators were to achieve their potential of being able to produce beams of particles of steadily increasing energy.

At the founding of Harwell, Cockcroft was naturally extremely keen to make the establishment the leading centre for particle accelerators in Britain. The pioneering work of Cockcroft and Walton was incidentally to be recognized by the joint award of the Nobel Prize for Physics in 1951. This may have been something of a surprise, as they had been overtaken in the Nobel stakes by Lawrence in 1939 and may have felt that their time had passed. By this time, Cockcroft had been 'Sir John' for some time, having been knighted in 1948 [16].

While for physicists the main lure of high energies has always been first for nuclear physics and then for the study of elementary particles, which are essentially the constituents of nuclei, the medical uses have always been recognized. Very early in the history of the cyclotron, Lawrence's brother, John, a doctor, had taken part in studies using the comparatively high energy neutrons produced by a cyclotron to fight cancer [28], and medical applications have continued and been developed ever since [24]. Industrial uses have also multiplied [24], so certainly Cockcroft had no difficulty in justifying the development of accelerators at Harwell.

At first he thought only a Van der Graaf machine might be required, but he soon decided that, in addition, a (by that stage) fairly routine 72-inch cyclotron was required, principally to produce radioactive isotopes and to accelerate fast neutrons to 50 MeV. With rapid technical development in the next few years, to be described below, the plan was itself developed; the machine would ultimately be called a 'monster', able to accelerate protons, deuterons, and alpha particles, the nuclei of helium atoms, to 175 MeV [30].

At the international level, the hopes of rapid development of accelerators were raised by two factors that became apparent at the end of the war. First, the intensive development of high-frequency radar techniques during the war provided oscillators of high power: klystrons, and cavity magnetrons, which became available as war surplus. Also, an understanding of waveguides had been developed [24]. A waveguide is a structure designed to transmit electromagnetic waves and, in an accelerator, the accelerated particle effectively 'surfs' on the wave. This made possible the building of linear accelerators of much higher energy than had been possible before, and the process was started by Alvarez, with the able assistance of Wolfgang ('Pief') Panofsky, at Berkeley; by 1947 they had accelerated protons to 32 MeV [5].

On the cyclotron side, a series of developments went some way towards solving the problem of relativistic change of mass and addressing concerns about focussing and the spiralling cost of increasing the energy of the accelerated particle [30]. The relativity problem was solved by the invention of the frequency-modulated cyclotron, or synchrocyclotron, in which the radio frequency is varied cyclically so that, as the particles gain energy, relativistic mass and frequency remain in step. Particles are accelerated in bursts rather than, as in the cyclotron, continuously.

An important principle that makes this type of operation possible is that of 'phase stability': a particle that is travelling a little faster than expected and thus gets ahead of its fellows therefore moves into a larger orbit than they are in and so takes longer than they do to perform the next orbit, thus falling back towards the average. The reverse happens to one travelling a little slower than the others. The idea was stated independently by Vladimir Veksler and Edwin McMillan towards the end of the war [5]. (It may be said that it had been known by James Clerk Maxwell nearly a hundred years earlier in connection with his work on the rings of Saturn [31].)

Focussing is also essential. While particle energy is, of course, crucial for any accelerator, the intensity of the particle beam is also extremely important and, to keep this intensity as high as possible, some type of focussing is essential. 'Weak focussing' works as follows. If a particle has a component of velocity perpendicular to the direction of its path, it would be expected to continue moving in that direction and hit one of the dees. This effect may be minimized by shaping the poles of the magnet to provide a restoring force; but this force and the resultant corrective effect must be relatively small, or instability is caused in the motion along its path. Thus, the focussing is indeed 'weak'; much better than that would be desired and was in fact forthcoming.

The last step, which reduces the costs enormously and changes the synchrocyclotron to a synchrotron [30], is to increase the magnetic field cyclically so that the particles have a constant radius. Thus, the massive dees can be replaced by localized accelerating electrodes. With the synchrotron replacing the cyclotron, and linear accelerators coming on stream, the end of the war meant that a new era for accelerators was commencing.

The organization of work on accelerators at Harwell was complicated by developments at the Telecommunications Research Establishment (TRE) at

Malvern. During the war, this had been the main centre for radar research in the United Kingdom, although its presence in Malvern was actually something of an accident. The establishment was originally based in Bawdsey [32] on the east coast of England but, by 1940, it was sited in Swanage roughly in the centre of the South Coast. However, at the end of February 1942, British commandos carried out the so-called Bruneval Raid on a German radar establishment on the northern coast of France, carrying away significant equipment.

It must have occurred to the powers that be that a retaliatory raid on the TRE at Swanage was highly likely and, in great haste, the whole operation decamped to Malvern, roughly 140 miles north of Swanage [33]. The buildings of Malvern College were occupied, the school itself having moved to Harrow [30].

At the end of the war, the buildings were returned to the school, and many staff returned to civilian life. Others, however, occupied barracks built on the college grounds and continued to work on radar as one component of the Radar Research Establishment (later Royal Radar Establishment), which remained in Malvern.

Still others were transferred to the Ministry of Supply to work at Malvern on 'spin-offs' from radar work, in particular, particle accelerators; and the accelerator group thus formed became effectively an outer substation of Harwell [21]. Skinner [19], who had studied under Rutherford at Cambridge before moving to Bristol University and who had worked at the TRE throughout the war, moved to Harwell as Head of the General Physics Division and deputy to Cockcroft, and Donald Fry [4] moved from Malvern to be in charge of accelerators. He would later be Deputy Director at Harwell and would end his career as Director of the Atomic Energy Authority Establishment at Winfrith, developing nuclear reactors. He was also to play a large part in Britain's relations with CERN.

Cockcroft appointed Thomas Gerald (Gerry) Pickavance [34], who had worked with Chadwick on the Liverpool cyclotron before the war, to lead the cyclotron group. Chadwick himself, for so long Rutherford's right-hand man in the Cavendish Laboratory, had been unable to persuade his leader to obtain a cyclotron for Cambridge and so had moved to become Professor of Physics at Liverpool, where he built a cyclotron of his own. Actually, as it happened, Cockcroft was allowed to build a cyclotron at Cambridge at about the same time. Both British cyclotrons were 36-inch ones and were completed in 1939; they were the only cyclotrons built in the United Kingdom before the war, as compared to 24 in the USA, and 9 in other countries—3 in Japan, 2 in Russia, and 1 each in France, Germany, Denmark, and Sweden [7].

Pickavance would later become Director of the UK Rutherford Laboratory and then Director of Nuclear Physics for the Science Research Council. Like Fry, for many years, he would play an important part in maintaining relations between the British government and CERN.

Others involved at TRE included Adams [30], Walkinshaw, and Frank Goward. Adams was an extraordinary man, an engineer without a degree, although admittedly with a Higher National Certificate in Electronics, who was chosen by Skinner to be in charge of construction of the cyclotron at Harwell. With fresh discoveries, the original plan for a 72-inch cyclotron became one for a 110-inch synchrocyclotron, the first machine to be built in Britain incorporating the principle of phase stability, and it produced protons with an energy of 175 MeV. It operated for the first time in December 1949, it was the second proton synchrocyclotron in the world behind only Berkeley, and it ran without problem until 1978. For two years, it was the largest operating accelerator in Europe.

Adams was to progress extremely fast, and was eventually to have spells as Director of Culham, the laboratory set up to study nuclear fusion; 'Controller', in the Ministry of Technology; and Director-General of CERN. In order to demonstrate the high quality of those involved, it may be mentioned that Skinner, Pickavance, and Adams were all to become Fellows of the Royal Society, while Pickavance and Adams would be knighted.

Walkinshaw [35], soon to be Bell's boss, was to play a large part in many of the theoretical developments in particle accelerators over a long period, and he later organized the computational effort to deal with all the data emerging from working accelerators, while Goward was to play a considerable part in the initial involvement of British scientists in CERN.

Among early achievements at Malvern was the construction in 1946 of the first microwave linear accelerator for electrons, powered by a wartime magnetron, using the waveguide principle as already explained, and producing a beam of 0.5 MeV. The following year a 4 MeV machine was produced, and such accelerators began to be used for the treatment of cancer and also for physics research. Walkinshaw was heavily involved in the design of this machine [30, 35].

Another first for Malvern followed the announcement of the theoretical ideas behind the synchrotron. In August 1946 Goward, together with his colleague D. E. Barnes, adapted an existing machine to produce the world's first synchrotron, which accelerated electrons to 8 MeV [30], and in the following year they constructed a 30 MeV machine of the same type with an 8-inch diameter.

It was to this thriving laboratory that Bell came at the end of 1949 to work with Walkinshaw. It was a great opportunity. The staff involved in accelerator work both at Harwell and Malvern were not only of the highest calibre but quite approachable. The field had made great strides since the war but clearly there was a very great need for further progress.

Walkinshaw had got his man and expected great things from him. He was certainly not disappointed. A considerable range of structures for the different types of accelerator were being considered and discussed, and Bell was very soon playing a full part in this discussion. Much later, Walkinshaw [22] was to say that 'I look back with great pleasure at the sharpness of John's mind and the challenge of keeping up with him. Here was a young man of high calibre [who] soon showed his independence on choice of project, with a special liking for particle dynamics. His mathematical talent was superb and elegant.'

Most of the work consisted of designing configurations of electric and magnetic fields through which the various beams of particles would pass, the aim being to achieve the maximum amount of focussing so as to keep the intensity of the beam as high as possible. Tracing the paths of particles through these fields was the centrepiece of the analysis—hence Walkinshaw's mention of 'particle dynamics'.

It must be remembered that this was, of course, well before the age of computers, and all the calculations had to be performed on the rather primitive desk calculators then available. Bell's talent in the task, as explained by Mary Bell [36], lay in his skill, based on his deep knowledge of electromagnetism and Newtonian kinematics, to simplify and approximate the rigorous mathematics, without losing any of the physical significance, to the level that the calculations could be performed on the equipment available. We shall see a particular example of this ability when we look at some of his work in detail.

Bell himself later told Bernstein [21, 37] that he found the work interesting and challenging as it required the use of Maxwell's equations of electromagnetism with varying and complicated boundary conditions, and relativistic mechanics with electromagnetic fields. It probably became clear to everybody that, while other members of the group were much more experienced than Bell in applications of electromagnetism, Bell had the clearest understanding of the fundamental physics. He was also practically alone in being competent in Hamiltonian mechanics, which was highly useful for these problems. Every so often it would occur to him that he would like to be tackling more fundamental areas of physics than these, but he was still very young and the future must have seemed very open.

It should be mentioned that in these years, although he did publish three papers in scientific journals, the great majority of his work was presented in AERE reports. There were 19 of these in all, 4 of which were published jointly with Walkinshaw, 1 with J. M. Bruce, and 1 with a certain Mary Ross, who was to play a much more important part in Bell's life. These papers are all listed in the collection of Bell's papers edited by Mary Bell, Kurt Gottfried, and Martinus Veltman [38].

From a career-building point of view, these reports did not really count as 'publications', but neither were they merely 'internal reports' for circulation only within Harwell. They were unclassified (that is, not secret) and were issued by His Majesty's Stationary Office; they were available for purchase by those interested at a cost of two shillings (equivalent perhaps to perhaps £5 today.) Many of Bell's reports were read extremely widely by all those concerned with accelerators.

We now discuss some of his work in a little more detail [22, 36]. When he arrived in Malvern, the work on small electron accelerators mentioned above had finished, and the group were looking at a number of possibilities for high energy machines. The first project that Bell worked on, together with Walkinshaw, was that of a dielectric disc loaded waveguide for a linear electron accelerator. The need for the dielectric disc is that, if it is not present, the electron will naturally fall behind the electromagnetic wave in the waveguide. The dielectric slows down the wave so that the electron can 'surf' on it. Bell and Walkinshaw studied the system in detail, analysing a number of waveguide structures, and they wrote two reports on the topic in 1950.

It was the spring of 1951 when Walkinshaw's group moved from Malvern to Harwell to become part of the Theory Group there and, at Harwell, Bell and Walkinshaw did a lot of work on the theory of the high energy proton linear accelerator. The favoured project had become this type of accelerator based on the Alvarez design, but focussing was a major problem.

From late 1950 through 1951 and into early 1952, Bell [38] analysed the situation thoroughly, considering quite a wide range of types of waveguides. He produced a substantial number of reports, mostly under just his own name, but two with Walkinshaw and one with Bruce, studying various geometries, including helical, semicircular, and spiral.

The fundamental difficulty in attempting to reduce divergence of the beam is the famous Earnshaw's theorem, produced by Samuel Earnshaw as early as 1842. Essentially, this theorem says that stability of the beam cannot be obtained by any arrangement of electric and magnetic fields. If a particular geometry causes the resultant force on particles in the beam to point (conveniently) inwards along one axis, there must be another axis in which it will point (inconveniently) outwards. As Bell was to say later, those thinking about these problems constantly came up with new forms of instability.

Yet Bell, and indeed a few other workers in different locations, came across hints that progress was possible. Bell's work on an accelerator with spiral orbits was exceptionally promising. Another effective scheme was to make use of a central control rod—at the position of the particle, the centrifugal and Coriolis forces could be made to cancel out, an observation suggesting a measure of stability. These hints were tantalizing but unfortunately there was no fundamental or general theory to explain what was happening.

Then, in the summer of 1952, the great discovery of *strong focussing* was announced by Ernest Courant, Stanley Livingston, and Hartland Snyder, who were working at the Brookhaven National Laboratory. A brief account of the various stages of their discovery is given by Livingston himself [27].

Strong focussing [27, 39] uses extremely large radial gradients in the magnetic field, but the crucial idea is that along the beam there are successive sectors in which the gradients are alternately radially inwards and outwards. (Another name for strong focussing is *alternate gradient focussing*.)

Each sector acts as a magnetic lens which is convergent in one transverse direction but divergent in the other. The perhaps rather surprising point is that the overall effect is to cause convergence in *both* directions. This result is analogous to the simple optical case where, if convex and concave lenses with focal lengths which are equal in magnitude, though, of course, opposite in sign, are situated along the optical axis some distance apart, the overall effect is one of convergence.

In the accelerator, the effect of strong focussing is that particles still perform oscillations about the direction of the beam, but the amplitude of these oscillations is very much reduced. This observation in turn implies that the sizes of the vacuum chamber and the magnet can be reduced by a substantial factor of maybe 5 or 10. Thus, the costs of the magnet and power supply are also very much reduced for the accelerator, as compared to one without strong focussing producing particles of the same energy. Alternatively, of course, the same financial outlay may provide an accelerator of much higher energy than one without strong focussing. Snyder was later able to show that strong focussing was an example of a very general principle, applicable in many mechanical, optical, and electrical systems, that rapidly alternating forces provide dynamical stability.

It should be mentioned that, when Courant, Livingstone, and Snyder announced their breakthrough, an unknown engineer called Nicholas Christofilos [40, 41], who had been born in the United States but was working on elevator maintenance in Greece, pointed out that he had discovered and patented strong focussing two years earlier, and the Americans were forced to admit his priority. Christofilos was a highly original thinker, who moved to Brookhaven National Laboratory in 1953 and then to Lawrence Livermore Laboratory where, among other fascinating projects, he developed and led the Astron controlled fusion experiment.

In 1952 Bell and Walkinshaw immediately saw the significance of strong focussing for their struggles with the linac, and Bell realized that his schemes with the spiral orbits and the central control rod were indeed nothing other than particular examples of strong focussing. His own understanding of the idea [37] was that, once alternating fields are allowed, the grip of Earnshaw's theorem is loosened. The first-order effects still obey the theorem, but there are important second-order effects for which the effective force may be arranged to point inwards from all directions.

Bell, perhaps a little irritated by his failure to come up with this important principle for himself, was determined to make himself one of the leading experts, if possible *the* leading expert, on strong focussing in the world. To do this, first he had to find out how to apply the method, which had been developed of course for the synchrotron, systematically to the linac, which was not at all obvious. The work of design of the accelerator had already been started, so it was essential for him to move swiftly, and he was indeed able to solve the theoretical and practical problems. Walkinshaw [22] was much later to say that this had been 'sterling work and his most important contribution to a real design.'

Bell's work was completed, and a paper [42] had been submitted to *Nature*, perhaps the most prestigious scientific journal in the world, by the beginning of November 1952; it was fast-tracked for publication in January of the following year. So little time had elapsed since the Brookhaven discovery that there was still no published paper on the topic, and Bell had to refer to a private communication from Brookhaven to Goward in Harwell.

The fundamental discovery is that, if electric rather than magnetic fields are used, the modifications to the original design are rather simple. However, Bell backed this up with a detailed geometry of the new arrangement, with dimensions of the apparatus, the magnitudes of the various fields, and the current and power requirements all calculated in detail. It was certainly a magnificent achievement in such a short time.

Just as important to Bell, though, was to develop the general theory [37]. In this task, he was totally alone in the accelerator theory group, the other members certainly being experts at design, but not experienced and really probably not interested in fundamental theory. His competitors included Ernest Courant and Gerhard Lüders, whom Bell was to meet up with in an entirely different context the following year.

The analysis involved such complicated mathematical topics as characteristic catonics and the stability of dynamical systems. Bell soon realized that the standard method of analysing accelerators, by the use of continuing differential equations, was far from ideal for studying strong focussing, making it difficult to visualize what was actually happening to the particles as they moved along the beam.

It was much more convenient to use matrices. For each stage in the physical process, the mathematical state of the system is multiplied by an appropriate matrix, and the physical meaning is very clear. As well as Bell, Courant saw the power of matrices, but Lüders worked with differential equations. It might be mentioned that, although the use of matrices in quantum theory had at this time been well known for a quarter of a century, they were not used as generally by physicists as they would be in a few years' time.

Indeed, Bell's great paper on strong focussing (Figure 2.8), titled *Basic Algebra of the Strong Focussing System* [43] and produced only as an AERE report, was basically mathematical. At a casual first reading, it might seem to be *all* mathematical, but there are a few translations to the underlying physics.

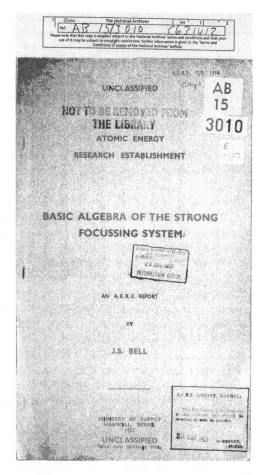

Figure 2.8 Bell's report on strong focussing. Courtesy National Archives.

The report deals with mathematical methods developed by Bell for the study of strong focussing in linear accelerators and for the study of oscillations in the proton synchrotron. He introduces a mathematical function that plays a role similar to that played by the energy of oscillation in simple harmonic motion. This quantity is usually known as the Courant–Snyder invariant, but, as has been said, it was discovered independently by Bell, who also in this paper studies the results of misalignments along the particle beam.

This paper was again produced remarkably quickly, in this case by January 1953. In their *Biographical Memoir*, Phil Burke and Ian Percival [44] write that this report was 'seminal ... [and was] read by all accelerator designers of the day.'

Bell also worked on strong focussing in types of accelerators used for accelerating electrons—the microtron and the betatron [45]—and one of his few published papers in this period was on the microtron. It is a substantial and much more physically oriented paper [46] than the internal report just discussed. Incidentally, papers sent to this journal, the *Proceedings of the Physical Society*, had to be submitted by a senior scientist, and in this case Cockcroft himself obliged. It was clear how involved he was with the detailed work of Harwell, but also perhaps how highly he regarded the work of the younger man.

During this period, Bell wrote another very important paper, which appeared only as an AERE report. This paper concerned circular accelerators, and the awkward topic of resonances [26]. Under the wrong conditions, a particle which is seemingly in a stable circular orbit may encounter a resonant frequency. It may then receive a kick at the same phase on each rotation and move to another orbit of different radius, and this process, of course, causes a diminution of the beam. Different types of resonances have been studied in some detail, one being known as the Walkinshaw resonance.

Bell's paper [47], which was written in October 1954, was related to this topic and was effectively a commentary on the work of three of the most distinguished American physicists outside the Nobel category. John C. Slater accomplished decades of highly significant work on the quantum theory of atoms, molecules, and solids, Robert Serber worked very closely with Robert Oppenheimer at Los Alamos and elsewhere, and 'Pief' Panofsky was one of the most distinguished accelerator physicists, being Director of the Stanford Linear Accelerator Center (SLAC) for many years. Alvarez later reported to the physicist and science historian Abraham Pais [5] that he rated the discovery of Panofsky among his finest achievements in science!

The question being discussed was which Fourier components (essentially, which component frequencies) of the accelerating field had to be taken into account in any discussion of their effect on the accelerated particle. In 1948 Slater had said that only the main component, that is, the one in phase with the particle, needed to be considered, as the effect of the other components, which might have caused resonances, would be oscillatory and so would average to zero. However, Serber in 1948 and Panofsky three years later claimed the opposite, arguing that the behaviour of small oscillations depended on these other components. However, Bell analysed the situation in detail and demonstrated agreement with Slater.

The mistake of Serber and Panofsky had lain in the way they handled the inevitable approximations. It is easy to chop the mathematics at a convenient point in the analysis, but it is essential that the approximated form remaining obeys the laws of physics. In particular, as Mary Bell points out, it must obey Hamiltonian dynamics. Serber and Panofsky failed to realize this, but Bell, skilled in manipulating the mathematics but also possessing a deep knowledge of fundamental physics, was successful.

He was justifiably extremely pleased with this paper but, when he submitted it to a journal, one referee said the manuscript was too short, and the other that it was too long. Since Bell was short of time, and was at this point in any case leaving the accelerator field, he did not submit the paper elsewhere, and Mary Bell [48] reports that many years later scientists were making the very mistake that he had pointed out. When Mary came to select papers for the volume of Bell's papers, *Quantum Mechanics, High Energy Physics and Accelerators* (QHA), space meant that only a selection could be published, and this was the one paper she chose from this period of his life.

Bell was to co-author one further paper [49] on accelerators in this period, although it did not appear until 1957, and then it was published in the *Proceedings of the Institution of Electrical Engineers* with five co-authors: R. B. R. Shersby-Harvey, L. B. Mullett, and B. G. Loach, as well as Bell and Walkinshaw. It was a detailed account, 17 pages long, of the work Bell and the others had performed on dielectric-loaded linacs, in Bell's case several years before.

A very important development in these years, important for the future of European science, and particularly for the eventual future of the Bells, was the formation of the *Conseil Européen pour la Recherche Nucléaire* (CERN; the European Organization for Nuclear Research) [4].

As we have seen, Britain was able and willing to develop a substantial programme for particle accelerators as part of its work on atomic energy. As well as the 175 MeV Harwell synchrocyclotron, there were plans for a 300 MeV electron synchrotron at Glasgow, a 400 MeV synchrocyclotron at Liverpool, and a 1,000 MeV (or 1 GeV) proton synchrotron at Birmingham. Other European nations recognized that individually they could not compete, almost certainly not with the British, and definitely not with the Americans. American building of accelerators was steady in the immediate post-war years, but the big statement of intent came with the founding of the US Atomic Energy Commission in 1948, which announced plans for two massive proton synchrotrons, one of 3 GeV at Berkeley and another of 6 GeV at the Brookhaven National Laboratory. It was clear that the American intention was to move to higher and higher energies.

The first idea for European collaboration came in 1949. Scientists were, of course, generally favourable, as it gave them the chance to build and use state-of-the-art accelerators. Fortunately, statesmen were also enthusiastic, despite the prospect of having to pick up quite large bills even to have a share in the scheme. Their enthusiasm was probably because it gave them the opportunity to look like good Europeans, which was regarded as positive at the outset of this period of peace. It was also convenient that work on accelerators was not expected to lead to commercial discoveries.

Britain, with its already large investment in its own accelerators, opted officially as late as 1951 not to take a full part in the project, although it expressed a willingness to give advice and assistance, and to assist in training CERN scientists, or to second its own scientists to CERN. However, Cockcroft was personally positive about the idea and sent Goward to a UNESCO meeting that would commence planning.

Thus in early 1952, British scientists were involved in some of the early planning of CERN, essentially as observers [30]. It was decided that the organization should be located near Geneva, and the initial plan was to build quite rapidly a fairly routine 600 MeV synchrocyclotron but also, much more excitingly, a weak focussing proton synchrotron of at least 10 GeV. This synchrotron would be more powerful than any machine already built or known to be planned. (Actually, a 10 GeV machine was being planned in Dubna in the USSR but, at the time, those planning CERN did not know about that.)

It was in August that strong focussing was discovered and the bold decision was made to work on a strong focussing proton synchrotron of energy between 20 and 30 GeV, rather than on the 10 GeV machine. In October, a conference of scientists from various European countries was held in Geneva to discuss the design of the new machine and, with his considerable expertise in strong focussing, Bell was called in as a consultant to the meeting, which actually took place in Paris. Thus, he gained an early exposure to the ethos of CERN, and also met Mervyn Hine and Kjell Johnsen for the first time—they were to become leading lights of CERN [37].

Then in November 1952 Lord Cherwell, scientific adviser to Churchill, who was back as Prime Minister, finally agreed that the United Kingdom should be a member of CERN. The inauguration ceremony was in 1954, and Bell was present for what was his first visit to Geneva (Figure 2.9). Adams [30] was seconded from Harwell to CERN and, by the end of 1954, he was head of the proton synchrotron group and so was in charge when at last, in 1959, a beam was produced, its energy being 25 GeV. Incidentally, with improvements in the performance of cyclotrons, the Harwell 600 MeV project was cancelled, but the design of sections of this accelerator formed the basis of the injector into the machine at CERN. As well as becoming a major centre for accelerator physics, during this decade CERN was also building up the leading European theoretical group working on field theory and the theory of elementary particles, a fact which was certainly not lost on Bell.

Bell had had a brilliant three years working in accelerators, particularly excelling in his grasp of strong focussing, and he certainly found the work challenging and fascinating. As has been said, he did on occasions think that he would like eventually to a more fundamental area of theoretical physics than that, but he could not have thought that the possibility would come so early in his career and so suddenly.

Figure 2.9 Removing the first earth at CERN on 17 May 1954. Courtesy CERN.

Since the demise of Fuchs, Peierls had been playing an important role at Harwell. It was probably his idea that a few of the leading theoretical physicists at Harwell, those who were extremely able but, for whatever reason, had not obtained research degrees, might spend a year performing research in a university. Walkinshaw [22] had reported to Peierls about the brilliant man who was working with him, so it was not surprising that Bell was selected for this project—although, actually, he was the second to be chosen.

The first was Brian Flowers (Figure 2.10). Flowers [50] was four years older than Bell and from rather a more privileged background. After obtaining a brilliant degree at Cambridge—in two years, because it was wartime—he went straight to the Anglo-Canadian atomic bomb project under Cockcroft. At the end of the war, he was recruited by Cockcroft to work at Harwell and, after a short spell in the Nuclear Division with Frisch, he moved to Theoretical Physics. His obituary writers suggest that his year with Peierls in Birmingham was a grooming to take over permanently from the disgraced Fuchs on his return and they stress that, when he did so, one of his main achievements was to encourage theoreticians and experimentalists to interact.

Flowers was to have a glittering career, although circumstances meant that more of his efforts were spent on administration and less on research than he might have liked. His main scientific achievement was in the shell theory of nuclei, while on the administrative side he became successively Professor of Theoretical

Figure 2.10 The Department of Physics Theory Group at the University of Manchester, 1959–60. In the front row: fourth from right, Sam Edwards; third from right, Brian Flowers; second from right, Franz Mandl. Courtesy University of Manchester.

Physics at Birmingham University at the age of 34, Fellow of the Royal Society at 37, Chairman of the Science Research Council, Rector of Imperial College London, Vice Chancellor of London University, and Chancellor of Manchester University. He was knighted in 1969 and became Baron Flowers in 1979.

Bell's career was to take a different route but he was equally delighted with the opportunity. It had not actually been stipulated that the work should be in Peierls's group at Birmingham, and Bell did indeed consider going to work under John Currie Gunn in Glasgow; but, in the end, like Flowers, he chose Birmingham and Peierls. He hoped to use his year in Birmingham to become an expert in quantum field theory.

Quantum theory 1: With particular attention to EPR

When he left Queen's in 1949, Bell had made a resolution, against his inclinations, to steer clear of worrying about quantum theory. He recognized that effort expended here would be at the expense of launching his career in more orthodox directions. Also, he could not ignore John von Neumann's 'proof', at that time only available in German, that hidden variables could not exist in quantum theory.

His resolution lasted only two, or at the most three, years, but there were extenuating circumstances, two of them in fact, and both of them were associated with the American physicist David Bohm.

It is probably fair to say that Bohm (Figure 2.11) was as able and inventive as any of the most famous theoretical physicists of the twentieth century. If his achievements strictly in the discipline itself are rather fewer than those of many others, that must be partly because of personal circumstances but mainly because his intellectual horizons were so much wider [51, 52].

Born in 1917, he had studied at Penn State University, a small institution, where, rather like Bell, he had the time and space to explore many aspects of physics and mathematics himself [51]. Then, from 1940 to 1947, he was at Berkeley, researching under Oppenheimer. Like many in the group, he

Figure 2.11 David Bohm. from F.D. Peat, Infinite Potential: The Life and Times of David Bohm (Addison-Wesley, Reading, Mass., 1997). Courtesy Addison-Wesley.

was involved in left-wing politics, and he did join the Communist party in November 1942 but left it after only a few months.

Because of this one-time membership, he was not allowed to move to Los Alamos with Oppenheimer; however, staying in Berkeley, he did exceptionally important work on developing the physics of *plasmas*. Plasmas are ionized gases in which some or most of the atoms have dissociated into electrons and positive ions. They are very important not only in astrophysics but also in attempts to utilize the power of nuclear fusion, because for this purpose very high temperatures are needed at which the substances involved are ionized.

In 1947 Bohm became Assistant Professor of Physics at Princeton University, where he interacted quite closely with Albert Einstein, who was at that time at the Institute for Advanced Study, also in Princeton. It was while he was at Princeton that Bohm wrote his influential book on quantum theory [53]; Bernstein [21] characterized this book as the only serious one on quantum theory that he had seen in which there were more words than equations. Indeed, in his preface, Bohm remarked that it had been his purpose to present the main ideas of quantum theory in non-mathematical terms. He did go on to admit, though, that for precision some mathematics was needed. There are, in fact, over 1,100 numbered equations in the book, as well as many more unnumbered ones!

Nevertheless, it is clear that Bohm made a supreme effort to *explain* the new concepts involved in quantum theory as well as presenting the calculations. The first part of the book is a substantial, broadly non-mathematical section comparing the wave and particle properties of matter, in which he also attempted to build a physical picture of the quantum theory of matter.

Then, towards the end of the book, he studied the relationship between quantum and classical concepts, and presented a sustained analysis of the quantum theory of measurement. Many of these topics would have been completely absent from, or, at the very most covered in a few paragraphs, in most of the more 'shut-up-and-calculate' school of quantum mechanical texts.

While Bohm presented all his arguments in a particularly lucid way, it may be said that, with one extremely important exception to be mentioned shortly, there may be little that was genuinely new in the book. In his account, he remarked that much of the general content of the book was based on lectures given by Oppenheimer at Berkeley, while for what he called the 'general philosophical basis' of the theory, he said that a series of lectures by Bohr [54] were of crucial importance.

Indeed, at the time of writing this book, Basil Hiley [51] reports that Bohm was quite a keen supporter of Bohr and the Copenhagen interpretation of quantum theory, because, with his left-wing beliefs, he considered Bohr's complementarity to be akin to Georg Wilhelm Friedrich Hegel's 'unity of opposites', which is the central category of dialectics. Thus, all the central strands of Bohr's position appear in the book.

In particular there is a 'proof that quantum theory is inconsistent with hidden variables'. Here Bohm takes the case of momentum and position both having exact values, in conflict, of course, with the Heisenberg principle, and claims that a simple thought experiment detecting the position of an electron by use of a proton beam demonstrates the falsehood of this idea.

From hindsight of 60 years or so, it seems a very unsatisfactory argument. At best, it appears to show that one cannot have *just* particles but must retain a wavelike aspect of the experiment as well. But that surely is quantum theory, not a demonstration of anything to do with hidden variables.

From a wider point of view than that, Bohm claims to suggest a clear experimental test between the orthodox interpretation and hidden variables. However, the real point of hidden variable arguments is that, at least in the very great majority of circumstances, they should reproduce the quantum mechanical results, and in any case it was an argument that Bohm would very shortly renounce.

We now come to the most interesting part of Bohm's book, for the reader of today, but particularly for Bell reading it in 1951, which is the discussion of the Einstein–Podolsky–Rosen (Figures 2.12, 2.13) or EPR paper. This argument

Figure 2.12 Boris Podolsky. Courtesy Xavier University Archives, Cincinnati, Ohio.

Figure 2.13 Nathan Rosen as the sole surviving member of EPR, at a conference held in 1985 at Joensuu in Finland to celebrate the fiftieth anniversary of the EPR paper. Far left, Rosen; third from the left, Constantin Piron; second from the right, Rudolf Peierls; and, far right, Max Jammer. Courtesy World Scientific.

from 1935 was the culmination of Einstein's efforts over the previous decade to demonstrate the inadequacy of the Copenhagen interpretation of quantum theory; his discussions with Bohr are often called the Bohr–Einstein debate [55].

Bohr quickly replied to the EPR paper, and, because the general presumption among physicists was that Bohr had already solved the conceptual problems of quantum theory in the previous decade, while Einstein had not been able to adjust to the demands of the new theory, it was generally taken for granted that Bohr had yet again won the day. In fact, there was very little discussion of the paper at all, and it was not at all surprising that, in his memoir of Bell, Kerr [23] should say that Bell had not seen it while in Belfast. In fact, it seems very likely that he met it for the first time in Bohm's book. He would have been enthralled!

So it was good that Bohm's book, following his general policy of discussing conceptual issues extensively, did include a substantial account of the EPR paper. What was even better than that was that Bohm had been able to simplify enormously the presentation of the argument. EPR themselves had actually put forward an example of their general argument; however, it was rather awkward, and the awkwardness did not help their presentation of their case. Bohm's example, in contrast, was exceptionally and brilliantly simple. In fact, ever since that day, anybody who discusses the idea (and it is now used throughout the study of quantum foundations and quantum information theory) automatically uses Bohm's example. The more punctilious call it the EPR–Bohm argument, but mostly, and unfairly to Bohm, it is just called the EPR argument.

Having sketched the history, we now look at the physics involved in a little more detail. First, we shall examine Einstein's views [56]. It is often said that he was 'against' quantum theory or even hated it. This is not really true, as he felt that it was a perfectly valid theory at the statistical level. Indeed, he felt that, even if his own vision, which will be described shortly, came into being, the physicist working, for example, on the physics of solids or elementary particles would carry on exactly as at the moment using our present theory, just as the engineer designing bridges or power stations today quite successfully uses the classical laws of Isaac Newton and James Maxwell and certainly does not need to consider quantum or relativistic ideas.

What Einstein disliked in Bohr's approach was partly the probabilistic nature of the theory. Einstein is famed, of course, for his saying: 'God does not play dice', and it is certainly true that he very much disliked the loss of determinism. However, at least in his later years, he was much more concerned with the loss of realism: the fact that, before a measurement was made, a particular physical quantity did not, according to the Copenhagen interpretation, have a distinct value. Einstein's description of the situation was that quantum theory as interpreted by Bohr and his allies was 'not complete' [55].

At first sight it would seem clear, then, that Einstein must have supported hidden variables, which would surely 'complete' the theory and, it would be hoped, restore determinism and realism. Bell certainly thought so [57, 58]. In 1976 he wrote: 'I have for long thought it quite conventional and uncontroversial to regard Einstein as a proponent of hidden variables, and indeed as "the most profound advocate of hidden variables".' However, on this point he is in disagreement with Max Jammer [59], the highly distinguished historian and philosopher of quantum theory; so let us proceed to examine the contrary position.

It is well known that Einstein's crowning achievement was the general theory of relativity, which he produced in 1916. The theory explained gravitation, and it was based around the type of mathematical object called the *tensor*. Much of the remaining 40 years of his life was spent in trying to extend the same type of mathematics to produce a *unified field theory*, which would include electromagnetism as well as gravitation. The mathematical complication was enormous; John Wheeler subsequently described Einstein as a 'wholesale dealer of equations' [21]. Einstein was, however, unsuccessful.

In retrospect, it is quite clear why this should have been the case. There are in fact four fundamental fields in physics: as well as gravitation and electromagnetism, there are strong and weak nuclear forces (or just strong and weak forces or interactions). At first, the strong nuclear force was described as binding nucleons (protons and neutrons) together to give nuclei, but, with the realization during the 1960s that nucleons were not fundamental but consisted of quarks, it is now described as binding quarks together to produce nucleons. The weak nuclear force is responsible for radioactivity (beta decay).

Some success has been had in unifying these forces. In 1968 a mechanism was found by Abdus Salam and Steven Weinberg for unifying the electromagnetic and weak nuclear force as the electroweak force [5]; for this discovery, they were awarded the 1979 Nobel Prize for Physics. There have also been bold attempts to join the strong nuclear force with the electroweak force, although these attempts have not been totally successful. However, the gravitational force has proved by far the least amenable to these machinations! It seems that Einstein really chose the wrong starting point for his labours.

However, Einstein hoped that, once this unified theory had been produced, while it would agree with quantum theory where that theory had been experimentally confirmed, at least to within experimental precision, it would automatically remove the conceptual problems of quantum theory—it would display determinism and realism. In his terms, it would be 'complete'. Jammer's argument is that this hope should definitely *not* be regarded as a commitment to hidden variables if, as is conventional, hidden variables are defined as relatively minor additions to the present structure of quantum theory. In the philosopher of science Arthur Fine's pithy phrase [60], for Einstein, completion of quantum theory of quantum theory will not come 'from within', which is to say, tinkering by use of hidden variables, but 'from without' by construction of an entirely new theoretical framework. Jammer's view may sound a little like arguing about the use of words but, as we shall see in the following section, it does help to explain Einstein's reaction to a second and hugely important publication of Bohm.

We will shortly move onto the EPR (or, strictly, EPR–Bohm) argument itself but, first, we shall briefly discuss what may be called the first two rounds of the Bohr–Einstein debate [55]. What has usually been thought of as the standard account of the debate was written by Bohr himself [61] in a volume published in 1949. Until quite recently, it was not been seriously questioned whether this account was accurate, although Bohr's claim to have won the argument in the third round, that of EPR, may have been disputed by some.

In the first two rounds, which were in discussions between the two men in the evenings of the Solvay Congresses of 1927 and 1930, Einstein claimed to show by thought experiments that measurements on a single system could be carried out to greater accuracy than allowed by the Heisenberg principle. In the first round, these were measurements of position and momentum; in the second, measurements of time and energy. In each case, though, again according to this standard account, Bohr was able, although not without a lot of thought, to defeat Einstein's strategy.

It is important to stress that Bohr did not *use* the ideas of complementarity in these arguments. From his point of view, of course, he could have done so, as he was convinced that complementarity was correct; but, obviously, he could not have expected his arguments to convince Einstein, who did not accept complementarity at all.

Rather, the great strength of Bohr's successes in defeating Einstein's arguments was that they were performed using only standard physics. Yet, they did appear to support, at least in a general way, the ideas behind complementarity. Quantum theory could not provide simultaneous values for, for example, position and momentum, and neither could these quantities *be simultaneously measured*. It might, though, be remarked that there was no proof that they could not *simultaneously exist*.

Although not mentioned explicitly in Bohr's discussion of Einstein's thought experiments, it could be suggested that there is a general truth behind the detailed arguments, which is that each measurement *disturbs* the other. This is, as it were, a physical argument that again agrees with the impossibility of simultaneous measurement.

One may imagine Einstein retreating after the 1930 Solvay Congress, licking his wounds and wondering if he could get round the seemingly intransigent difficulties in evading the Heisenberg principle.

(Such, as we have said, was the universal view of the situation for a long period. More recently, though, the philosopher Don Howard [62] has assembled evidence that Bohr had misunderstood Einstein's position in the second round of the debate. He argues rather convincingly that Einstein's real focus was on *entanglement*, which was, as we shall shortly see, the centre of the *third* round of the debate. However, this does not affect our argument which particularly concerns beliefs in the 1950s and 1960s.)

Now we turn to the EPR argument, or in fact the EPR–Bohm argument; to explain it, we need a few facts about the *spin* of an electron. In a simple Bohr model of the atom, the atom moves in an *orbit* about the nucleus and so has *orbital angular momentum* about the nucleus, in a way that is rather similar to the way the earth has orbital angular momentum about the sun. And, just as the earth also spins about its own axis, we picture the electron spinning about a particular axis, and so having *spin angular momentum*; we usually just use the term *spin*.

The spinning may be in the xy-plane about the z-axis; we call this the z-component of spin. Similarly, we may consider spinning about the x- or y-axes; we call these the x- and y-components of spin. These definitions apply both classically and in quantum theory, and in the classical case we may talk about 'how the particle is spinning'.

However, in quantum theory, as always, we cannot discuss the behaviour in the absence of measurement; all we can do is explain the results of measurements. The first important statement is that, if we measure the z-component of the spin of an electron, we must get one of two possible answers. We may just call them $(+)$ or $(-)$, although technically, the $+$ or $-$ signs should be multiplied by a constant equal to Planck's constant divided by 4π.

What is perhaps a little surprising is that, if we measure the y-component of spin, or the x-component, or indeed the component in any other direction, we will have exactly the same possible results of the measurement. If we find this result unexpected, it may be because we are thinking in terms of a spinning system before any measurement, and trying to relate results of measurements of different components; however, this is just what quantum theory tells us not to do.

The other important point according to quantum theory is that, just as if you know the momentum of a particle precisely, you can know absolutely nothing about its position, if you know the z-component of spin of an electron exactly, you can know nothing about the x- or y-components. Indeed, according to the Copenhagen interpretation, in each case, if you know the first, a value of the second just does not exist.

Now let us look at the EPR–Bohm argument. This thought experiment deals with an extremely cleverly dreamed-up type of system, which, shortly after the original EPR paper in 1935, Erwin Schrödinger [63, 64] (Figure 2.14) called *entangled*. The system consists of two electrons, but it has only a single wave function for both electrons. The combined wave function does not provide a value for any component of spin for either electron, but the values for, say, the z-component of each are intimately connected.

To be specific, we assume that the two electrons have been produced by the decay of a different type of particle which has zero spin; so, by conservation of angular momentum, the sum of the spins of any component of the two electrons must be zero.

So, again concentrating on the z-components, we may have two possibilities. The z-component of the spin of the first electron may be found to be $(+)$, and that of second $(-)$; alternatively, that of the first spin may be found to be $(-)$ and the second $(+)$. This result would be true classically, in which case we would have an easily understood case of one situation *or* the other. But, of course, quantum mechanically we do not have one or the other. We have a state for which, at least until a measurement is made, both possibilities exist together. Mathematically, we may say that the wave functions for the two possible cases are added together to give a linear combination, or *superposition*, of different states.

So we have to say that the wave function of the system does not give us the value of the z-component of either spin. Actually the same goes for the y- or x-components as well and indeed for any other component of spin for each particle. And it must be stressed yet again that the Copenhagen interpretation does not allow any hidden variables giving extra information. They might, for example, give us the values of the z-components of spin for each of the two

Figure 2.14 Erwin Schrödinger, by the River Liffey in Dublin. Courtesy Cambridge University Press.

particles, which would certainly make things a lot simpler conceptually! But they are prohibited.

Now, we may imagine that the two electrons move apart as far as we wish. If we like, they may be right across the universe from each other. Next, we measure, say, the z-component of spin of the first electron. We may, of course, get the result $(+)$ or $(-)$. But the very surprising point is that, instantaneously, the z-component of spin of the second electron must become fixed. If the result for electron 1 is $(+)$, that for electron 2 must now be $(-)$, and vice versa. Essential in the argument is the fact that, for this system, the measurement on the first spin cannot in any way *disturb* the second spin.

EPR then would say the following. (Bohm, at least when he wrote his book, would have dissented since, although he presented the argument and simplified its form, he did not agree with it.) There are two ways of explaining this. The first is that the Copenhagen interpretation is wrong to consider that the components of spin do not have values before any measurement; in fact, they do have them. The second is to say that a message of some kind travels instantaneously from one electron to the other, essentially instructing the second electron what value to take up.

Since the first explanation is ruled out by orthodox Copenhagen theory, we may add a third area where quantum theory presents conceptual problems, which we call *locality*. The fundamental meaning of this concept is that cause and immediate effect must be at the same point. A football may break a window when it hits it but not when it is somewhere else! The concept is generalized, though, to the recognition that cause and effect may be separated in space, but in that case a signal must pass from cause to effect, and this signal cannot go faster than c, the speed of light.

Thus, the second explanation violates the law of locality and, since, as has been said, the Copenhagen interpretation does not allow the first explanation, we may conclude, as a final statement of the conceptual problems of the Copenhagen interpretation, that there are three: realism, determinism, and locality.

Einstein had no such hang-ups about considering the first explanation, since he did not subscribe to the Copenhagen vision. Thus, he had the choice to accept realism with the first explanation, or reject locality with the second and, since he had a very strong belief that physics should be both real *and* local, for him there was no choice. He opted for the first explanation, and this was the conclusion of the original EPR paper. His conclusion was that the universe demonstrated *local realism*. (As already explained, this pithy statement might suggest that Einstein was explicitly endorsing hidden variables, but his vision was actually wider than that.)

A few variants of the EPR argument will be sketched. An extremely common alternative approach, actually used by Bohm [52], argues that measurement of the z-component of spin for the first electron may be regarded as a measurement of the same component of spin for the second electron. Thus, the possibility of performing this measurement, together with the lack of interaction between the electrons, means that the value of this component of spin for the second electron must have been fixed in advance.

But the argument may be repeated with x-components instead of z-components. This implies that *both* components have fixed values in the absence of any measurements—in conflict with the generalized Heisenberg principle. The general conclusions are the same as for the previous approach.

This approach is very much analogous to Einstein's intentions in the first two rounds of the Bohr–Einstein debate—two quantities, such as momentum and position, energy and time, or, in the EPR case, different components of spin, could be measured simultaneously, in contradiction to the laws of quantum theory. There was a major particular difference, though, between Einstein's previous challenges and EPR's challenges, which will be discussed shortly.

First, we include a little historical background. The original EPR argument of 1935 did not deal with different components of spin of the two entangled particles, but their momentum and position. The combined wave function was an extremely complicated function of the coordinates of both particles. Essentially, where the EPR–Bohm argument had two terms, $(+)(-)$ and $(-)(+)$, the original EPR argument had an infinite number.

Einstein was dissatisfied with the presentation of this paper; the paper had been written by Podolsky and the way he did so made it appear to be, as much as anything, an exercise in logic chopping. As a result, at various times in the late 1940s, Einstein produced an argument that was much simpler than that presented in the paper. His argument is rather general, and Bohm's own argument could be regarded as a special case of Einstein's.

Now, we turn to the responses to the original EPR paper and, of course, the one people waited for was Bohr's. Certainly, the EPR argument took Bohr by surprise. Rosenfeld [65] has written that 'This [EPR] onslaught came down upon us like a bolt from the blue. Its effect on us was remarkable.' Bohr and Rosenfeld worked tirelessly at their analysis of the argument.

Yet, strangely and, one might say, almost smugly, in his response in 1935, Bohr [66] wrote that 'the special problem of [the EPR argument] ... does not actually involve any greater intricacies than the simple examples discussed above.' These 'simple examples' are precisely what we have called the first two rounds of the Bohr–Einstein debate, and it will be remembered that Bohr defeated Einstein's arguments in these rounds through standard physics without using complementarity.

Indeed, his argument in the EPR case does contain elements of 'normal' physics—a diaphragm and slits—but it is also centred round *complementarity*, the central doctrine of the Copenhagen interpretation. But Bohr knew, of course, that Einstein did not accept complementarity. From this point of view, it seems that Bohr's response to EPR was not saying that their argument was flawed but merely that it did not damage his own interpretation. However, this is not the rhetoric of his response, which definitely claims to show that the EPR paper failed in its objectives.

We shall say little about the actual argument, which is frankly rather obscure. Howard Wiseman [67] says that 'Bohr's reply was a quagmire from which even his supporters had difficulty extracting any clear meaning.'

Wiseman adds, perhaps a little maliciously, that, when the paper was reprinted in the collection edited by John Wheeler and Wojciech Zurek [64], the pages were printed out of order—but nobody noticed!

Many authors have discussed Bohr's response to EPR. The historian Mara Beller [68], for example, has given a fairly destructive criticism. The present author [69] has also expressed rather a negative view; however, the philosopher Jens Hebor expressed full agreement with Bohr's position [70].

Actually, the central point in Bohr's argument was that the choice of one or other of the two possible measurements, of momentum or position in the original EPR paper, has 'an influence on the very conditions which define the possible types of predictions regarding the future behaviour of the system.' This statement would seem to be just a way of using complementarity—we must choose one of the two experimental arrangements to discuss; we may discuss one or other of the physical quantities involved but not both.

Yet, there was one enormous admission by Bohr. He admits that there is 'no question of a mechanical disturbance' of the second particle as a result of the measurement of the first one. This admission was the hugely important difference between this round of the Bohr–Einstein debate and the previous two. In each case, there was the possibility of two measurements; however, in the first two rounds, Bohr could justifiably claim that the first measurement disturbed the system. In this case, he did not try to do so.

This point was not related solely to the EPR discussion. As mentioned before, up to this point, Bohr had used the disturbance argument to buttress his conceptual argument centred on complementarity. It was perhaps an argument that appealed to the physicists, while the conceptual arguments perhaps appealed more to philosophers. From now on, the disturbance argument was not available to him and, although it would certainly not happen for some time, eventually it would damage his standing with at least some physicists.

At the time, the only person who responded positively to the EPR paper was indeed the only person who broadly sympathized with Einstein's positon—Schrödinger. The two men corresponded over the issue, and Schrödinger [63] also published a three-part paper titled 'The Present Situation in Quantum Mechanics,' as well as two other papers. It was in the trilogy that, as already mentioned, he invented the term *entanglement* for the type of situation dreamed up by EPR. This is an extremely useful term that has been used ever since. He also said the following about entanglement: 'I would call it not one but the characteristic trait of quantum mechanics, the one that enforces its entire departure from classical lines of thought.'

As is well known, he also introduced the famous *Schrödinger's cat*, which has as its wave function a sum of terms corresponding to *dead* and *alive*, its fate only determined when its box is opened and its state is observed. This creation has, of course, captured the attention of the public far more than the EPR paper ever did. As a conceptual problem, on the other hand, it is moderately interesting but of much less depth than the EPR argument.

Incidentally, it seems that Bell never discussed Schrödinger's cat in print. He did, however, refer to it in his lectures; however, rather than having the cat poisoned so that it was dead or alive (or, more accurately, dead *and* alive), the cat was fed so that the states were 'fed' and 'unfed'. Even in a thought experiment, he did not approve of cruelty to animals!

Unfortunately, and quite inadvertently, Schrödinger also did something at this time that may have made the general perception of the EPR paper among physicists even less positive than it might have been anyway. He christened the argument *the EPR paradox*. Now, for Schrödinger, the word 'paradox' meant only a surprising or puzzling result. He used the same word in other places to mean just that.

However, probably a more commonly held meaning of the word 'paradox' is very different—an argument that leads to an absurd result, perhaps that $1 = 2$, or that all angles are acute. A 'paradoxer' is often regarded as somebody who irritatingly tries to create problems in perfectly satisfactory sets of ideas. This is precisely how at least some physicists viewed Einstein's ideas at this time. A slightly different understanding of the word was an assumption that Einstein had genuinely reached a contradiction in his arguments and was confused. Much later, some 'kind' people helped him out! As stressed by Wiseman [67], far from presenting a 'paradox', EPR had presented a 'logical argument'; but, for a quarter of a century, it was rarely seen that way.

It will be admitted, by the way, that both Einstein and Bell did use the 'paradox' terminology: Einstein in the Schilpp volume [71], in which a range of physicists and philosophers discussed his work and Einstein himself replied, and Bell in his most famous paper on EPR [72]. It is probable that they were merely using what was by then the conventional wording rather than expressing any deep-felt belief. In later papers, Bell avoided the word, although whether deliberately or not it is impossible to say.

Discussion of EPR and of Bohr's response to it was rather limited for a long period, so we shall move ahead to Bell's views. Entanglement and EPR were, of course, to comprise the major component of his most important work in quantum theory. In many cases, he reproduced the argument as part of his own analysis. There is no evidence that he had any concerns about its truth and its relevance to the physical situations he was discussing.

His view of Bohr's responses was completely the opposite. His first comment was made in a paper from 1976 [57]. He characterizes Bohr's remarks in the Schilpp volume as suggesting that we should not worry about the situation and that to try to analyse nature at any level deeper than that of observation would be a waste of time and against the lessons learned at the very beginning of the theory, lessons which had to be understood in order that the theory could actually be constructed.

He also picks up on a development in Bohr's work that had occurred after 1935 and which to some extent was a replacement for the 'disturbance' idea. This development was the novel use of the word *phenomenon* to encompass, in the EPR case, *both* particles, treating them as single system. According to this idea, it is strictly illegitimate to discuss possible interactions between the two particles of the EPR system, and the EPR analysis is irrelevant. This approach, Bell says, is what he believes is the orthodox view, and he says that many people are content with it (although it is to be understood that he is not included in that group).

By 1981 he was a lot more combative [73]; he discusses Bohr's 1935 paper, noting that the section he is discussing was repeated in 1949 [61]. He comments that while Bohr agrees that there is no *mechanical* disturbance of the second particle, he still insists that there is an *influence* that may affect the future behaviour of the system. However Bell say that he does not understand in what sense the word 'mechanical' is used for the disturbances that he is not contemplating, in contrast to those that he is.

Bell now addresses Bohr's claim that the analysis of the experiment using complementarity 'may be characterized as a rational utilization of all possibilities of unambiguous interpretation of measurements compatible with the finite and uncontrollable interaction between the objects and the measuring instruments.' Bell wonders whether Bohr is saying any more than that different experiments on the first particle give different information on the second and, if so, whether Bohr is actually ignoring the point of the EPR argument, which was that only the first particle was supposed to be disturbed by the measurement, and yet predictions could now be made about the second.

We shall have much more to say about Einstein, Bohr, and Bell in the next chapter and the rest of the book.

Quantum theory 2: With particular attention to Bohm and to hidden variables

Let us now turn back to Bohm's book [53] and briefly look at his own statement of what he felt was wrong with the EPR argument. Like that of Bohr, it does not really address the point that a measurement on one particle disturbs a second spatially separated particle. In any case, we need not delay to discuss Bohm's arguments in that book, because his views would swiftly change in a dramatic way.

The cause for this change of belief is not quite clear. It is certainly the case that Bohm [74] discussed the book with Einstein, who said that it was the best presentation of Bohr's views, to which, of course, he was opposed, that he had seen. Nevertheless, Einstein persuaded Bohm that the approach to quantum mechanics in his book was not 'complete' in the terms it has been defined above and, according to Bernstein [75], he asked Bohm to perform something 'fairly specific'.

Bohm later suggested to Jammer [59] that, at the time, he had been influenced by a Russian paper, the author of which he could not remember, and perhaps as well as by Einstein, while Hiley [51] reports that Bohm said that Einstein had not influenced him at all! All these statements, of course, were made some decades after the event!

Actually, Bohm was not the only person to write an authoritative text on quantum theory from an orthodox (Copenhagen) position and then almost immediately change his mind and become a keen critic of this position. The same thing happened to Alfred Landé, who in the 1920s made important contributions to the application of quantum theory to atoms—students of physics will remember the well-known Landé interval rule. In 1951 he wrote a textbook supporting the Copenhagen interpretation [76] but thereafter wrote voluminously criticizing it.

However, Bohm [51, 52] now had other problems—his political past was catching up with him. These were the days of the House Un-American Affairs Committee, and Joseph McCarthy. Bohm had been called in for questioning about his Communist past as early as 1949, although David Pines [77], who collaborated with Bohm highly effectively at the time, remarks that the purpose of the committee was to get at Oppenheimer through his contacts with Bohm and others.

Bohm was arrested and charged for refusal to answer the questions of the committee. He pleaded the US Constitution's Fifth Amendment, which allows a refusal to incriminate oneself, and was eventually acquitted; however, Princeton refused to consider him for tenure and paid him for the last year of his contract only on the condition that he did not enter the campus. He was effectively unable to gain employment in the United States; Pines [77] reckoned that the failure to employ Bohm on nuclear fusion, a topic on which he was probably the most knowledgeable person in the world, must have cost the country many millions of dollars.

Bohm was forced to move to Brazil to obtain employment, and he was appointed Professor of Physics at the University of São Paulo from 1951. It was here that he completed his volte-face over quantum theory by writing a pair of papers [78] demonstrating in detail a hidden variable model of the theory. This change of heart could only be seen as an insult by Niels Bohr, Werner Heisenberg, and Wolfgang Pauli!

To give some understanding of Bohm's papers, it is helpful to remember that two of the most important discoveries in the early years of quantum theory were, at least in retrospect, very much related [55]. In 1905 it had been known for around 200 years that light had wavelike properties—interference, diffraction, and polarization. Indeed, at that time, such language would have seemed pedantic in the extreme; it would just have been said that light *was* a wave; but in that year Einstein showed that, in other types of experiment, in particular the photoelectric effect, light behaved as particles.

Similarly, from its discovery at the end of the nineteenth century, experiment showed the electron behaving as a particle. However, in 1923 Louis de Broglie (Figure 2.15) suggested that, in other circumstances, it might behave as a wave and, although he had no evidence for this suggestion, and indeed it seemed a very strange idea, in fact very soon it was found that electrons did exhibit the wavelike property of diffraction. (So, in fact, do any of the entities we normally call particles—protons, neutrons, etc.)

We may sum this up by saying that both light and electrons behave as waves when they travel or *propagate* but as particles when they collide or *interact* with similar objects. This behaviour is often called *wave–particle duality* but, in the

Figure 2.15 Louis de Broglie. Courtesy Cambridge University Press.

context of Bohr's ideas, it may be termed *wave–particle complementarity*: rather, as complementarity says that we may discuss momentum *or* position but not both simultaneously, it also says that we need to use particle-like ideas *and* wavelike ideas at different times, but always one or the other, not both at the same time.

In 1952 Bohm broke this rule. Rather than having wave *or* particle, he had wave *and* particle. There are different ways of thinking about the theory but one interesting and popular one is to regard the wave as dictating the motion of the particle, and so it is often called a *pilot-wave* theory.

In this approach, we have a wave function in the usual way, and, as usual, it is the square of this function that is physically significant. However, in the conventional account, it gives the *probability* of finding the particle at a particular position in a measurement—indeed, it is called the *probability density*; however, in Bohm's version, it gives an actual density of particles. Indeed, as well as the wave function, we have a particle with a precise position and a precise momentum at all times, independent of any measurement. It is the position that is the *hidden variable*.

Bell [79] pointed out that this terminology was 'a piece of historical silliness'. If we perform a measurement, it is the hidden variable for the individual particle that will provide the result. (We shall see the details shortly.) The wave function itself will show up only in statistics across many particles.

The particle follows a trajectory, and its speed may be obtained from the wave function. Technically, the speed is equal to the ratio of the *probability current* to the probability density. The probability current itself relates to the flow of probability and again is obtained from the wave function, and the momentum of the particle is, of course, obtained from the speed. Altogether, it is a relatively simple attractive scheme.

It is important to recognize, though, that while the particle does have a position at all times, a measurement of position does not just record this value. Rather, the measurement process must be described as a physical interaction between the particle and the system that is performing the measurement and, during this process, the value of the position of the particle may change deterministically.

Bell [79] stressed this point, as it demonstrated that measurement should never be regarded as a simple registration of pre-existing properties of the system. Rather, measurement results are the product of the complete set-up of system plus the measuring apparatus. This argument was certainly a central point for Bohr and, while very often disagreeing strongly with Bohr, Bell in this case was in full agreement.

It should be pointed out, though, that in this analysis the apparatus and the observer are given no special status distinct from the system being observed. Neither is there any process in which, from a number of potential results, one is somehow 'selected' and becomes an actual result.

The dramatic point about the Bohm theory is that it reproduces every experimental aspect of regular quantum theory—it is essentially a new interpretation of the same theory, and yet, completely contrary to the claims of the Copenhagen interpretation, it is realist and deterministic and, completely contrary to the argument of von Neumann, it is a hidden variable theory.

It is not, however, local, a point of which Bohm was well aware. To explain this point, it is easiest to sketch the mathematical background of the theory a little more closely than we have done so far [80]. If we take a conventional formalism of classical mechanics, all we need to do to transform it to quantum mechanics is to add an extra term to the classical potential. Not surprisingly, this term is called the *quantum potential*.

This statement sounds very straightforward, but the quantum potential has some properties which are totally different from those of any classical potential. Rather than depending on position only, like a classical potential, it depends on the wave function itself in rather a complicated way. Furthermore, the quantum potential does not fall off with distance, as any classical potential would. The quantum potential, and thus the force, between two particles may remain large even when the particles are a very long distance apart. Indeed, directly from the equations of motion it can be seen that, in the Bohm scheme, the behaviour of any particular particle depends on the positions of *all* the other particles at that time, however far away from the first particle they may be. Clearly, from any classical point of view, this fact is unpleasant! This argument would be very important for Bell.

Bohm himself was wary of one point. His analysis was designed to reproduce the same results that the usual set of ideas would produce, and he felt open to the complaint that his work produced nothing new and was therefore pointless. He tried to defend himself against this charge by suggesting that, while there was agreement between his results and the current ones in areas already investigated, there could be differences for inter-particle distances of about 10^{-15} m, which is much smaller than the size of nuclear radii and at which distance the present theory was unsatisfactory. But, without actual proof, the argument seemed unconvincing. Bell was also to suggest the possibility of flaws in orthodox quantum theory a decade later, but his area of potential weakness was chosen to be much more specific than Bohm's.

We shall make one last point about Bohm's ideas. Since his theory was based on hidden variables, it was in itself an implicit rebuttal of von Neumann's famous argument. Bohm actually went further and claimed to show where von Neumann's argument was in error. He suggested that violation of von Neumann's prohibition required hidden variables in the measuring apparatus as well as the measured system. Bohm's own argument was not very clear or convincing and, indeed, Bell was later to show that it was wrong.

We now move onto the response to Bohm's work. An immediate point is that many of the older people remembered that Bohm's approach was along exactly the same lines as that of de Broglie a quarter of a century earlier. The only thing generally known about de Broglie's ideas during the 1920s is, of course, the wavelike properties of 'particles', but this idea was actually just one component of a substantial body of work carried out from 1923 and culminating in the 1927 Solvay Congress, where three approaches to quantum theory were presented: those of de Broglie himself, Schrödinger, and a combined presentation of Born's and Heisenberg's ideas.

De Broglie presented his pilot-wave theory, which, as already stated, was broadly the same as Bohm's, which has just been sketched. Born and Heisenberg presented an account of quantum theory according to the Copenhagen interpretation, while the presentation given by Schrödinger stressed the wave picture of quantum theory in a mathematical context.

Guido Bacciagaluppi and Anthony Valentini [81] have recently published a full transcription and discussion of this important meeting, which they call *Quantum Theory at the Crossroads*. By the end of the meeting, it might be said that the die was (practically) cast. Schrödinger's physics—the Schrödinger equation—was quite acceptable and indeed greatly prized, but his wavelike interpretation would have to be abandoned; this result was perhaps inevitable. Much more problematically, the de Broglie interpretation was jettisoned, and the Copenhagen interpretation would reign supreme for the foreseeable future.

Why did this happen? James Cushing [82] asserts that the de Broglie theory has 'nothing incoherent or logically inconsistent about it', and, in a chapter titled 'An Alternative Scenario?' in his book *Quantum Mechanics: Historical Contingency and the Copenhagen Hegemony*, suggests that there was no logical reason why de Broglie could not have swept the board. In real life, though, the intellectual power and aggressive approach of Pauli was too much for de Broglie, who voluntarily renounced his own ideas and actually became quite an advocate for the Copenhagen interpretation; in 1952 he even criticized Bohm strongly before reverting after a time to his own original ideas.

When Bohm presented his interpretation, it goes without saying that the high priests of Copenhagen were the first to respond, although their arguments were mostly dismissive rather than substantive. Born wrote to Einstein [83] that 'Pauli has come up with an idea ... which slays Bohm not only philosophically but physically as well'; however, Pauli's actual complaint, that Bohm provided no explanation of the relationship between wave function and probability density, seems tame enough, and Bohm, together with his assistant Jean-Paul Vigier, was able to answer the objection.

Neither Pauli nor Heisenberg was impressed with the suggestion that Bohm might be able to modify the quantum formalism to tackle areas of physics where the present theory was not yet established, Heisenberg [84] suggesting that this was rather like a wish that occasionally 2 + 2 would equal 5.

As Bohm expected and feared, he was criticized because his results only reproduced those of the present theory. Heisenberg called the particle trajectories 'superfluous "ideological superstructure" ', while Pauli [85] described his work as 'artificial metaphysics'.

Bohm might have hoped that he would get a more favourable response from Einstein, but this was not to be. It was Einstein [83] who wrote to Born saying that 'Bohm believes (as de Broglie did, by the way, 25 years ago) that he is able to interpret the quantum theory in deterministic terms? That way seems too cheap to me.'

Born expressed surprise at Einstein's reaction, and Bell, commenting on the exchange, and of course imagining Einstein to be a supporter of hidden variables, was surprised too. However, remembering that Einstein was hoping that all the conceptual dilemmas of quantum theory would be solved as part of a grandiose unification theory, we are not at all surprised that he thought Bohm's clever but basically quite simple scheme was indeed 'too cheap'. If, as Bernstein [75] suggested, Einstein had given Bohm a particular task, then obviously he had not performed it!

Bohm was justifiably disappointed by Einstein's response. Much later [74] he was to write: 'It was important that the whole idea did not appeal to Einstein, probably mainly because it involved non-locality. I felt this response of Einstein was particularly unfortunate as it certainly "put off" some of those who might otherwise have been interested in the approach.' Bohm's suggestion that the problem for Einstein was non-locality is certainly plausible, although of course it was not the point that he made to Born.

Bohm also picked up the same point as Cushing about the 1927 Solvay Congress. He argued that 'if de Broglie's ideas had won the day, they might have become the accepted interpretation: then, if someone else had come along to propose the current interpretation, one could have said that since, after all, it gave no new experimental results, there would be no point in considering it seriously. In other words, I felt that the adoption of the current interpretation was a somewhat fortuitous affair.'

In a recent book, Peter Holland [80] has suggested that Einstein's rebuttal of Bohm's ideas must have been a 'tactical mistake'. Whatever the perceived inadequacies of the theory, 'some model such as that of de Broglie and Bohm was better than none at all in countering the prevailing vagueness of interpretation, at least as a makeshift before a more satisfactory foundation could be found.' But Holland did not think Einstein's support would have helped Bohm, because nobody was taking Einstein's views at all seriously at the time either.

Fortunately there was at least one person (although perhaps only one) for whom Bohm's work was of immense interest, bringing great satisfaction and, one might almost say, joy. This was, of course, Bell. It is not hyperbole to suggest that for him it was intellectually a liberation.

It must be remembered that, from his days as a student, Bell had been extremely dismissive of the reigning Copenhagen orthodoxy. He had been very much attracted by the idea of an approach which would be closer to realism, and it seemed to him that hidden variables was the obvious way of pursuing this ideal. It had only been the von Neumann argument as described by Born that dissuaded him from this belief. Now it seemed clear that von Neumann must have been wrong all along! It could be said that Bell was released from his intellectual shackles; he would take full advantage of this freedom for the rest of his life. Much later, he would write [79]:

> In 1952 I saw the impossible done. It was in papers by David Bohm. Bohm showed explicitly how parameters could indeed be introduced, into nonrelativistic wave mechanics, with the help of which the indeterministic description could be transformed into a deterministic one. More importantly, in my opinion, the subjectivity of the orthodox version, the necessary reference to the 'observer', could be eliminated.

Bell noted that the essential idea had been put forward by de Broglie a quarter of a century earlier, and asked why Born had not pointed this out in his book, and indeed why von Neumann himself had not recognized its significance in his book of 1932.

Bell was to have many arguments in later years over quantum theory, and although he must frequently have been irritated by what he would have seen as the logical failures or evasiveness of his opponents, his reaction would have been polite, if determined.

But one must suspect that he was actually very angry with the response of the leading advocates of the Copenhagen interpretation to the Bohm papers. As they had insisted for several decades that the existence of hidden variables in quantum theory was completely impossible and indeed that this prohibition was an integral part of the logical structure of their own ideas, Bell must have considered their response to Bohm's demonstration of hidden variables intellectually dishonest.

Rather than admit their rather drastic mistake, or even having the humility to search for their error, they merely ignored the real and important significance of Bohm's argument, and concentrated their fire on the fact that his physical predictions did not differ from theirs (apart from his rather unwise suggestion that there might be a difference for small distances).

Heisenberg [84] provided a good example of this in a chapter titled 'Criticism and Counterproposals to the Copenhagen Interpretation of Quantum Theory'. He wrote that a particular group of critics, among which he included Bohm, 'does not want to change the Copenhagen interpretation so far as predictions of experimental results are concerned but it tries to change the language of this interpretation in order to get a closer resemblance to classical physics'. He added that 'one may even say that we are concerned not with counterproposals to the Copenhagen interpretation but with its exact repetition in a different language.'

It is scarcely to be wondered that this statement would make Bell extremely angry! The argument links the experimental results so uniquely with the Copenhagen interpretation that any attempt to question the interpretation is almost by definition futile. As Bohm and Bell (and Cushing [82]) would insist, the reasoning is entirely fallacious. The experimental results exist, and may be interpreted in various ways—via the Copenhagen interpretation, the pilot wave, or, 50 years after Heisenberg's words, many others. Advocates of the Copenhagen interpretation should not assume inherent superiority—it is one interpretation among many, and, at least from Bell's point of view, a rather unsatisfactory one at that.

Heisenberg's implication that Bohm was trying to make the language of quantum theory resemble that of classical physics was also something of a loaded statement. The idea was that the pioneers of quantum theory had had the vision and the courage to attain an entirely new understanding of nature and that this understanding was more subtle and richer conceptually than that provided by classical physics. However, it would be admitted that assimilating and accepting this new set of ideas could be a difficult and perhaps lengthy process.

To accuse Bohm, just as Einstein was accused, of wishing to revert to classical physics is to suggest that neither of them had the flexibility of mind to come to terms with the conceptual revolution required in quantum theory. Thus, they were unable to address the numerous conceptual difficulties that the theory presents.

Now let us compare Bell's ideas as presented in 1982 [79]. He by no means pushed the Bohm scheme as 'the only way' but did ask why it is ignored in the textbooks. Surely, he said, it should be taught alongside other interpretations 'as an antidote to the prevailing complacency'.

As for the idea that a retreat to some aspects of classicality is necessarily a backward step, Bell argued that this would only be the case if the 'forward' steps taken by Bohr and colleagues were either clearly positive in terms of tackling the difficult issues successfully, or even necessary. Bell would vehemently deny both. Bohm's approach, on the other hand, showed that 'vagueness, subjectivity and indeterminism, are not forced on us by experimental facts, but by deliberate theoretical choice.'

Bell did not keep his excitement about the Bohm papers to himself. Mary Bell [48], who was not yet married to Bell at that time, later recalled that 'when the David Bohm papers appeared in 1952, I can remember how excited he [Bell] was. In his own words, "the papers were for me a revelation". When he had digested them, he gave a talk about them to the Theory Division.' Walkinshaw [22] remarked that 'I remember his taking a liking to Bohm's work in particular. When Bohm came to give a lecture it was John [Bell] who shone at question time and it was clear that he had studied Bohm's work with some care [probably an understatement!].'

At Harwell the person with whom Bell discussed Bohm's work most intensely was Franz Mandl (Figure 2.10). Mandl himself [86] was five years older than Bell and had come to England as a refugee from Germany in 1936. He had interrupted his degree in the last years of the war to work on the atomic bomb project at Birmingham University and, in 1950, had moved to Harwell, where he worked for eight years. In 1958 Bell and Mandl published a pair of joint papers, which will be discussed later in this chapter, on general aspects of scattering experiments—the so-called polarization–asymmetry equality.

At this time, Bell's main concern in quantum theory was to get to grips with von Neumann's argument. To Bell it was quite clear that Bohm had shown that von Neumann was wrong—hidden variables *could* exist inside quantum theory. The EPR argument pointed in the same direction, so definitely the next step was to detect the error in von Neumann's argument. However the book in which von Neumann had published his argument had been written in German, a language which Bell didn't know, and the book would not be translated into English until 1955.

In these circumstances, Mandl was an ideal person for discussion on the topic. Of course he knew German, but also he had rather strong views on von Neumann's theorem—he was convinced it was right! He and Bell had good arguments, with no holds barred; these arguments, of course, were hugely beneficial for John, helping him with the formulation of his slowly emerging ideas and also giving him practice in discussion and argument with highly intelligent people who strongly disagreed with his ideas. Mary Bell later remarked that, when Bell gave his seminar on Bohm's 1952 papers, 'There were interruptions, of course, from Franz Mandl, with whom he had many fierce arguments.'

Fifty years later, these discussions remained vivid in Mandl's mind. The two men talked about many things, both inside and outside physics, and Mandl always found Bell's conversation stimulating; however, for him, the high-lights were in the discussions of von Neumann's proof. These discussions went on with some intensity for several months and then more sporadically until Mandl left Harwell. Mandl felt that he was lucky to have been in the right place at the right time.

For Mandl [87], four features stood out in Bell's approach: (i) the simplicity of his language, reflecting, as Mandl felt, the clarity of Bell's thinking; (ii) the originality and depth of his probing of von Neumann's arguments; (iii) Bell's modesty and open-mindedness—Mandl commented that Bell was a good listener; and (iv) Bell's conviction that there was something missing or wrong with quantum mechanics, and his determination to find out what.

Mandl was highly impressed by Bell's personality:

He was a quiet rather private person. At the same time he had an extremely warm personality, confidence inspiring and friendly. He had a dry subtle sense of humour, at times a little bizarre but always without malice. He was able to laugh at people and things. He was clearly a profound thinker. In speaking with him, an answer was not always instantaneous, but it was thought through and carefully expressed. Although he spoke very quietly, he often conveyed the feeling of great conviction and intensity.

Mandl had occasionally seen Bell angry; when angry, Bell would still talk quietly but the intensity in his voice would double. One example concerned the establishment of the state of Israel; he did not feel that you could take the land away from a people. Mandl added that, of course, Bell was in no way anti-Semitic. The other thing that Mandl mentioned as making Bell angry was the treatment of the work of de Broglie and Bohm, already discussed at length in this section. Overall, Mandl considered Bell one of the most impressive people he had known. His admiration of him as a thinker and person was boundless.

Much later, Bell was asked by Paul Davies [88] how long his famous results published in the 1960s had taken to obtain. He replied that, for many years, the ideas had been at the back of his mind. It is fairly clear that, from 1952 and Bohm's papers, his prior resolution to leave the basic ideas of quantum theory on one side had been overtaken by events. Mandl himself later said that, while he did not think Bell had made a breakthrough on the von Neumann issue by 1958, at which point Mandl had left Harwell, he had wondered why it had taken Bell 14 years between reading the Bohm papers and formulating his own response.

It may be surmised that Bell was prepared to take his time in thinking out the issues as thoroughly as conceivably possible. He was in no hurry—there was little possibility of his being scooped (although one can, of course, never be too sure of that). Still a young man in 1952, he must have felt it would seem presumptuous for him to dive in where Einstein, Bohr, and von Neumann had, at least to an extent in his opinion, floundered. He certainly did not want to state his case and be found in any way wanting. He was prepared to bide his time.

When Mandl left Harwell in 1958, it was to take up a lectureship at Manchester University. Although he was an able research worker—he ended up as Reader—it probably became clear to him that he was even more interested in teaching and in writing books for physics students. Even while he was at Harwell, he was writing a book on quantum field theory, and he enjoyed discussing the book with Bell.

At Manchester, from 1970, he became the instigator and Chief Editor of the Manchester Physics Series [89], an extremely popular series of student texts of moderate length. He himself wrote the books *Quantum Mechanics* and *Statistical Physics*, which gained an enviable reputation and have remained in print, in successive editions, for several decades. He retired in 1984 but was working at his editorial task until a few days before his death in 2009.

We here may also briefly review the future of Bohm and his theories. Bell and Bohm incidentally interacted very little (although there was one rather crucial interaction in 1964) because Bohm's ideas veered well away from traditional theoretical physics. In 1984, though, Bell did write a paper in the collection of essays in honour of Bohm [74] and he dedicated the paper to him, presumably in gratitude to him and in recognition of the impetus given to Bell by the 1952 papers.

Bohm [51, 52] himself stayed in Brazil until 1955 and then moved to Israel, where he worked until 1957. Actually, as his American passport had been confiscated by the American consul in Brazil, Bohm had to take out Brazilian citizenship in order to leave the country. From Israel he moved to England, where he was to stay for the remainder of his life; he had a spell at Bristol University and then, from 1961 until his death in 1992, he was Professor of Theoretical Physics at Birkbeck College in London.

Birkbeck College was definitely not one of the grand colleges of London University. In fact, it was the only college which provided lectures in the evening only; they were primarily for part-time students, and the workmanlike approach to study may have appealed to Bohm with his egalitarian views. His interests encompassed philosophy, mathematics, psychology, biology, and art, as well as physics, and this fact is demonstrated by the wide range of contributions to the essays in his honour [74].

Personal life in the 1950s

While Bell would certainly have missed his home and his family—he was to make reasonably lengthy visits to Belfast at Easter and Christmas and in the summer, at least until he was married—it must have been invigorating to have, as well as a challenging and interesting job, more freedom than he had been used to and also a little money in the pocket.

Figure 2.16 John Bell, on his new motorbike at Harwell. Courtesy Ruby McConkey.

At Harwell he lived in the 'A' hostel, the former officers' mess, now occupied by the scientists. Then, when he moved to Malvern, he lived in the Geraldine Road hostel. Mary Bell [48] has said that he was without the famous beard and was one of a group of young men, all with motorbikes, which they regularly took to pieces. For Bell, the motorbike (Figure 2.16) was probably his first (moderate) extravagance and he would have enjoyed excursions (Figure 2.17)

Leslie Kerr [23], who had known Bell from Belfast, spent quite a lot of time working in places reasonably close to Malvern and Harwell between September 1949 and 1958, and the two met up quite frequently. Kerr began a PhD in Belfast, working on electron/atom collisions but, in the summer of 1950, he was sent for a two-month course in electronics at TRE in Malvern. He was met by Bell at the railway station and he remembers the rather hair-raising journey on the pillion of Bell's old Velocette 350 to Geraldine Road. Bell had no goggles and was in any case short-sighted; since the road was full of pot-holes, Kerr was relieved to reach the hostel in one piece!

Perhaps Bell may have been rather a reckless motorcyclist or maybe just inexperienced, but at least two hazardous incidents have been reported. In his memoir of Bell, Kerr said that, on one occasion, Bell was travelling at a fast speed when a mishap allowed the valve to fall into the cylinder, where it went straight through the ascending piston. Had the engine seized, Bell would probably have been killed; but, fortunately, he was able to coast to a standstill.

Figure 2.17 John, on the beach.

Annie Bell [90] tells of an even more serious incident. Bell and a friend were travelling on separate machines, again at a reasonably high speed, when they came across a sharp change of surface. Bell was thrown off the bike and suffered a severe cut around the mouth. According to Annie, this incident underlies the origin of the famous beard, although Bell's brother Robert [91] tells a slightly different story—a nurse told him that his brother's injuries would have been more severe *if he had not already had the beard*. Perhaps at that time, the beard appeared and disappeared. It may not be particularly surprising that, when Bell graduated from motorbikes to cars, he settled for sedate models, such as Volvos!

In his memoir of Bell, Kerr recalls many pleasant times spent with Bell during the period when they were both in Malvern. They spent several evenings testing extrasensory perception. One of them would think of a number, and the other would write down a guess. Rather surprisingly, agreement was significantly more than chance could account for, but then, Kerr says, 'with his typical simplicity and elegance', Bell suggested that one of the lists should be inverted. A new comparison gave just as good agreement as before. Presumably, the distribution of numbers on the two lists was not random.

Kerr also recalls going to a demonstration of hypnotism with Bell, and also to the Malvern Festival Theatre to attend either a Shaw play or an Elgar concert. Kerr cannot remember which, but remarks that Bell had told him that the only aspect of music that appealed to him was the rhythm, a fact that surprised Mary Bell when Kerr told her after Bell's death. After the performance, Bell and Kerr had gone walking in the Malvern Hills, where again they were at least in a little danger—the possibility that, in the pitch-black darkness, they might tumble into a quarry.

Another amusing evening was spent studying the well-known '12 ball problem', which Bell had brought back from work. In this problem, one is given 12 numbered balls and a pair of scales; all of the balls have the same weight, apart from one, which may be heavier or lighter than the others. The task is to determine which is the odd ball, and whether it is heavier or lighter, by three independent weightings. Bell enjoyed solving the problem, but found demonstrating the proof that no solution is possible for 13 balls even more interesting than finding the original solution.

In that summer, the two friends went on a camping and cycling trip to the Lake District. This expedition entailed the not-unusual delights of a British summer holiday on the cheap—pitching tent supposedly in open country and waking up in the middle of a council estate; becoming so saturated that they tried to get into a couple of hotels but were turned away for being unshaven and scruffy; and being shown pity by a pensioner couple and allowed to stay the night but then having to leave before daybreak because such activity violated the by-laws. They managed two good walks: one around Derwentwater in excellent weather, and one near Borrowdale in atrocious weather.

Two other memories of Kerr may show a little more about Bell, his thoughts, and the impression he gave to others at the time. At lunch in the Geraldine Road hostel one day, a topic came up and Bell started to pronounce on it, rather as if he were addressing a meeting. One of the group at lunch imitated winding up a gramophone under the table. As Kerr says, this 'expounding' style was totally at odds with the relaxed lecturing style for which he would be famous later. Another comment from the time is that of John Perring [92], who remarked that Bell was chiefly noticeable for his 'vegetarianism and generally ascetic air'.

Kerr's other memory is that once, when the pair were walking near Malvern Girls' College, Bell remarked: 'Women are a different species.' When Kerr challenged this statement, Bell became heated and insisted on the truth of his aphorism, much to Kerr's surprise. Kerr was not sure whether the remark referred to Mary or some other girl of his acquaintance.

In 1952 Kerr took up a Warren Fellowship of the Royal Society at the University of Birmingham and, in 1957, he was appointed to a lectureship there, resigning in 1958 to move back to Queen's University Belfast. During this period, he visited Malvern on occasion and, indeed, Bell himself worked in Birmingham for a year; however, with the arrival of Mary on the scene, naturally relations between Bell and Kerr became less close than they had been. Even during Bell's period in Birmingham, he was anxious at weekends to get back to Harwell and Mary. Kerr remembered from that time only a few conversations, mostly on physics.

Then, when the Bells moved to Geneva, interactions between Bell and Kerr were reduced to occasional meetings in Belfast when Bell was 'home'. Nevertheless, they remained good friends, and Kerr was delighted when, as already mentioned, he was invited to the dinner following the 1991 CERN symposium in honour of Bell and, indeed, he was placed in a prestigious place there.

Now we must turn to the meeting of John and Mary Bell, as theirs was obviously by far the most important relationship of either of their lives.

Mary Bell, née Mary Ross [37], was born in Scotland and has retained a pronounced and pleasant Scottish accent all her life. Her parents were a little higher up the social scale than John's, and the family environment was also more bookish. Mary's father started work as a clerk in the shipyard, moving up to become a commercial manager specializing in the sale of wood, while her mother was an elementary school teacher; the family was vegetarian.

From the time of her childhood, Mary had enjoyed arithmetic problems, and it was lucky that the school she attended was co-educational, the presence of the boys ensuring that there was a quorum for a physics course, which Mary took. She performed extremely well at school, but her family's slight affluence compared to that of her husband-to-be did not mean that she would have been able to attend university without financial support. Fortunately, she obtained a bursary to attend the University of Glasgow, where she studied mathematics and physics. The Scottish course took four years but, after three, since it was 1944 and wartime, she was seconded to TRE at Malvern to work on radar, although she has claimed that she did not achieve a great deal there.

At the end of the war, she returned to Glasgow to finish her degree and then, like John, took up employment at Harwell, where she worked on neutron physics, specializing in computational work. During John's brief period in Harwell before he moved to Malvern, she must have come to his notice, although whether it was through meeting socially or through discussions on physics is uncertain!

Some time after John had joined the Walkinshaw group in Malvern, the amount of work requiring numerical calculations there had risen considerably, and Walkinshaw was intending to ask Jack Howlett, who was head of computing at Harwell, for 'one of his girls' to be moved to Malvern. As to which one to choose, since John had been working at Harwell recently, Walkinshaw

[22] asked his advice. John suggested Mary Ross and, when Walkinshaw asked, 'Is that the fat one, the tall one …?', replied 'No. She's the pretty one.'

John's motives may have been mixed, but certainly there is no doubt that Mary was excellent at her job, which was always on the computational aspects of research. First at Malvern and Harwell and later at CERN, her work was of the highest quality, and, at CERN, both John and Mary were to obtain permanent positions, which required work of very high calibre.

She did move to Malvern and, clearly, the relationship progressed suitably. It was not so very long afterwards, in May 1954, by which time the Walkinshaw group had moved to Harwell, that Mary Ross became Mary Bell.

Mary's parents were a little older than John's and not in particularly good health, so they did not wish to travel to Malvern for the wedding [90] (Figure 2.18); in addition, John's father was also a reluctant traveller (in complete contrast to his mother, who would willingly have travelled anywhere!) It was therefore decided that the wedding would be in Harwell and would be extremely quiet and that the newly married couple would make trips to Belfast and Glasgow to meet parents and in-laws in the next few months.

This approach to matrimony was thus convenient for all, but one must suspect that a low-key ceremony in any case suited both parties. Perring [92], who was told the story by one of the witnesses, later described how the day went. The two participants, with the two witnesses, walked a mile and a half to Rowstock Corner, from where they caught the bus to Wantage. (At least they did not arrive on motorcycle and pillion!) They were married in the registry office there, had lunch at 'The Bear', and returned by the same route to continue with their work for the rest of the afternoon.

The newly married couple moved into a Harwell prefab (Figure 2.19), although actually at this time and for a few more months John's weeks were spent in Birmingham, the marriage being restricted to the weekend [37]. John had to tell his parents that his visits to Belfast would not be as frequent, which of course they understood, and Annie made up for this by visiting Harwell when she could. She was very impressed by the prefab [90]. Mary did travel to Belfast with John, but fairly naturally later when they lived in Geneva and as the Troubles developed in Belfast, she was a little reluctant to come, so sometimes John went on his own. Nevertheless, John's relatives were certain that Mary was a perfect wife for John (Figure 2.20).

Indeed, it is clear that the marriage was ideal for both Mary and John. It gave them a perfect space in which to practice their mutual passion for studying the physical world. During John's accelerator period, they produced one report together and when, towards the end of his life he returned to working on accelerators, at a time when in fact quantum theory was needed to solve some of the problems, they happily collaborated together.

(a)

(b)

Richard Ridley

William Watkinshaw

Margaret Walkinshaw

Mary Bell

John Stewart Bell

Sheila Roberts.

Figure 2.18 (a) Mary and John Bell on their wedding day, May 1954. From left: Richard Ridley, William and Margaret Walkinshaw, Mary and John Stewart Bell, and Sheila Roberts. Courtesy Ruby McConkey. (b) Back of the photo, with a list of the names of the people in the photo. Courtesy Ruby McConkey.

Figure 2.19 John Bell, in front of the prefabs at Harwell. Courtesy Ruby McConkey.

Between these periods, of course, their main research interests were very different. However, John's work on quantum theory was perhaps somewhat different in general. This was his 'hobby', and much of the thought must have taken place at home when Mary was around. His tribute was heartfelt at the end of the preface to *Speakable and Unspeakable in Quantum Mechanics*: 'I here renew my warm thanks to Mary Bell. When I look through these papers again I see her everywhere.'

Outside work, they both enjoyed the outside—walking and later skiing, although, as mentioned earlier, they did not take part in active sports. They enjoyed the hills of Berkshire, although admittedly not as much as they would the mountains of Switzerland [37].

Figure 2.20 Mary and John at Stonehenge. Courtesy Ruby McConkey.

In character, they were both private people. This certainly did not mean that they were not polite and good company. At work and in their social lives, they would, of course, interact with others in a positive and pleasant way. Yet, with one possible exception in the 1980s, it might be suggested that they never actually *needed* others. Perhaps they were never happier than when walking the hills or mountains together, or closing the door on their Harwell prefab or their Geneva flat and being able to concentrate on their physics and each other.

Birmingham, Peierls, and CPT

Bell's leave of absence at Birmingham University began in October 1953. Technically, he was on a year's leave from Harwell. Although, as has been said, he had contemplated going to Glasgow for this year, there can be no doubt that Birmingham was the best place for him. In 16 years as Professor of Mathematical Physics (later Theoretical Physics), Peierls [12, 13] (Figure 2.21) had built up an exceptionally strong research group. Freeman Dyson, who had been a teaching fellow at Birmingham between 1949 and 1951 [93] when he had already made an extremely important contribution to quantum electrodynamics, wrote that the department had 'left Oxford and Cambridge far behind'.

Figure 2.21 Rudolf Peierls. Courtesy Royal Society.

Peierls himself was one of the very last physicists who contributed enormously to virtually every branch of theoretical physics. As we saw earlier in this chapter, he had kick-started the atom bomb project with Frisch, and he was to remain highly influential right through the whole process. While he had obviously been concerned that Hitler should not get the bomb first, it should not be thought that Peierls was at all a warmonger; after the war he played an important role in the Pugwash Movement, the physicists' attempt to contribute to world peace.

He made other important contributions to areas as diverse as solid state physics, nuclear physics, quantum field theory, statistical mechanics, and superconductivity. Undoubtedly, he achieved far more than many Nobel Laureates, but it could be said that he missed the one crucial contribution that would have sent him to Stockholm.

Unlike many brilliant researchers, Peierls was also deeply concerned with teaching, and both at Birmingham and, from 1963, at Oxford, he built up departments that combined strong research with an equally passionate interest in students and their progress. Among the promising young people in the

Figure 2.22 Paul Matthews. Courtesy Royal Society.

Birmingham department were Sam Edwards [94] (Figure 2.10), Paul Matthews [95] (Figure 2.22), and Walter Marshall [96] (Figure 2.23). All three not only carried out research of the highest calibre but reached important positions in universities and beyond. Edwards was Cavendish Professor at Cambridge from 1984 to 1995, Matthews became Professor at Imperial College London from 1952 and was later Vice Chancellor at Bath University, and Marshall became Chair of the United Kingdom Atomic Energy Authority and later of the Central Electricity Generating Board. Edwards was later knighted and Marshall became a baron. All three became Fellows of the Royal Society.

In some ways, Peierls' wife Genia, who came from Russia, was almost as well known in physics circles as her husband. A woman of tremendous personality, she had had four children of her own but welcomed many of Birmingham's research workers to stay in her house [97]. Dyson, for example, has said that Peierls himself was always supremely busy and it was Genia's hospitality that made his stay in Birmingham most memorable. Bell, though, while admitting Genia's many good points, confessed to Bernstein [37] that he found her rather overwhelming, and could not have stood actually living in the Peierls house.

Figure 2.23 Walter Marshall. Courtesy Royal Society.

He was, of course, returning to Harwell at weekends to see Mary, and it was during his period in Birmingham that they were married.

At the outset, Peierls told Bell that, recognizing Bell's slightly advanced age compared with most research students, and particularly his maturity and considerable research accomplishments, he did not intend to treat him as a beginner [37]. Bell was asked to give a short talk on some area of research to introduce himself to the group, and he suggested two topics to Peierls. The first was the theoretical aspects of accelerators, on which, of course, he was already an acknowledged expert. However, since the publication of Bohm's papers, Bell had been intensely involved with the foundations of quantum theory, and his second proposed topic was the work of Bohm.

Peierls considered himself to be a devoted follower of the Copenhagen doctrine, to the extent that he did approve of the term 'Copenhagen interpretation' because he felt that it implied that there were other valid (or even invalid) interpretations [98]. For him, what was called the Copenhagen interpretation was just quantum mechanics! Thus he had no wish to hear Bell expound on the merits of Bohm and asked him to talk on accelerators; doubtless, Bell's account of them would have been excellent.

Over the next decades, the interactions between Bell and Peierls over the foundations of quantum theory would be quite interesting. Peierls never accepted Bell's view that the Copenhagen approach was flawed in any way, and that new interpretations could be required. However, he did come to appreciate very much the actual mathematics of Bell's theorem. And, in the last decade of Bell's life—technically, probably just after he died—it appeared that Peierls' views were in any case rather different from those normally associated with the Copenhagen interpretation (see Chapter 5). What is pleasant to record, though, is that throughout the two men retained a high regard and personal friendship for each other [99].

With the talk over, Bell had to choose or be given a topic of research. Actually, while Peierls was nominally supervisor of all research students in the department, Bell was to be guided on a day-to-day basis by Matthews. Bell and Matthews were to become very close and, when Bell was proposed as a Fellow of the Royal Society, Matthews was the main sponsor.

However, it was Peierls that chose the topic of research, and here, in his memoir of Bell, Kerr [23] has maybe muddied the waters with his memories of half a century before. To explain his suggestion, it is necessary to give a brief elementary account of quantum electrodynamics (also known as QED), which is essentially quantum field theory for the electromagnetic field [5]. Whereas, in quantum theory itself, photons (which are the particle aspect of light) appear and disappear as atoms make transitions, quantum electrodynamics is a theory that encompasses the photons as well as the atoms.

However, in the years before the 1939, the theory met seemingly insuperable obstacles. Virtually every calculation of even the simplest physical quantities led to infinite results. Tantalizingly, if one took what should have been the most important term in an expansion, the answers were good; however, when other terms were included, which one would have hoped would cause only small changes to the initial results, the mathematical result, as physicists say, 'blew up'.

After the war came the breakthrough [5]. In 1948 two Americans, Julian Schwinger and Richard Feynman, and the Japanese physicist, Sin-Itiro Tomonaga, independently came up with solutions. Feynman's approach was very pictorial, involving the famous and attractive *Feynman diagrams*, which actually implied deep mathematical truths. The approaches of the three men could not, at least at first sight, have been more different, but each used fundamentally the same 'trick' to remove the infinities: essentially, one infinity was subtracted from another, leaving a finite result. The process was known as *renormalization*. The contribution of Dyson was to understand the three approaches and to be able to relate them.

Some giants from the previous generation, in particular Paul Dirac (Figure 2.24), denounced the procedure as mathematical nonsense, as

Figure 2.24 Paul Dirac (right) and Wolfgang Pauli in 1938. Courtesy CERN.

subtracting infinites without strict criteria must always be worrying; but the method took hold and the three founders were to share the Nobel Prize for Physics in 1965. Now a principal requirement for any quantum field theory is that it should be renormalizable.

Bell's choice of topic was occurring, of course, only five years after the original work, and Kerr's idea that Bell wanted to work out what 'was wrong' with quantum field theory, while Peierls wanted him to calculate results of practical interest, makes sense in its own terms. However, Kerr also suggests that this disagreement led to Bell giving up working with Peierls. Since the latter is, at the least misleading—Bell's day-to-day guide was Matthews, but under the general supervision of Peierls—one may be fairly confident that the former is not true either. It seems likely that Kerr's memory is of Bell saying that he followed Dirac in being totally unconvinced by renormalization and that he might like to 'put it to rights' someday—that is not at all unlikely, and he might well have seen the 'fudge' in quantum electrodynamics as not dissimilar to those he hated in quantum theory itself.

He must have realized, though, that Peierls would never have let him work on such a fundamental and probably intractable problem, and he would certainly have had no wish to fall out with the older man. It must also be said that later in his career, Bell took renormalization for granted and never had any inclination to question it.

The problem that Peierls did suggest was actually extremely interesting, and it again requires some basic information on the physics involved [5].

The original discoveries of quantum theory by Heisenberg and Schrödinger in 1925–6 were non-relativistic, and it was Dirac who was first able to bring in relativity, although only for the hydrogen atom. This finding was a great discovery and gave Dirac a share with Schrödinger of the 1933 Nobel Prize for Physics, Heisenberg being awarded the whole 1932 one for himself.

In every way but one, the work of Dirac was immediately seen to be a great theory. In particular it demonstrated the existence of the spin of an electron, with all the correct properties, from first principles; until then, the spin had been tacked on to the basic formalism in an ad hoc and rather unsatisfactory way.

However, the seemingly crippling snag was that the theory predicted the existence of electrons with negative energy and mass as well as those with the expected positive energy and mass. That such potentially supremely important work seemed to be so flawed was a great disappointment; Heisenberg called it 'the saddest chapter in modern physics'.

Dirac was eventually able to come up with an idea that could explain his theoretical results. The negative energy electrons, he surmised, formed an infinite sea which was itself unobservable, and it was only when an electron was able to jump from this sea to a positive energy state that the *lack* of this electron in the sea would be observable. The lack of an electron with negative mass and negative charge is a particle with positive mass and positive charge, and the existence of such a particle was Dirac's new claim. A perhaps even more startling suggestion was that this particle could be nothing other than the proton and, if this were true, far from the theory being a failure, it was a magnificent success, explaining not only the electron but the proton as well!

It could be said that the chief aim of atomic physics and later elementary particle physics is to explain the physical universe in terms of the smallest number of basic particles. From this point of view, Dirac's theory led to the most successful moment in the history of physics—everything could be explained in terms of two particles: the electron and the photon.

Unfortunately, it didn't work! In 1930 Oppenheimer was able to point out that such a theory would make the hydrogen atom unstable, as the proton and the electron could decay to two photons (two rather than one being needed to satisfy conservation of momentum). It was thus clear that *if* Dirac's particles of positive charge existed, they were definitely *not* protons.

Then, in 1932, Carl Anderson found such particles: studying cosmic rays with the aid of a cloud chamber and a very strong magnetic field, he saw particles travelling for which the size of the curvature in the field was the same as that for electrons but the curvature was in the opposite direction. This observation showed that the particles had a positive charge but the same mass as the electron. These were the particles Dirac had predicted, at least before his flirtation with the proton. The new particle was called the *positron*, and it was termed the *antiparticle* of the electron, and the first example of *antimatter*. When electron and positron meet, they *annihilate* each other. For this important discovery, Anderson was awarded a share of the Nobel Prize for Physics in 1937.

The change in the landscape of theoretical physics was enormous. It could be assumed that every particle would have an antiparticle, though in a few cases antiparticle would not be distinct from particle. For any charged particle, particle and antiparticle clearly differ as they have opposite signs of charge. For example, the antiproton has a negative charge.

The antiparticles of many uncharged particles also differ from the corresponding particles. For example, neutron and antineutron differ in the relative orientations of spin and magnetic moment. (Of course, once it was realized in the late 1960s that neutrons were composed of quarks, it would be said that the difference was that antineutrons were composed of antiquarks.) Similarly, neutrino and antineutrino have been said to differ in the relative directions of spin and velocity, although, now that it is known that neutrinos have a small but non-zero mass, this distinction is less clear. The important example of an antiparticle which is identical to the particle is the photon.

As has been said, theory suggested all this practically at once, but experimental confirmation of some aspects was necessarily slow, because of the large amount of energy required to produce an antiproton or antineutron. They were not detected until the mid-to-late 1950s.

We may now introduce the *charge conjugation operator*, which is always written as 'C'. This operator replaces all particles by their antiparticles so, for example, electron becomes positron, antiproton becomes proton, and photon remains unchanged.

Now we reach our main point. We start from an event that takes place physically, and we are essentially thinking of an event with a small number of elementary particles taking part, a decay of a particle or an interaction of two particles. We then apply C, changing particle to antiparticle. The important assumption is that the result caused by this procedure will be another possible physical event, that is to say, an event that is allowed by the laws of physics and so may take place. This assumption was made fairly universally in the early 1950s, on general grounds but with no actual proof, theoretical or experimental. It was expected to apply if the event were governed by any of three of the four fundamental forces of nature—electromagnetism; the weak nuclear force, which causes radioactivity; and the strong nuclear force, which was originally regarded as binding nucleons (protons and neutrons) together to form nuclei but is now regarded as binding quarks together to form the nucleons themselves. The fourth force is that of gravity, which we do not bring in here since it has an extremely small effect at the level of fundamental particles.

We may call this assumption a belief in the conservation of charge conjugation, or just the conservation of C. Alternatively, we say it assumes that C is a good symmetry.

A second operation that may be applied to a physically allowed event is that of *parity*, always written as 'P'. This operation relates to reflection of the event. In principle, the reflection should be about each of three perpendicular planes—mathematically, we could say the *xy-*, *xz-*, and *yz-*planes—but in practice it is usually simpler, for any particular event, to use just one strategically chosen plane of reflection.

Just as for C, we may say that, at the beginning of the 1950s, it was generally assumed, again on general grounds but without any proof, that the event thus obtained would be a perfectly proper physical event. This may be called conservation of parity, or conservation of P, or symmetry under P.

One way of discussing conservation of P is that, once the *x-* and the *y-*axes are defined for a physical event and pictured as usual on, say, a piece of paper, the physical system cannot 'prefer' a third direction coming out of the paper rather than going into it, or the other way round. An even simpler way is to say that physics cannot 'choose' its right hand rather than its left.

Anybody who has taken a course in basic electricity and magnetism will probably be up in arms about this last statement, for it must be admitted that such courses are full of references to right and left hands, such as the left-hand motor rule, the right-hand dynamo rule, the right-hand grip rule, and so on.

In fact, a little study will demonstrate that all these rules are related to the explicit presence of a magnetic field in the analysis. If the magnetic field is produced by an electrical phenomenon and results in a mechanical effect and if one is able to 'short-circuit' the magnetic field in the analysis by going directly from the electrical phenomenon to the mechanical effect, no use of right or left hands is required.

The fundamental mathematical point is that, although in many ways behaving as vectors, the magnetic field and also the quantities associated with rotation, such as angular velocity and angular momentum, behave in a different way under reflection. What we may call normal vectors, technically called *polar vectors*, include quantities such as force, velocity, momentum, and electric field. Their direction is absolutely clear physically. It is obvious that, when they are reflected, their direction will change.

The magnetic field and the rotational quantities behave differently. Indeed, for the rotational quantities in particular, the student is always quite justifiably surprised to find that these quantities may be treated as vectors and has to be told what direction should be assumed. Nevertheless they, and the magnetic field also, can highly conveniently be treated as vectors in every way except, as we have been discussing, that they do not change direction under reflection. They are called *pseudovectors* or *axial vectors*.

As a technical point for those who have studied basic ideas of vectors, we may mention the *vector product*, often called the *cross product* of two vectors. The details of the relationship are very fundamental in the subject, of course, but are a little complicated and are not needed here. The very idea of the product, though, implies that, if the two constituent vectors are polar vectors and so change sign under reflection, the vector product itself cannot change sign and so must be a pseudovector.

Anyway, the point of introducing this material on electromagnetism is to reassure the reader that, in the early 1950s, as indeed today, there was little or no doubt that P is conserved in electromagnetism.

We now move onto a third operation: time reversal, or 'T'. To study this, we may, for example, take a film of a physical event and then project it backwards in time. Conservation of T would imply that the physical event thus shown would still be a perfectly proper physical event; it would obey the laws of physics. As for C and P, in the early 1950s it was generally assumed, although again with no proof, that T was conserved in all interactions, or we might say, in interactions governed by the three fundamental forces we are considering. We may alternatively say that physics is time-reversal symmetric or time-reversal invariant.

To clarify this last operation, we must remember that we are discussing an event containing a small number of particles. For a system of many particles, the second law of thermodynamics tells us that there is an *arrow of time* and that processes naturally move in a particular direction and not in the opposite direction [100]. The conceptual problem of how time reversibility for a small number of particles (and Newton's laws certainly obey time reversibility) seems to change to irreversibility for many-particle systems is, of course, a major one for thermodynamics and statistical mechanics but is not relevant here.

To sum up this topic, there are three symmetries: C, P, and T; at the time when Bell went to Birmingham, it was generally assumed that each of them held in all interactions, although there was no proof. In the 1950s and 1960s things were to change dramatically and Bell would contribute substantially to this process.

We now though come to Peierls' suggestion for Bell's research. Just at this time there was news of experiments that seemed to give evidence of a negatively charged particle which was stable but had a mass less than that of the proton. It was the fact that the mass was comparatively low that meant that it could be produced by the accelerators of the day.

The experimenters asked Peierls whether it was possible that this particle could be the antiproton. It seemed very unlikely, since it was generally assumed that particle and antiparticle should have the same mass, like the electron and the positron. But just possibly, as was to turn out for the conservation of C, P, and T, what was generally assumed to be the case was not actually so.

Peierls suggested that Bell think about this, and, as he says [99], this was a 'problem after his [Bell's] own heart'. Bell hated to assume that views were necessarily correct just because they were commonly held. He always wanted to ask: 'How do you know?'

To answer this problem, in due course Bell came up with what remains an extremely important theorem in quantum field theory: the CPT theorem. This theorem said that, while it was only an assumption that any of the three operations individually on a physical situation produced another allowable physical situation (an assumption that would in fact soon be found wrong), the strict laws of quantum theory *demanded* that the combined operation of all three act in this way [101]. This theorem confirmed that particle and antiparticle must have the same mass, and indeed the experiments suggesting otherwise were soon to be found wrong.

We shall come back to Bell's own work and presentation of the theory shortly, but first it must be mentioned that unfortunately it soon emerged that Lüders [102], with whom Bell had already competed over the analysis of strong focussing in accelerators, had already produced an independent proof of the same theorem. Indeed, this proof reached Birmingham not as a *preprint*, a manuscript distributed at the same time as the paper was submitted for publication, but as a *reprint*, so the paper had actually been through the refereeing stage and had already been published. Lüders was a whole year ahead! And Bell could not even rule out the possibility that Peierls had heard some garbled report of Lüders' work so, in a sense, Bell's may not have been actually totally independent [21].

In fact, though, Bell's work was so different from that of Lüders that he was able to publish his own work independently, and indeed it was published in the very prestigious *Proceedings of the Royal Society*. And, since his derivation was much simpler and more straightforward than Lüders's, it is actually more accessible for many readers.

Despite this fact, Bell is extremely rarely credited with this independent discovery of the theorem. The chronology might seem to make this reasonable, but Roman Jackiw and Abner Shimony [103] believe that it is 'presumably because he was not in the circle of formal field theorists (Wolfgang Pauli, Eugene Wigner, Julian Schwinger, Res Jost, and others) who appropriated and dominated the topic.' In fact, because of independent work by Pauli [104], the theorem is nearly always referred to as the 'Pauli–Lüders' theorem.

It may be fair to comment, though, and this will not be for the last time, that perhaps Bell did make it rather easy for others to ignore or just fail to notice his work. In his paper, he mentioned the work of Lüders, of course, and also papers of Wigner, Schwinger, and Pauli from Jackiw and Shimony's list, as well as a few other papers that had been published around the same time. He remarked, though, that 'there still seems room for an elementary treatment starting from first principles.' In comparing his approach to that of Lüders, he remarked that his own approach differed mainly in stressing the rather classical analogue. It must be said that none of this actually demands attention!

The actual paper is, as one would expect, beautifully written, the main content being absolutely clear and the mathematical details relegated to appendices. It reads just like a well-constructed textbook, but a research paper should probably be somewhat different if it is to attract attention, making the most of the newest and most exciting aspect of the work!

Interestingly, at the very end of the paper, Bell mentioned the idea of renormalization, pointing out that his work explicitly dealt only with systems without renormalization. He suggested though that his conclusions should also apply to the renormalized case. His stress on this point perhaps suggests that Kerr's memories may be correct to some extent and that Bell was indeed concerned with this point.

Could the paper become a doctoral thesis? While obviously the quality was there, either Bell himself or Peierls must have decided that the quantity of work needed to be greater, and performing the extra would have to wait until the return to Harwell. However, the important thing for Bell was that Peierls [99] had been as impressed with Bell as Walkinshaw had predicted he would be. Bell had become popular with the members of the department in Birmingham, as they were impressed with his ability and also the clarity of his thoughts. Rather, as in Harwell, at first they had thought he was a little pedantic but soon realized that he was merely taking care to get the important points across accurately. Peierls also thought that Bell's experience with the 'concrete' problems of accelerator design influenced his later style; faced with an abstract problem, he would always try to find a simple tangible example to test his general ideas.

So Peierls arranged that, when Bell returned to Harwell, he returned to the Theoretical Physics Division. Much as Bell had enjoyed working with accelerators, he would certainly appreciate the opportunity to work on more fundamental topics—quantum field theory, nuclear physics, and elementary particle physics. These areas of physics would take the greatest amount of his time for the rest of his life, although towards the end of his life he would return to perform some excellent and important work in accelerator physics. He would also, of course, think a great deal about the foundations of quantum theory, although, since he felt that he was not employed to do such work, it was, at least for many years, kept firmly in the background.

His return to Harwell must also have been appreciated by both John and Mary Bell as, up to this time, their marriage had been 'a thing of weekends'. It could now begin in earnest!

Back to Harwell and to theoretical physics

Back in Harwell, Bell joined the group of Tony Skyrme [105], which was working mainly on nuclear physics [99], since this subject was, of course, the official and main interest at Harwell, and indeed the members of the group tackled several of the experimental problems that came up in the work at Harwell. Bell would probably have preferred it if he had been able to have the freedom to devote more time to quantum field theory and elementary particle physics rather than nuclear physics and, indeed, the work he carried out with Skyrme himself was more general in nature than that of the others in Skyrme's group. However, this period has been described as 'the golden age of nuclear physics', and Harwell played a big part in the work.

Skyrme (Figure 2.25), who had been born in 1922, studied first at Eton, and then at Cambridge, where he showed that he was clearly an outstanding mathematician, and in June 1943 he was assigned to work on the atomic bomb project under Peierls in Birmingham. He soon showed great skill in answering problems of the experimentalists, in particular in clarifying the effect of the restriction to finite dimensions of calculations originally carried out assuming free space, and also dealing with neutron scattering; in addition, he studied the properties of shock waves in air moving outwards from an explosion.

Skyrme's reputation grew so fast that he was asked to follow Peierls to the United States, first to New York, where he was by far the youngest member of the British Mission, and then on to Los Angeles. Here he worked on the study of implosion; this study was necessary to develop a technique for detonating the plutonium bomb because the gun method, by which the two halves of the fissile material in the bomb are brought together to reach critical mass (the method used in the uranium bomb), was much too slow to detonate a plutonium bomb. An implosion is essentially the opposite of an explosion; thus, because of his previous work on shock waves, Skyrme was able to make an enormous contribution, combining theoretical ingenuity with brilliant use of the modest computational facilities available. Margaret Gowing [11] describes Skyrme and Fuchs as 'Peierls' assistants' in this period.

Figure 2.25 Tony Skyrme. from G. E. Brown, Selected Papers with Commentary of Tony Hilton Royle Skyrme (World Scientific, Singapore, 1994). Courtesy Dorothy Skyrme.

On Skyrme's return to England in 1946, he was awarded a research fellowship at Trinity College, Cambridge, but in fact spent the next two years working with Peierls in Birmingham, where his interest turned to nuclear physics, although he developed an unfortunate trait of obtaining important results but being extremely reluctant to publish them. After a year at MIT and another year at the Institute of Advanced Studies at Princeton, he moved in 1950 to Harwell, where he was Head of the Nuclear Physics Group in the Theoretical Physics Division. He was graded Senior Principal Scientific Officer, a level perhaps equivalent to that of a university professor; at that time, he was only 28— indeed, only six years older than Bell—but, like Flowers and others, his war service gave him a veteran status and an effective seniority beyond his years.

In the early days in this post, he published a number of important papers, including those on the photodisintegration of a carbon nucleus to three alpha particles, on the so-called double Compton effect, in which a high energy X-ray incident on a nucleus gives rise to two photons, and studies of the effect of neutrons from the Harwell synchrocyclotron on various metallic targets.

Figure 2.26 John Perring (left), John Bell (right), and Mrs Waste (centre), who was the wife of one of Bell's friends in Oxford, 1988. *Europhysics News*, Volume 22, April 1991, p. 71. Courtesy European Physical Society.

Publication was guaranteed by the existence of co-authors who wished to see their work in print, despite the lack of similar desire on the part of Skyrme himself; much of this work was carried out with Mandl.

After Fuchs's arrest, Maurice Pryce, who was Wykeham Professor of Physics at Oxford, acted as the temporary head of the Theoretical Physics Division; however, in 1952, Flowers returned to take up this position permanently. He confessed that one of his main duties was to persuade Skyrme to publish the fascinating results that were stuck in his filing cabinet! In the division, there was a great wealth of talent, including Flowers, Skyrme, and Mandl themselves, Phil Elliott, Tony Lane, John Soper, John Perring (Figure 2.26), Roger Phillips, Bill Thompson, and John Hubbard. Marshall, whom Bell had met at Birmingham, was also at Harwell at this period. They were all highly able physicists, but here we shall pick out those who became best known.

Elliott [106], who was actually a year younger than Bell, was to become a towering figure in nuclear physics. Together with Flowers, Elliot carried out work on nuclear structure; one of his most important pieces of work was to relate mathematically the motion of nucleons (protons and nucleons) inside the nucleus to the collective nuclear rotations. This work was done during a period when it was difficult to relate two very different models of the nucleus.

One of the models was the shell model, in which each nucleon is described by a set of quantum numbers, just as in the shell model of the atom. The support for this model had received a boost in 1949 with the realization by Maria Goeppert-Mayer and Hans Jensen that, if one included a strong coupling between spin and orbital angular momentum of the nucleons, agreement with experiment was greatly improved. For this discovery, Goeppert-Mayer and Jensen were to share the Nobel Prize for Physics 14 years later.

The second model was the liquid drop model, which had originally been formulated by Niels Bohr in the 1930s; in this model, the nucleus is treated as a dense, liquid drop, with no reference to individual nucleons but with behaviour characterized by rotation and vibration of the whole drop. During the 1950s, Bohr's son, Aage, together with Ben Mottelson and James Rainwater, further developed this model, which became known as the collective model of the nucleus; for this work, the three were awarded the Nobel Prize in 1975.

Although Elliott worked from the shell model side, his sophisticated mathematical techniques, based on group theory, helped to bridge the gap between the two models. He subsequently left Harwell in 1962 and was at the University of Sussex from 1964.

Incidentally, Perring and Skyrme also helped to clarify the relationship between rather different nuclear models. In a paper published in 1956 [107], they showed that each successful nuclear model could illuminate the validity and use of other successful models. In an article about Skyrme, Richard Dalitz [105] said, 'The paper became well-known as the first major step in the coordinating and reconciliation of different nuclear models.'

Lane [108] was another extremely important figure, definitely one of the leading nuclear physicists of his generation. He worked mainly on the theory of nuclear reactions and the interpretation of nuclear data, and he also specialized in the study of photonuclear reactions. Later he turned to work on atomic theory, including the study of the separation of isotopes by lasers and the feasibility of muon-catalysed fusion. Unlike Elliott, he was to stay at Harwell all his working life.

Both Elliott and Lane became Fellows of the Royal Society, and Hubbard [109] certainly would have done so had he lived a little longer: although he was three years younger than Bell, he died ten years before him, in 1980. Hubbard studied the behaviour of interacting electrons in narrow energy bands, and his highly influential *Hubbard model* was the simplest mathematical model that could yield both insulating and metallic behaviour in appropriate limits. Hubbard was to leave Harwell in 1976, and he worked for the remainder of his life at IBM San Jose in California.

It was to this division that Bell came in 1954, and it must have been exactly to his taste. While he was working with accelerators before his time in

Birmingham, he may still have been regarded as something of an 'apprentice', although by all means an apprentice of both enormous ability and, by the end of the period, actual achievement, as well as a person from whom his 'masters' were happy to learn much.

Back at Harwell in the Theoretical Division, though, he was definitely regarded as at least an equal among a group of scientists of the highest ability and enthusiasm. Some of this group were around his own age, and a few were not much older in years but had the experience of war work behind them. There was cooperation and mutual support but also, as they were young, highly talented men all anxious for success, a decided element of competition.

On the cooperation side, other members of the division were delighted with the arrival of Bell as, for one thing, it seemed that he was the only one of them that could understand Skyrme. And, as another example of cooperation, in what is probably Hubbard's most important paper, he thanked Bell for 'helpful advice and criticism'.

The regular meetings of the Nuclear Physics Group were perhaps a little more competitive than co-operative. Bell, with typical humour, christened these meetings 'skyrmishes', and they mostly consisted of rather intense confrontations between Skyrme himself and individual members of the group on progress made on their current research topic. It seems that the only other person who would become involved in these two-body interactions was Bell, who had the nerve to put forward opinions and ideas of his own.

Seminars in the Nuclear Physics Group were also occasions of some excitement. With Skyrme and Bell in the front row, together with perhaps Marshall and Thompson, speakers had to expect a hard time. Charles Clement [110] commented that only at Princeton, where Oppenheimer and Wigner played similar roles, did he experience the speakers' work being subject to such penetrating criticism. He particularly remembered one seminar when a speaker from the Meteorological Office was talking about numerical methods in weather forecasting, but was unable to present all his material because of the heated argument between Skyrme and Bell about the origins and validity of the equations the speaker was using.

We now turn to Bell's own work and his completion of his PhD thesis. The thesis would eventually be submitted in April 1956, around 18 months after his return to Harwell from Birmingham. In fact, Skyrme had made it clear that he would be quite happy for Bell to complete the thesis with work carried out at Harwell. Indeed, while the work of Skyrme's group was, of course, centred on nuclear physics, Skyrme gave Bell the opportunity to ensure that the thesis would be relatively homogeneous by allowing him to work on quantum field theory for the second part of the thesis, so that it would broadly match the first.

The paper on Bell's work on CPT invariance [101] was ready for submission to the *Proceedings of the Royal Society* in April 1955. In the paper, Bell stated that conversations had with Skyrme after Bell had returned to Harwell had resulted in substantial improvements in the presentation of the work. The manuscript had to be communicated to the Royal Society by a Fellow, and Peierls took on this role. The publication of the paper was a major event for Bell, as it was his first important paper in the area of general theoretical physics. In addition, this paper was to constitute the first part of Bell's thesis. Only the most minor changes were made to the text of the published paper when it was presented as the thesis: a footnote on units was assimilated into the text, a short portion became italicized (or, actually, underlined in the typing of the thesis), and an in-text equation was given a full line to itself.

The remainder of the thesis appears to have come from a suggestion by Skyrme. In March 1955, Skyrme had published a paper [111] presenting a novel approach to quantum field theory, broadly based on the formalism of Feynman but also with some connection to the previous work of Peierls, Edwards, and Matthews. In this approach, he introduced a quantity analogous to the scattering matrix, which is an important quantity in quantum theory, into classical field theory. Quantization can then be introduced in a way that is rather analogous to the passage from Newtonian mechanics to statistical mechanics. This process is very straightforward mathematically, although Skyrme admitted that the physical significance is unclear.

He was then able to introduce a relativistic analogue of the non-relativistic variational principle; using this analogue in the derivation of the Schrödinger equation, he was able to deduce some values for the nuclear moments of the neutron and the proton, although these values were in rather poor agreement with those known from experiment.

Skyrme must have suggested that Bell should complete his thesis by developing this method further, analysing its relation to other approaches to quantum field theory, and investigating some applications. This work was to constitute the second part of Bell's thesis and was titled 'Some Functional Methods in Field Theory'; Bell modestly described this part of the thesis as little more than an account of Skyrme's own methods from a different viewpoint. This part of the thesis is actually much longer than the first part, occupying more than three-quarters of the total length of the thesis.

In it, Bell began by arguing to Skyrme's general conclusions, using Schwinger's method, which was based on canonical quantum theory, rather than Feynman's method, which was based on the quantization of classical fields. He applied the technique to the case of a single nucleon and was able to obtain correct results in certain limiting cases. In addition, like Skyrme, he obtained values for the magnetic moments of the nucleons (neutrons and protons); using these, he deduced a value for the so-called neutron–electron interaction.

The thesis (Figure 2.27) was typed by Mary in the evenings and finally submitted in April 1956. It was certainly a highly impressive piece of work.

(a)

1. TIME REVERSAL IN FIELD THEORY.
2. SOME FUNCTIONAL METHODS IN
FIELD THEORY.

J. S. BELL.

Thesis submitted under General Regulations
at the University of Birmingham
for the degree of Doctor of Philosophy.

Division of
Theoretical Physics
A.E.R.E., Harwell.   April, 1956.

(b)

2

ACKNOWLEDGEMENTS.

The work of Part 1. was performed while the author
was on leave of absence from A.E.R.E. at the University of
Birmingham. It is a pleasure to thank Dr. P.T. Matthews
for much instruction and encouragement during this period.
Several other members of the Department of Mathematical
Physics are thanked for helpful discussions, and particularly Professor R.E. Peierls, F.R.S., who suggested this
topic for study.

In connection with Part 2. I have had valuable conversations with Professor Peierls and Dr. S.F. Edwards. I
am of course very greatly indebted to Dr. T.H.R. Skyrme;
this work is little more than an account of his methods
from a different viewpoint; he is thanked for much assistance at all stages.

Figure 2.27 (a) Title page, (b) acknowledgements, and (c) synopsis of Bell's PhD thesis,
April 1956. Courtesy University of Birmingham and British Library.

(c)

                                                                        3

SYNOPSIS.

       Part 1. examines the reversibility in time of quan-
tised fields. It is shown that reversibility of a certain
kind (that of Schwinger) is a concomitant of Lorentz invariance, rather than an independent symmetry to be imposed on the theory.

       Part 2. is an account, with some new applications,
of work of T.H.R. Skyrme. First we derive his closed
forms for propagators from canonical quantum theory, rather
than from the Feynman quantization of classical fields.
The discussion is then specialised to the one nucleon
propagator and a variational method developed. With a
very simple form of trial function this is applied to the
no-recoil neutral scalar theory, where it is shown to give
accurately that part of the propagator which describes the
real nucleon. A similar approximation is then used in the
pseudoscalar symmetric no-recoil theory, and found to give
correctly the electrical properties of the nucleon in the
weak and strong coupling limits. Finally a closely related trial function is used in the relativistic symmetric
(P.S.,P.S.) theory, where we extend Skyrme's work on the
nucleon anomalous magnetic moments to obtain also the so-
called "neutron-electron interaction".

Figure 2.27 (Continued)

The first part contained an important theorem, while the second presented a highly elaborate and mathematically sophisticated account of a new approach to quantum field theory. As implied, agreement with experiment was only moderate, but this was to be expected; at this stage of development of the theory, setting up a consistent formalism was the principal aim. There could certainly be no doubt that Bell would be awarded his doctorate!

In the acknowledgements section of his thesis, as well as thanking Skyrme of course, Bell acknowledged valuable conversations with Peierls and Edwards.

Bell and Skyrme continued to work on their approach to quantum field theory and, by May of the following year, 1957, two papers [112, 113] were submitted, again to the *Proceedings of the Royal Society* and again communicated by Peierls. The first paper, authored by Bell alone and seven pages long, was much shorter than either the 1955 paper or the second paper of this pair, and contained some of the findings reported in the second part of Bell's thesis.

The second paper, in which both Bell and Skyrme were named as authors, although still based on Skyrme's previous paper and Bell's thesis, contained much more new material than the first paper, and was specifically addressed to the problem to which both Skyrme and Bell had paid comparatively little attention in their previous work: the 'anomalous [magnetic] moments of nucleons'.

The meaning of the term 'anomalous' derives from the idea of the 'nuclear magneton', which itself is derived from that of the 'Bohr magneton'. The Bohr magneton may be described as the natural unit of magnetic moment for electron systems. It is a result of Bohr's famous theory of the hydrogen atom of 1913, and it showed that the basic unit of angular momentum was $\hbar$, which is Planck's constant divided by 2π, and that the basic unit of the magnetic moment was $\hbar$ multiplied by the electronic charge and divided by twice the electronic mass. In fact, the magnetic moment of the orbital motion of the electrons is always an integral multiple of the Bohr magneton, and that of the spin motion is very close to one Bohr magneton.

It would therefore be predicted that the natural unit for nuclear magnetic moment would be equivalent to the Bohr magneton, but with the mass of the nucleon replacing that of the electron. In other words, the value of the nuclear magneton would be that of the Bohr magneton divided by the ratio of the nucleon and electron masses. Simple theory would also predict that, since the charge on the proton is equal in magnitude to that of the electron, its magnetic moment would be 1 nuclear magneton while, since the neutron has no charge, its magnetic moment would be 0.

Experiment shows that this formula does indeed give the scale correctly, but the actual values are stubbornly different, that of the proton being 2.8 and that of the neutron being −1.9 nuclear magnetons.

Bell and Skyrme were, of course, going beyond the simple theory by working within the realm of quantum field theory, where the work, even using rather simple approximations, was extremely complex mathematically. Their final predictions were that the values of the proton magnetic moment and the neutron magnetic moment would be 1.8 and −1.3 nuclear magnetons, respectively. These values were maybe not too bad for a first shot but obviously there was a long way to go! More recent analyses, of course, work in terms of the quark structure of the nucleons.

Skyrme meanwhile had become interested in the ideas of Keith Brueckner [114]. Brueckner, an American physicist who was just four years older than Bell and so was much of an age with most of the members of the Harwell group, had in 1955 produced a novel approach to the many-body problem in the nucleus. Up to this time, a straightforward treatment involving each nucleon interacting with every other nucleon had broadly led to nonsense. Brueckner showed how to reorganize the calculations so that the nonsensical terms vanished and meaningful results were obtained. In fact, he was able to show that strong interactions between many particles could effectively be shown to be equivalent to the system behaving as individual particles, with each particle moving almost freely. Skyrme found these ideas extremely attractive, as they emphasized the mean coherent properties of the nuclear medium in which each nucleon moved, rather than the particulate structure of the nucleus.

In 1956 Bell and Skyrme together [115] used the Brueckner method to analyse spin–orbit coupling. As has been said earlier in this section, it had recently been shown that this effect in the nucleus as a whole had been particularly important in the resurgence of the shell model in nuclei, although the actual calculations were complicated. Bell and Skyrme noted that the interaction between two nucleons had recently been determined from data on scattering, and, following Brueckner, they were able to use individual nucleon–nucleon calculations to derive the all-important coupling for the full nucleus.

Together with Richard Eden, who was visiting Harwell from the University of Manchester, Bell and Skyrme [116] also used Brueckner's method to investigate the magnetic moment of nuclei. In the general approach, the behaviour of each nucleon could be discussed in terms of a simple potential; however mathematically this potential depends on the momentum of the particle. In turn, in some applications, the effect of this momentum dependence may be represented by the mass of a single nucleon being replaced by an effective mass equal to roughly half the real mass of the nucleon.

The three authors investigated the situation rigorously, considering in turn the different causes of the momentum dependence of the potential. Their conclusion was that, as the various interactions increased in strength, different mathematical terms with different characteristics came into play. The naïve use of the effective mass was quite unjustifiable. This paper was an important contribution to the study of Brueckner's ideas.

In the following year, Bell [117] wrote a further paper in which he used a simple model to give a numerical estimate of how the results obtained from shell theory were caused by the exchange forces due to mesons. As in all this work, the 'corrections' turned out to be rather uncomfortably large.

In those years, most of the work that Bell carried out on his own or with colleagues other than Skyrme was on nuclear physics. As Peierls [99] later pointed out, Bell was never satisfied with routine applications of standard methods and always preferred to go back to the foundations of the topic. As mentioned earlier in this chapter, Bell would probably have preferred to concentrate on quantum field theory or move on to the emerging field of elementary particle physics rather than study nuclear physics; however, Peierls later stated that he believed that this work put Bell in a strong position in the 1960s, when he tackled problems which straddled the borderline between nuclear and particle physics, such as muon capture in heavy nuclei, and the nuclear optical modes for pions.

Throughout his career, Bell retained an interest in problems in the area of CPT, particularly those involving time reversibility. Before we look at a paper he wrote in 1957 in this area, we must mention by far the most important development of the 1950s in terms of nuclear physics and possibly the whole of physics [5]: the discovery of non-conservation of parity in the weak interaction. (Parity is conserved in the strong and electromagnetic interactions.)

The possibility that parity might not be conserved in the weak interaction was raised by the so-called tau–theta puzzle. In order to discuss this topic, it should be mentioned that, while parity was initially introduced in this chapter as the symmetry of a physical event, individual elementary particles can also be assigned parity—either positive or negative. The proton, neutron, and electron are assigned positive parity by convention, and the parities of other particles may then be deduced through a variety of arguments. In particular, particles called *pions*, or *pi-mesons*, may be shown to have negative parity.

The pion should be briefly introduced. It had long been known that the correct description of the electromagnetic interaction was that it was carried from one charged particle to another by the *exchange* of photons. It may be said that the interaction is less like boxing, where forces are applied *directly* from one boxer to the other, and more like tennis, where each player may apply a force *indirectly* to the other with the tennis ball as an intermediary.

In 1934 Hideki Yukawa [5] suggested that the strong interaction could similarly be *mediated* by a new particle, now called the pion, or pi-meson, or often just meson. We shall see a little about the discovery of this particle in Chapter 3. It is natural to assume that the other two interactions would similarly be mediated by *exchange particles*. The gravitational force is so weak that the hypothecated particle, the *graviton*, must interact with matter in a manner that is far too weak to be detected, at the very least for the foreseeable future. The exchange particles underlying the weak interaction will be discussed in later chapters, as will the changes in our ideas with the coming of the quark theory.

In 1954 among the data emerging from accelerator experiments were the signs, it seemed, of two particles, τ (tau) and θ (theta), that had masses that were (at least) very similar, and other properties that were identical. However, it was taken for granted that the particles themselves could not be identical because, while τ decayed to three pions and so must have negative parity (because the parity of pions combines multiplicatively), θ decayed to two pions and so must have positive parity. The decay, it should be noted, occurred via the weak interaction.

Then, in 1956, Tsung Dao Lee and Chen Ning Yang, young Chinese theoretical physicists working in the United States, suggested what seemed to many to be heretical—that, in fact, the particles were actually identical. The implication was that this one-and-only particle could decay either to a state of even parity or to one of odd parity, thus in implying in turn that parity was not conserved in the weak interaction.

Lee and Yang made a thorough search of previous experimental results but could find no other evidence of this breakdown of parity conservation; so, to test their controversial suggestion, they suggested an experiment in a completely different area of nuclear physics: the beta decay of cobalt-90 nuclei. This experiment would require the magnetic moments of the cobalt nuclei to be pointing preferentially in one particular direction, a situation that would only occur at extremely low temperatures.

Then, non-conservation of parity would imply that there might be a correlation between the preferred direction of the nuclear magnetic moments and that of the beta decay electrons emerging asymmetrically. This situation would indeed violate parity conservation, because momentum is a 'normal' or polar vector but magnetic moment, which is directly related to angular momentum, is a pseudovector, or axial vector. Thus, if such a correlation existed, when the physical event was reflected, the momentum vector would change direction but the magnetic moment pseudovector would not! What would result would not be a physical event.

Lee and Yang persuaded their compatriot, Chien-Shiung Wu, often known as Madame Wu, to carry out the experiment, and a team including Ernest Ambler and others was gathered together. The extremely low temperature required for the polarization of the magnetic moments of the cobalt nuclei meant that the experiment was a tour de force when it was carried out in 1957 at the National Bureau of Standards at Washington. It gave the result that Lee and Yang had predicted—parity was not conserved in the weak interaction.

The conclusion was shocking—Pauli for one had been dismissive of the very idea of the experiment, believing that conservation of parity had to be a fundamental feature of physics. The importance of the discovery was seen in the behaviour of the committee for awarding the Nobel Prize in Physics. Many eventual winners of the prize have had to wait years or decades for its award. In contrast, Lee and Yang shared the prize in 1957, the same year in which the experiment had been carried out.

From the point of view of Bell's work, the discovery served to magnify the importance of the CPT theorem. For, when it was taken for granted that C, P, and T were each conserved individually, the mathematical proof that CPT was conserved was important conceptually but added little to physicists' view of their subject. But, once it was clear that P was not conserved, the CPT theorem made it clear that the conservation of at least one of C and T had to be abandoned as well.

In fact, the type of experiment discussed by Lee and Yang and carried out by Wu, while making it clear that P was not conserved, suggested that it was highly likely that CP was conserved. This finding implied that, if one started with a physical event, carried out charge conjugation, *and* reflected the event, one would end up with another physical event. So, while C was not conserved, T, or time-reversible symmetry, could be maintained. This conclusion was the general belief until 1964.

Bell's paper [118] in 1957 was a response to the work of Lee and Yang. From the acknowledgements to this paper, it seems that both Tony Skyrme and David Candlin had posed the following question. Spatial symmetry, they said, or in other words, conservation of P, implies restrictions on the angular distribution of electrons in beta decay. It was, of course, the failure of nature in Wu's experiment to obey these restrictions that meant that conservation of P had to be abandoned. Skyrme and Candlin asked: did time reversibility, or the conservation of T, enforce conclusions on beta decay analogous to those enforced by the conservation of P?

Bell first pointed out that it was scarcely possible to consider the reverse of a process in which a nucleus decayed to three particles: the new nucleus, the electron, and the antineutrino. One would have to imagine two incoming spherical waves and then the reassembly of the initial nucleus.

However, Bell tackled the problem mathematically and was able to show that, to first approximation or, one would say, to the first order of perturbation theory, an equation was produced which showed what Bell called *spatial irreversibility*. This result implied that a transition to a particular state is as equally probable as one in which the directions of both momentum *and* angular momentum have been reversed. As Bell says, useful restrictions are obtained only because of the adequacy of first-order perturbation theory.

In another paper [119], he investigated why experimental values for the rate of beta decay were sometimes considerably reduced, as compared with those predicted by the simple shell model. He calculated the results of various additional effects due to the other electrons; these moved the rate in the right direction, although not as far as the experimental values.

Bell also collaborated with Roger Blin-Stoyle (Figure 2.28), who was carrying out research at Oxford. Harwell and Oxford were close and there was much joint work; it will be remembered that Cockcroft was keen to site Harwell near either Oxford or Cambridge and, in the post-war period, physics at Oxford certainly improved dramatically.

Bell and Blin-Stoyle [120] attempted to discover whether the presence of mesons would affect interaction and hence lifetime in beta decay. Experimental data did show disagreements with values calculated from the shell model, especially for so-called mirror nuclei, where, for example, hydrogen-3, which has 1 proton and 2 neutrons, decays to helium-3, which has 2 protons and

Figure 2.28   Roger Blin-Stoyle. Courtesy Royal Society.

1 neutron; or calcium-39, which has 20 protons and 19 neutrons, decays to potassium-39, which has 19 protons and 20 protons. The authors investigated various possible deficiencies of the theory but tentatively suggested that they were all too small to explain the differences between theory and experiment and concluded that rather substantial mesonic effects might indeed be the actual cause.

Blin-Stoyle later became highly influential in the setting up of the new University of Sussex, which insisted on providing a broader education for commencing science students than was normal in British universities. He also became a Fellow of the Royal Society. In Phil Elliott's 'Biographical Memoir' of Blin-Stoyle [121], Elliot reports that Blin-Stoyle described Bell as 'a physicist and philosopher of great charm and depth'.

In 1958 Bell wrote a couple of short papers with Mandl on the so-called polarization–asymmetry equality. This equality relates the two simplest experiments involving scattering of electrons, and it equates the spin polarization produced in a collision process with the angular distribution produced by the inverse process. The previous proofs assumed conservation of parity, so Bell and Mandl [122] produced a proof which they stated did not rely on this assumption. In some ways, this proof might have seemed unnecessary, since the scattering process uses the strong interaction, where it was (and is) assumed that parity was conserved, but, for a full understanding of the issue, it did seem worthwhile.

Unfortunately there was a mistake in the argument, since they neglected the possibility that, without conservation of parity, the beam might be polarized along its direction of travel—this would imply correlation of momentum and angular momentum, a situation which is not allowed if parity is conserved but is permissible if parity is not conserved. The authors did their best to retrieve the situation [123] but with perhaps questionable success. As it is said, even Homer nods! Incidentally, as well as Skyrme, Euan Squires was thanked for 'valuable discussions' on this second paper. He had recently joined Harwell as a postdoctoral fellow and was to carry out important work with Bell.

Another area where Bell [124] succeeded in solving problems raised by very senior physicists was his study of radiation from high energy electrons undergoing multiple scattering. A group of Russian physicists, including Nobel Prize winner Lev Landau, had suggested that the loss of energy by radiation of these particles actually decreased with the energy of the electrons, owing to the destructive interference of the radiation. Thus, the electrons were predicted to be more energetic than might be expected and thus highly penetrating. As Bell pointed out, such a prediction conflicted with the standard result from classical physics, and he was able to show that the amount by which the level of radiation was reduced at low frequencies was exactly equal to the amount by which it increased at high frequencies. He did have to admit, though, that this increase occurred in a region where quantum effects must be important, so further study was necessary.

In another short note, Bell debunked a suggestion made by a surprising number of the great and good of physics—the Nobel Prize winners Leon Cooper, Aage Bohr, and Ben Mottelson, as well as Kurt Gottfried and David Pines. Using a complicated formalism to study a two-body interaction, they had claimed to find a mathematical singularity—the interaction becomes infinite. Bell instead [125] used a system that was much simpler than theirs and for which the analysis was so simple that it was absolutely clear that there could be no singularity. However, when he studied this system with the complicated formalism used by Cooper and the others, he found exactly the same type of singularity that they discovered. He thus deduced that the singularity was due to an unfortunate selection of higher-order terms in the first approximation and so had no physical meaning.

Another very imaginative paper [126], published in 1959, obtained many of the standard results of the shell model, including the celebrated formulae of Giulio Racah, by using only the symmetry between particles and holes.

(A hole is essentially an absence of an electron in an otherwise full system of electrons.) Bell's methods were much quicker and more elegant than those in use up to that point.

In 1959 Bell returned to the analysis of nuclear moments with Soper [127]; this time, the analysis was in terms of correlations with the hard core. The influence of these correlations on several parameters of the shell model was considered, and the corrections were all found to be very small and, in a number of cases, identically zero.

Also in 1959, Bell and Squires [128] published a paper on the nuclear optical model. This model, which had been first discussed towards the beginning of the decade, considered the nucleus to be analogous to a cloudy crystal ball in optics. Thus, the model predicted that, when a beam of particles struck the nucleus, the beam would be partially absorbed, partially scattered, and partially transmitted. A wide range of scattering data could be accounted for by the model, and the corresponding wave functions were used to obtain information on nuclear structure. The paper by Bell and Squires was short but very significant, and it attracted a lot of attention.

Around this time, Bell also lectured on the nuclear optical model and several other topics in nuclear and many-body physics at several conferences and schools on physics. These are included in the list in QHA [38] and in a bibliography prepared by the Royal Society [129].

Bell and Squires also wrote an important and lengthy review [130] of the theory of nuclear matter, studying the application of the methods of Brueckner and also of Robert Jastrow, who later became known as an astronomer. This review has also been widely used.

Farewell to Harwell

The end of the 1950s was certainly Skyrme's most productive period in research [105]. He published a series of fundamental papers dealing with non-linear meson theories and their application to nuclear physics, and his ideas promoted a nuclear theory based entirely on meson fields without actual nucleons. His theory involved the existence of a so-called *skyrmion*, which may be regarded as a representation of a nucleon emerging from the nuclear field. Mathematically, a skyrmion has the nature of a *soliton*, a solitary wave that maintains its shape as it travels. At this time, skyrmions attracted interest but little enthusiasm.

Skyrme had spent a year away from Harwell in 1958, using part of the time to work with Brueckner in the University of Pennsylvania at Philadelphia; but when he returned, it was to a changing Harwell. It could be said that the task for which it had been set up had been achieved—nuclear power stations had been built, and it seemed that the Harwell scientists would now be encouraged to earn their keep by obtaining contracts to carry out work for industry and other government bodies, and fundamental research would be squeezed. Flowers moved to Manchester, being succeeded as Head of Division by William Lomer and then Marshall; Elliott went to Sussex, and Bell, as we shall see, to CERN. Squires went to another fellowship in Cambridge, then to Berkeley, and on to his long stay in Durham.

In 1961 Skyrme himself left Harwell, spending a period at the University of Malaya before becoming Peierls's successor at Birmingham. The latter was obviously a highly prestigious appointment, but perhaps not especially successful or happy from Skyrme's point of view. As was mentioned earlier in the chapter, he was not a very good teacher and, although he gave substantial help to anybody—physicists, mathematicians, statisticians, engineers—seeking it, his unwillingness to publish his own work became almost total. His Department of Mathematical Physics was forced, because of its small number of students, to amalgamate with Mathematics. This amalgamation was not necessarily against Skyrme's tastes, but, like the malaise at Harwell, perhaps an indication that a highly fruitful period in British physics was coming to an end.

Then, in the late 1970s, interest soared in skyrmions [131]. Considerable interest developed in particle physics and, although initial hopes that skyrmions might play a role, normally assigned to quarks, in building up baryons were dashed, skyrmions are still mentioned in many discussions and theories in particle physics. They are also highly important in many areas of solid state physics, particularly the quantum Hall effect, where the application of a magnetic field causes a highly precise increase in current, but also in many aspects of electronic and magnetic behaviour, and possibly even quantum computation. John Polkinghorne [132] has suggested that, with its applications in topology, Skyrme's work has led to 'the richest and most mutually satisfactory interaction between mathematics and physics since the nineteenth century'.

Skyrme himself took little part in the work at this time, but he did give one historical talk. He was awarded the Hughes Medal of the Royal Society in 1985 but died rather suddenly in 1987.

Incidentally, the award of this medal was rather unusual. Becoming Fellow of the Royal Society is in itself a high honour and a small number of physicists might be elected each year, but a further step up for a Fellow broadly in the area of physics is the Hughes Medal awarded each year. The next step would be a Royal Medal; one of the three awarded each year is in the general area of physical science, including chemistry. The highest accolade is the Copley Medal, of which one is awarded each year to any branch of science; this medal could be considered to have rather the same status as a Nobel Prize.

For example, Nevill Mott, one of the most respected physicists of his time, was awarded Hughes, Royal, and Copley Medals in 1941, 1953, and 1972, respectively, before becoming a Nobel Laureate in 1977. Surprisingly, although Cockcroft, who was awarded the Nobel Prize in 1951, had been awarded the Hughes Medal in 1938 and would receive a Royal Medal in 1954, he never received the Copley Medal.

As we shall see, Bell would be awarded the Hughes Medal in 1989. Doubtless, had he lived even a little longer than he did, he would have climbed the other two rungs of the ladder. The unusual point about Skyrme's award is that he never actually became a Fellow of the Royal Society, so the award of the Hughes Medal was interesting. Presumably, at this time, his work still divided opinion.

However, let us return to John and Mary Bell. Like others, they were very concerned about the potential loss of freedom to carry out fundamental research [37]. They had also become somewhat concerned that work on weapons was being carried out against the original purpose of the institution, and certainly to the disapproval of the Bells. While this situation had probably not changed over the decade, it may be that what had seemed allowable only a few years after the war was much less so ten years later.

But where could they go? Both had good research records and so either could have hoped to obtain a lectureship or higher position in a university reasonably easily, but it was rather less likely that they could obtain positions in the *same* university or even in the same city. But the problem was greater than that. Mary wanted to continue research in accelerator design, and John in theoretical physics, preferably quantum field theory and elementary particle physics; it seemed unlikely that any university could cater for both interests.

The fairly obvious possibility was to go to CERN. As we saw earlier in the chapter, John had been involved in the early days of the institution and must have been pleased to see it become established and start to carry out respectable research, even although it was not as yet quite competing with the leading American laboratories in the major discoveries. Certainly, both John and Mary's scientific interests would easily have been met at CERN, and indeed John would have been happy to leave behind the moral obligation to concentrate most of his efforts on nuclear physics.

The 'fly in the ointment' might have been the fact that, in making this change, permanent civil service jobs were being relinquished for short-term contracts. The general policy at CERN was to have the great majority of the scientific work carried out by visiting teams or fellows seconded to CERN for a year or so. On joining CERN, the Bells would be on a three-year contract, with the understanding that a further three-year contract would probably be offered, but that, at the end of the second contract, they could only stay on if they were offered a permanent position, and these were very few in number—at one stage only 20 per cent of the total number of the positions in the laboratory. John was later to tell Bernstein that, in retrospect, he was amazed at how readily they left permanent jobs for uncertainty. Mary's parents were a little concerned, but the couple themselves not at all. So, CERN it was, a decision that would turn out to be a permanent one and which was, of course, to play an important part in the rest of their careers and lives.

It is interesting to consider what might have befallen the Bells had they remained in Harwell, and in this context a book edited by Cockcroft [133] and published in the mid-1960s is of interest. It discusses the organization of research establishments and contains chapters on Harwell, the Rutherford High Energy Laboratory, and CERN, with the chapter on CERN having been written by Adams.

The Harwell chapter, written by Arthur Vick, who had been until shortly before that date its Director, stressed 'the changing nature of Harwell's work'. Work on defence had practically disappeared (which was good from the Bell's point of view) but state secrets had been replaced by commercial ones—very much as the Bells had feared. The theme of the chapter was, essentially, coping with change. Harwell has moved essentially in the same direction ever since.

However, the chapter on the Rutherford, written by Pickavance, it's director, was more hopeful than the Harwell chapter. The Rutherford was set up adjacent to Harwell, although outside Harwell's security wire, by the National Institute for Research in Nuclear Science. In 1960 many of the staff at the Rutherford had transferred there from Harwell, a 7 GeV proton synchrotron called Nimrod had been constructed, and work was continuing using a 50 MeV proton linear accelerator which had been handed over, partly built, by Harwell. It seems that the Bells (although perhaps just Mary) could have been happy here.

As for CERN, this, of course, had its own problems, which we shall survey in the next chapter.

For John Bell, the 1950s had been an excellent decade, both personally and professionally. On the professional side, he had certainly established himself as an extremely able physicist—it was generally acknowledged that, among an exceptional group at Harwell, he and Skyrme were the outstanding talents. His achievements across the areas of accelerator physics, quantum field theory, and nuclear physics were substantial, and they showed both tenacity and imagination. His friends and admirers would certainly have hoped that, in the future decades, he would make outstanding contributions to his subject—and they would not be disappointed!

3

The 1960s

The Decade of Greatest Success

CERN

Chapter 2 gave an account of the 1952 involvement of John Bell and some of the other Harwell scientists with CERN, at the time when strong focussing was discovered and it was built into the plans for the CERN proton synchrotron, while this chapter provides a general discussion of the genesis of the institution, those who were responsible for promoting it, and how they got others to pay for it!

Another interesting question is why a very strong theory group was built up at CERN. In their article on Bell's physics, Jackiw and Shimony [1] commented that CERN 'houses Europe's preeminent particle theory group', and John Iliopoulos [2], in his historical account of the Theory Division at CERN, said that 'CERN in general and the Theory Division in particular, were the undisputable scientific centres for high energy physics in Europe' at that time and that CERN probably had 'the largest theoretical high energy physics department in the world'. In contrast, he noted that Brookhaven, the nearest analogue and rival to CERN, never developed a large theoretical group.

The successful promotion of the concept of CERN [3] depended in a general way on the scientific situation and political circumstances following the Second World War, but particularly on the efforts of quite a small number of scientists, especially the Italian Edoardo Amaldi (Figure 3.1) and the Frenchman Pierre Auger (Figures 3.2 and 3.3).

Amaldi [4], born in 1908, had been one of the physicists of the 'School of Rome' centred on Enrico Fermi in the 1930s. The group included Emilio Segrè, Franco Rasetti, and Bruno Pontecorvo as well as Fermi and Amaldi, and their crucial work involved the discovery of many new radioactive nuclei via the bombardment of targets with neutrons. In 1938 the group dispersed because of the introduction of regulations against Jewish people; Fermi and Segrè went to the United States, Rasetti to Canada, and Pontecorvo to France. Amaldi was able to stay in Rome, where he practically kept research alive during much of the war and in the years afterwards. From 1947 onwards, his research was in cosmic rays, of which more will be said shortly. In addition, he was Vice

Figure 3.1 Edoardo Amaldi. Courtesy Royal Society.

President (from 1948 to 1954) and later in the 1950s, President of IUPAP, the International Union of Pure and Applied Physics; these positions were useful for promoting CERN.

In 1968 he was elected a foreign member of the Royal Society, a high distinction normally offered only to a few chosen Nobel Prize winners. His citation said that he was 'distinguished for his important experimental work on the properties of neutrons and for making the first observation of an event attributed to the creation or annihilation of an antiproton.' It is true that he was involved with the discovery of the antiproton [5], working initially with cosmic rays; however, the Nobel Prize for that discovery went to Emilio Segré and Owen Chamberlain for their work using the Berkeley accelerator, while Clyde Wiegand and Thomas Ypsilantis, who co-authored the paper with Segré and Chamberlain, were disregarded, and Oreste Piccioni [6] resorted to law to get credit for his contribution. Assigning responsibility for scientific advances can sometimes be extremely tricky!

Pontecorvo [7] also had an interesting life. During the war, he worked for the Anglo-Canadian project in Montreal. After the war, he was employed at

Figure 3.2 John Cockcroft (centre), with Pierre Auger (far right) and Lew Kowarski (second from the right), and two other French scientists who worked on nuclear physics at Montreal during the war: Jules Guéron (far left) and Bertrand Goldschmidt (second from the left). ACME Newspapers.

Harwell and was indeed due to take up a position as Professor of Physics at Liverpool University in January 1951 when, in August 1950, he disappeared, turning up shortly later in Moscow. At first it was feared he was another Fuchs, but it turned out that Pontecorvo's work had been in reactors rather than bombs.

Auger [3], born in 1899, worked first in atomic physics, where he discovered the 'Auger effect'—atoms in a state of energy higher than the lowest state may lose the energy by emitting an electron rather than (or as well as) a photon. Then, he moved onto the study of cosmic rays [8], where he discovered 'Auger showers' or large numbers of particles resulting from a single cosmic event. During the war, like Pontecorvo, he worked on the Anglo-Canadian project, after which he became Director of Higher Education in France, and then, from 1948 to 1959, Director of UNESCO's Department of Exact and Natural Sciences, which was another good place to promote the idea of CERN. Later in his life, he was instrumental in the founding of the European Space Research Organisation, of which he was Director from 1962 to 1967.

Figure 3.3 Pierre Auger (left), with Gustavo Colonnetti and Miss Thorneycroft at the meeting of 15 February 1952 to set up the provisional CERN. Courtesy CERN.

Also mentioned should be a second Frenchman, Lew Kowarski [3], who had been involved with Frédéric Joliot-Curie and Hans von Halban before the war on the study of nuclear chain reactions and power production. In 1940 Halban and Kowarski came to Britain, bringing the world's entire stock of heavy water for use in nuclear reactors. During the war, Kowarski, like Auger, worked on the Anglo-Canadian atomic energy project. Then, after the war, he supervised the construction of the first two French reactors. Unlike Auger and Amaldi, once CERN was founded, Kowarski would spend the remainder of his career working there.

We now come back to cosmic rays [9] and their part in the physics of elementary particles. Cosmic rays are particles of extremely high energy, reaching the earth from other parts of the universe; they are mostly protons but can also be nuclei of atoms larger than hydrogen, or high energy photons. Cosmic rays may be studied from two points of view. The first is to study their origins. It is extremely difficult to determine the origin of the charged particles, since they swirl about in the strong magnetic fields of interstellar space. In contrast, since photons are uncharged, they may be traced to various locations in our own galaxy, such as the remnants of supernovae, as well as to sources in other galaxies, such as black holes, and it may be assumed that the charged particles have the same sources.

The second approach is to study the particles formed by collisions of the cosmic ray particles in the upper atmosphere. The first particle of antimatter, the positron, was discovered in this way in 1932, and the muon in 1937, both by Carl Anderson. In 1947 positively and negatively charged pions were found by Cecil Powell and Giuseppe Occhialini; in the same year, George Rochester and Clifford Butler also used cosmic rays to discover the first two of the particles soon to be known as *strange*.

All these discoveries were certainly excellent for cosmic ray research, but a warning shot came in 1950. As well as positively and negatively charged pions, there is a neutral variety; however, this type had not been found using cosmic rays, because it was much harder to find than its charged brothers, for two reasons. The first was that, fairly obviously, charged particles interact more readily than neutral particles do; the second is that charged pion decay takes longer than neutral pion decay because a neutral pion can decay directly into a pair of photons, while the decay route for charged pions involves muons. In 1950, the neutral pion was detected with the synchrocyclotron at Berkeley—the first time that a new particle was discovered using an accelerator [5, 10]. This discovery did not mark the end of cosmic ray physics but it may have indicated the beginning of the end.

For example, after Rochester and Butler's initial discovery of strange particles via cosmic rays, they had to wait for two rather embarrassing years before they saw more examples [3, 10]. Cosmic ray research at the forefront of expanding knowledge was always a question of making the best of an almost vanishingly small number of examples. For example, the Powell group struggled for 47 days to produce cases demonstrating the existence of charged pions; once the Berkeley synchrocyclotron was put on the problem in 1948 [3], it produced a hundred times as many examples in 30 s!

In 1953 Powell remarked at a cosmic ray conference: 'Gentlemen we have been invaded ... the accelerators are here', while, at the important Rochester Conference of 1956, Robert Leighton said that, at the following year's conference, those studying strange particles using cosmic rays would have to 'hold a rump session somewhere else'—they would not be allowed in the hall.

It is quite true that cosmic rays do contain particles of energies up to 10^8 TeV (or 10^{20} eV), and these energies are higher than those produced by any accelerator constructed on earth, by a factor of at least ten million. But such particles are extremely few and far between in cosmic rays, so what little work is done with them is extremely specialized. Once the Cosmotron at Brookhaven and the Bevatron at Berkeley were producing particles of energy well over 10^9 eV in the 1950s, the die was cast; in *History of CERN* [3], John Krige wrote of 'the general eclipse of cosmic ray investigations by accelerators which was more or less total by 1955'.

At the founding of CERN in 1954, Cockcroft and Powell did argue for the setting up of some cosmic ray work. A limited pocket of money was put into this, and two groups were funded, but by 1958 all such work had been abandoned. Accelerators were to be the (unique) future for particle physics.

Despite their own background in cosmic rays, Amaldi and Auger were well aware of this, certainly as early as 1945 or 1946, when the large American accelerators were planned. They realized clearly that the Americans would be able to move ahead speedily in particle physics, Britain *might* be able to follow on their coat tails, but, unless steps were taken, the remaining European countries would flounder and it would certainly be impossible to arrest the brain drain to the United States The state of physics in Auger and Amaldi's own countries was perhaps worse than that in, say, Scandinavia, so there was a particular reason for concern from them. Action was needed at a European level.

It should be emphasized that their vision as it developed in the next five or six years centred on the building of an accelerator. Of course, it would be built for a purpose, although, as Polkinghorne says [10], through the 1950s there was often in effect a policy 'to build the machine and then begin to think what to do with it'. At least to an extent, this statement does seem to have been true of the proton synchrotron, CERN's first big machine [11]. In any case, it was probably assumed that users and projects would come from universities in a conventional way. Again, while theory would be required for building and using the accelerator, it was not obvious that the employment of theoretical physicists needed to be tied to the machine itself. It would probably never have occurred to Auger and Amaldi in these years that there would be a group working on fundamental theory at CERN.

While the vision of Auger and Amaldi was straightforward, they were able to take advantage of a number of interconnected 'currents of history', despite the fact that many of those involved in these 'currents' had no interest whatsoever in particle accelerators! One important trend, as we saw in the previous chapter, was the support of science, in response to the triumphs of science during the war. For example, in 1946, the newly founded United Nations called for the founding of UN research laboratories [11]. However, these laboratories would be designed, in a totally praiseworthy fashion, 'to improve the living conditions of mankind'. CERN, as it emerged, would not qualify on this criterion!

Although the UN Atomic Energy Commission that was also set up in 1946 was aimed primarily at preventing nuclear proliferation, discussion among the physicists and also the diplomats present would certainly have included research in the atomic area. Again, though, if accelerators were included in the suggestions, they almost invariably came along with the 'useful' reactors. Of course, in the long run, it would be clear that not every European politician had a clear idea of the distinction between reactors and accelerators!

Also at the European level, cooperation was under way. The Organisation for European Economic Cooperation was formed in 1948, and the Congress of Europe met later in the same year. In the following year, the European Movement was established and its establishment led to the creation of the Council of Europe, which aimed to establish common standards in Europe in such matters as human rights and the rule of law. Then, in 1951, the Treaty of Paris set up the European Coal and Steel Community, which would in few years lead to the Treaty of Rome and today's European Community.

The motivation for such projects was twofold. First, cooperation was seen as essential for economic progress. But perhaps more important was the sense that, during the twentieth century, two world wars had begun in Europe and that, if a spirit of cooperation or even unity could prevent a third, that would certainly be extremely worthwhile. Both these worthy objectives had the advantage, for the proponents of the future CERN, that they, and the methods of achieving them, were a little vague. Cooperation over accelerators would provide an excellent example for other fields, while nuclear physics, although admittedly not necessarily the accelerators themselves, would surely provide a stimulus to the national economies.

It might be mentioned here that, just as the physics of France and Italy was particularly weak after the war, thus making those countries enthusiastic for projects like CERN, so also, admittedly in a very considerable generalization, these countries were 'better Europeans' than, for example, the Scandinavian countries, and, of course, Britain, which had the same effect of pushing Auger and the others in the CERN direction.

The first specific discussions took place in a number of different places and between very different groups of people at the end of 1949 and through the first half of 1950. The physicists involved, particularly Kowarski, were naturally keen on the development of accelerators, while the civil servants and diplomats, although generally supportive, were still confused about how the key word 'nuclear' should be interpreted and developed in practice.

Gradually, it became clear to everybody that not only was application to military matters completely ruled out on the grounds of security, but application to reactors had also to be ruled out on the grounds of industrial secrecy. The work at any future institution had to be purely scientific. There is an irony in the change from the UN's desire to improve living conditions in 1946, to CERN in the end being supported not *although* it promised no immediate applications, but *because* it had no immediate applications.

Progress in gathering support was slow until the General Conference of UNESCO in Florence in June 1950. At this conference, Isidor Rabi, a member of the American delegation and the 1944 winner of the Nobel Prize in Physics, vigorously suggested the establishment of European 'regional research centres', including one with a powerful accelerator. The suggestion of support—technical and perhaps even financial—from the United States certainly helped to sway opinions in Europe. In the debate following Rabi's speech, Auger was given responsibility for getting the project moving, and he was determined to follow it through.

His first step was to organize an international meeting in Geneva in December 1950, although even before that meeting he had received the first national contributions to add to the small sums he himself could bring from UNESCO. These sums, from France and Italy, were small but useful; Auger's tactic was to ask for small contributions to kick-start the process, in the hope that, once inside, the various countries would feel an obligation to provide the very large sums that would eventually be required to build the actual accelerator.

At Geneva, Auger obtained the recommendations he sought. It was decided that the project should be for a laboratory based around a particle accelerator, which should be of an energy greater than those being constructed at Brookhaven (3 GeV) and Berkeley (6 GeV), certainly an ambitious undertaking. Its location was yet to be decided, although Geneva was always probably the most likely choice. Finance would be obtained from the countries involved, France being asked for the largest contribution—30%—with Italy and Germany being asked for the next largest, and other countries asked for smaller sums. These were, of course, just suggestions, but some actual money did follow, first from Italy, and a little later from Belgium and France. The British response was negative and one practically of ridicule.

Through the first eight months of 1951, the project made steady progress through waters that were, admittedly, sometimes choppy—it was decided, rather than outdoing Berkeley in the energy stakes, to simplify things by copying their machine. Dominique Pestre called this chapter of the *History of CERN* 'The Period of Informed Optimism' [3], but his next, covering the remainder of the year is called 'The Period of Conflict: The Gradual Emergence of an Alternative Programme'; the conflict referred to was centred on Niels Bohr.

Bohr was, of course, by far the most prestigious European physicist, and Amaldi had not been above 'taking his name in vain' in publicity for the proposed accelerator. However, the last thing he and Auger wanted was for Bohr to take over their project and derail it from its sole aim of providing a high energy accelerator.

Something of this nature seemed to be happening in July 1951, when Bohr and his ally, the Dutchman Hendrik Kramers, made two suggestions— suggestions that certainly turned out to be extremely constructive. The first was that, rather than starting solely with the building of the large accelerator, it would be prudent to put into construction also a small machine that could be constructed quickly, so that scientific work could begin as soon as possible. The second was that, since education was as important as facilities, an *Institute for Advanced Studies in Nuclear Physics* should be set up for courses and colloquia. Both suggestions were broadly agreeable to Auger and Amaldi.

However, in the following months, Bohr and Kramers went much further, effectively challenging the whole case for action that had been made so far. Their argument was that the new laboratory should be built around an existing research centre, and that the natural choice was Bohr's institute in Copenhagen as, they said, it already had a strong theory group and good experimental facilities. In this way, they suggested, money would be saved. Furthermore, they said that the programme so far suggested was entirely wrong. Work should start with a thorough survey of the theoretical and experimental background, and only then should a decision be made on the building of accelerators. In Pestre's words [3] and in Bohr's view, 'the project had taken a wrong turning and everything had to be cast back into the melting pot.'

Britain was at this stage on the sidelines of the general discussion, at best regarding the plans as good for Europeans other than themselves, but did come down broadly on the side of Bohr; this particularly applied to Chadwick and Thomson, although much less to Cockcroft. Those on Bohr's side in the argument were perhaps older than those supporting Auger and Amaldi, largely removed from personal involvement in research, and uninvolved in large-scale or nuclear research; Bohr and Kramers, of course, were theoreticians. They were unaccustomed to expect or even to ask for large sums of money and, indeed, despite his immense prestige, Bohr had spent much of his time before the war soliciting money from charitable institutions for his own work [12].

Those on Auger's side were just the opposite: young, involved in nuclear physics, aware of the new financial requirements to carry out experimental physics, and not afraid to ask for large sums of money.

At the national level, and again making sweeping generalizations, the countries supporting Auger were those which acknowledged their poor infrastructure for research in physics and were strongly involved in European affairs—France, Italy, Belgium, and (eventually, and with the exception of Kramers) the Netherlands. Against were the countries more self-confident in nuclear matters and less committed to European projects than the former group—the Scandinavian countries and, a fortiori, England. Germany's main preoccupation was to be centrally involved. The Swiss were probably most concerned with making sure that the laboratory was located in their own country.

Auger's approach to this intense conflict was to use the all-important intergovernmental conferences that took place in Paris in December 1951 and in Geneva in February 1952, to present his own ideas as the working document on the process, to allow Bohr success in certain compromises, but to use his own allies to achieve confirmation of the central points of the Auger–Amaldi position.

As a response to Bohr's idea for using existing facilities, British and Swedish offers to use the upcoming Liverpool and Uppsala synchrocyclotrons were accepted. Also a theoretical study group would be set up in Copenhagen. However, this study group would be broadly of the same status as three others which would be set up to plan the two accelerators to be built as well as the laboratory itself. These study groups would inevitably be based initially across Europe but, once the main laboratory had been constructed, they would obviously move there, and there was, at the very least, no guarantee that theoretical work would remain in Copenhagen.

Auger's side of the argument was scarcely dented by these 'concessions'. He had what he wanted—the three study groups on the experimental side, and the statement that the organization was to be a 'laboratory' actually doing things, not just a 'centre' around which things might be done in other places. And his side of the argument gained the agreement to set up a Council of Representatives from the participating countries; this Council would be essential if the organization were to develop as Auger wished, and particularly essential for paying the bills.

So February 1952 may be called the beginning of the *Provisional CERN*, which was actually to last longer than expected. In fact, an intergovernmental convention was signed in July 1953, but it was not until October 1954 that the full CERN came into being The cause of the lack of speed was that, however anxious those involved were to make a good start on the project, to get governments to make the formal commitments was still rather a lengthy and bureaucratic process.

However, an enormous amount was achieved by the Provisional CERN during this period. On the technical side, of course, the great event was the discovery of strong focussing. The Norwegian Odd Dahl, who was the group leader for the synchrotron, and the CERN Council had to decide either to play it safe and stick by their previous plans or to dare to opt for strong focussing, with all the risks that would entail but also all the tremendous gains if they were successful. They made the latter choice, so the intended energy of the accelerator was increased from 10 GeV to 30 GeV, later reduced to 25 GeV for financial reasons, while the construction time was reduced to six years. The proposed time for building the small machine, a 500 MeV synchrocyclotron, was between two and three years.

For CERN, the exciting point about the coming of strong focussing was that Europe and America were essentially back together at the starting point of a new race. Indeed, the various timescales meant that, not long after lagging far behind the Americans, for a period CERN would have the most powerful accelerator in the world.

The choice for the location of CERN was bound to be contentious but it was probably always likely that it would be in Geneva. In retrospect, Polkinghorne [10] felt that this may have been an unfortunate decision, as Geneva was 'an expensive milieu with an ethos of diplomatic-style loving'. Indeed, some of the tensions that were perhaps bound to emerge between those with the cachet of being based in Geneva, and those from outside wishing to use the facility, might have been dissolved sooner if CERN had been in a less prestigious location than Geneva.

Copenhagen had been vigorously pushed by the Scandinavians but had little support elsewhere—it was too far north for the southern Europeans, too near Russia for Lord Cherwell of Britain, and too likely to create a CERN dominated by Bohr.

Once defeated on that point, the Scandinavians devoted themselves, in Krige's phrase [3], to 'winning something for Copenhagen'. The Swedes, backed up by Germany, Norway, and Yugoslavia, proposed that two theory groups should be established—one in Geneva and the other in Copenhagen— and only after five years should a decision be taken over the long-term future of the Copenhagen group. However, the Council's decision was that the theory group should stay in Copenhagen for at most three years. Effectively, like the other groups, they should move to Geneva as soon as accommodation was available for them there.

Bohr must have bitterly disappointed and commenced discussions with CERN Council members from Denmark, Sweden, and Norway, as well as representatives from Iceland and Finland, to set up a Scandinavian institute for theoretical physics. These discussions led to the establishment of NORDITA, the Nordic Institute for Theoretical Atomic Physics, which was founded in Copenhagen in 1957 and moved to Sweden in 2007. We met this institute twice in Chapter 1.

For the eventual CERN career of John Bell, the most interesting point is that, as a result of the politicking of Bohr and certainly far from the original plans of Auger and Amaldi, theory became, one might almost say, central in the development of the project. Indeed [2], the first fully operating research group was the Theory Group, and the special role of the theorists was recognized: they were encouraged to act as consultants to the experimentalists and to take part in the establishment of the experimental programme, although the latter was never an obligation. What *was* demanded was excellent research and, for some members of the group, this research was to be abstract and mathematical.

In Copenhagen, the group was centred on four Danes: Bohr himself, who was Director until September 1954, Christian Møller, who took over at that date, Jacob Jacobsen, and Stefan Rozental. However, by 1952, there were in addition nine fellows from different European countries and, by 1954, the total complement of the group was around 25 and included Aage Bohr and Ben Mottelson, whose nuclear physics work, which would give them shares in the Nobel Prize in 20 years' time and which was mentioned in the previous chapter, had already been performed in the United States.

In these early years, the work of the Theory Group was the only scientific activity of CERN and also the main item in the budget, and thus it was closely discussed at Council meetings. This situation was, of course, in total contrast to the one that developed within a few years, when the cost of theory was a miniscule fraction of the total cost of CERN. However CERN management realized that the organization needed to win Nobel Prizes, and theoreticians might stand as much chance as experimentalists of achieving that distinction. Certainly, a theory prize would come vastly more cheaply than an experimental one. As from so many points of view, John Bell should have lived longer than he did!

Another contentious issue in these years was the appointment of the first Director-General. Of the six members of the appointment panel, three—Bohr, Heisenberg, and Cockcroft—had Nobel Prizes and, acting in consort in a way in which seemed rather arrogant to other Council members, insisted that, in order to convince the world of the status of the organization, it was essential that the Director-General also be a Nobel Prize winner. Their choice was Felix Bloch, who was certainly a wonderful physicist—he had shared the Nobel Prize in 1952 for his invention of the technique of nuclear magnetic resonance, but had many other achievements to his name, in particular, his work on the quantum theory of solids.

However, there were a number of difficulties associated with this choice. Although born in Switzerland, he had been working in Stanford since 1934 and had taken out American citizenship, a fact which might not have pleased countries such as France and Italy, which had strong Communist parties. He also had little experience and probably little interest in accelerator physics. That may not have been too important if he was to be regarded merely as a totem but, in addition, he wished to avoid actual administration as much as possible.

Bloch was appointed in April 1954 and arrived at CERN in September. He had insisted on retaining his Stanford salary, which was not far off double what had been budgeted for at CERN and, when he arrived, he found that CERN was little more than a building site, so he asked for costs for two assistants to assist in theoretical physics. These were both excellent physicists. Anatole Abragam was employed to work with Bloch on nuclear magnetic resonance; he was subsequently to write two hugely important books (one with a co-author) on magnetic resonance [13, 14]. Bernard d'Espagnat was a particle physicist but, as we shall see later in the book, he became particularly well known for sharing John Bell's interest in the foundations of quantum theory and for working effectively in this area.

However, Bloch was still decidedly unhappy. Amaldi had been appointed his deputy, but he realized that Bloch was not willing to undertake very much administration, and, having no wish to stay at CERN as an administrative dogsbody, Amaldi announced that he would soon leave. In turn, as early as February 1955, Bloch announced his own departure at the end of August. He was to be replaced by Cornelis Bakker, a Dutch physicist who had previously been leader of the synchrocyclotron division and representative of the group leaders, effectively third in command, behind Bloch and Amaldi. He was a good physicist, although not of the calibre of Bloch or even Amaldi, but he was prepared to devote himself to CERN. Sadly, he was killed in an aeroplane accident in April 1960.

Another important development in the years of the Provisional CERN was Britain joining the project. In fact, since the Geneva meeting of February 1952, British attitudes had been changing. The offer of the use of the Liverpool cyclotron, meant to imply standing rather aloof, seemed, rather paradoxically, to be taken effectively as a token of membership. As the Harwell people became involved with the project, giving advice that was certainly much appreciated since, in Europe, they were the experts, they found the approach interesting and highly likely to lead to very successful results. The fact that the Brookhaven 3 GeV Cosmotron, the first accelerator in the world delivering much over 500 MeV, had been completed satisfactorily added to this positivity.

In June 1952, Jim Cassels from Harwell argued that Britain must be competitive. He listed the then ten member states of the Provisional CERN. It would be one thing, he said, to be second to the United States, but quite another to be twelfth behind Yugoslavia! Then, the discovery of strong focussing and its planned use at CERN made the prospect of joining even more enticing for those actually involved with accelerators. It would also be important that Britain joined by Christmas 1952 if she were to join at all, in order that she might get her share of important posts in the organization.

Wheels, both within the community of physicists and within the government, particularly the Treasury, still moved slowly, but the decision to join was announced at the very end of 1952 and, from the start of 1953, Britain contributed financially and was treated as a member of the council, although she never officially joined the Provisional organization. Then, in July 1953, she signed the Convention.

Operations in Geneva began as early as October 1953, when part of the proton synchrotron group took up operations, first in the University of Geneva, and then in an assortment of temporary accommodation, which eventually housed experimental equipment, stores, a workshop, and offices. Then, on 17 May 1954, building commenced on the main site on the Franco-Swiss border, and it might be said that CERN was really under way.

One important appointment took place around this time. As the head of the proton synchrotron group, actually the key position in the organization since so much was reliant on this accelerator, Dahl felt he could not spend all his time at CERN, so Frank Goward from Harwell was put in charge in Geneva. However, Goward died suddenly in March 1954, so Adams [15] took over the responsibilities of both Dahl and Goward in October. He was to play an enormous part in the development of CERN.

We shall briefly survey the development of CERN (Figure 3.4) up to the time of the arrival of the Bells [11]. First, we shall follow the progress made with the 600

Figure 3.4 A view of CERN, with the proton synchrotron ring in the right foreground, March 1959. Courtesy CERN.

MeV synchrocyclotron. The point of constructing this small accelerator was that it should be based on well-understood ideas, so the construction should be reasonably quick and easy. This proved to be the case. Design took from 1952 to 1955, and then construction a further two years. By early 1958, the machine was available for use, producing pions and muons for experimental purposes and, even by the end of the decade, some good research had already been performed with it.

The most important achievement was certainly the identification of the decay of a pion to an electron and an antineutrino. While it had been generally assumed that this mode of decay was possible, although admittedly rare in comparison with the usual decay to a muon and an antineutrino, it had escaped detection. The CERN scientists were able to identify it and to compare the likelihood of this mode of decay to that predicted by theory. This was an important achievement and thus an indication, if there had been any doubt, that CERN would be a serious player in particle physics.

Over these early years, the time available on the machine was split roughly between CERN groups and outside teams, the total number of experiments each year being around 15. The outside teams came from such locations as Harwell, Liverpool, Utrecht, Saclay, and Trieste, and among their most important results was obtaining a value of the magnetic moment of the muon; this was in excellent agreement with theory.

So, by 1960, the vision primarily of Bohr a decade before of constructing a small machine as well as the proton synchrotron had been shown to be excellent. At this time, the synchrocyclotron had a staff of well over 200, not including visitors, and it would continue to good work up to 1991, although in later years it was largely 'relegated' to nuclear physics rather than particle physics.

Constructing the proton synchrotron (Figure 3.5), with all the novelty of strong focussing was, of course, a far more challenging task than the

Figure 3.5 The completed proton synchrotron ring, 1959. Courtesy CERN.

synchrocyclotron and took from the earliest ideas at the end of 1952 to November 1959, when the first beam of 24 GeV protons was produced, to be increased to 28.3 GeV a little later. By 1960 the number working in this group was around 350. Also, just as important as accelerators for the laboratory were detectors, and much effort was expended in obtaining suitable bubble chambers.

Thus, it seemed that the Bells were joining CERN at an exceptionally good time. In the *History of CERN* [2], 1958 is called 'the year when the synchro-cyclotron yielded its first important results', 1959 is 'the year when the proton synchrotron gave its first beam, the year when the future of CERN was no longer in doubt', and 1960 is 'the year when all seemed possible'. By 1960 the total workforce, including those providing scientific, technical, and general services, was over 1000, with over 150 visitors, a fair number from outside Europe, on site at any one time. All that was required, it might be said, was to be awarded a few Nobel Prizes.

John Bell at CERN, and the neutrinos

Mary and John settled into Geneva very easily [16]. They got to know people from CERN and perhaps did not feel unduly concerned to interact with other local people, so the language issue was not especially important. Indeed, while they were far from unfriendly—they were charming company when it was required or with close friends—they liked to think of themselves as 'dull' and not in need of stimulation from others. In truth, they enjoyed best each other's company and getting on with their work!

They did enjoy the mountains. As well as their flat in Geneva, they were to own a modest apartment in Champéry, which was a ski resort near the city, and they came to enjoy greatly downhill skiing. As a theoretical physicist, John always liked to get to grips with the ideas of a practical discipline before attempting it in practice [17]. For example, when he was a boy, he studied a book titled *Every Boy and Girl a Swimmer* before daring to enter the water. He learned dancing from a book by the famous dance-band leader Victor Sylvester prior to launching himself on the dance floor, and he studied the theory of skiing from a French instruction book that, happily for a physicist, was based on dynamics.

At CERN, after introductions had been made, John was a little lonely—in the Theory Division there was no real tradition of getting newcomers involved with the ongoing projects. However, he soon became more involved with the work of CERN and indeed, even much later, when frankly he might have preferred to spend time on his studies of quantum theory, he was to feel morally obliged to concentrate his work on subjects that were clearly related directly to the work of the institution—the physics of elementary particles (which was also becoming known as high energy physics), and field theory.

Even within that range of topics, he felt a responsibility to spend some part of his time on subject specifically connected with activities at CERN. However Martinus Veltman said that Bell was easily persuaded to join in discussion on any subject in physics [18]. In his early years at CERN, Bell was particularly involved with the work of the neutrino group, and especially with Veltman, and indeed it may be said that the two were the 'quasi-official theorists' or 'house theorists' of the experimental work on neutrinos.

Veltman (Figure 3.6), a Dutch theoretical physicist, spent considerable periods of time working at CERN before moving to the United States in 1981. At CERN, Veltman and Bell were close colleagues, and Veltman [19] later recognized Bell's important contribution to his own work on the weak interaction in the 1970s; this work led to Veltman sharing the Nobel Prize for Physics with Gerard 't Hooft in 1999.

Although Veltman was a great admirer of Bell—indeed [18], he became joint editor of his collected papers—it is interesting that he had practically no sympathy with Bell's work on quantum theory [20], regarding it as 'foolishness not good for anything to do with the real world'. Strangely, he quotes Bell as saying that 'he intended to do away definitely with this nonsense of hidden variables' and remarks that he actually did so. These statements seem to be almost the direct opposite of Bell's actual beliefs and programme! Veltman was to be slightly appeased by the fact that Bell's work could be verified by experiment.

Figure 3.6 John Bell (right) and Martinus Veltman (left). Courtesy CERN.

In the early 1960s, it could be said that neutrino physics was the central concern of CERN [11]. To give a little background, we will remember from Chapter 2 that the weak nuclear interaction was responsible for beta decay, in which a nucleus changes its nature to one with either one more proton and one fewer neutron, or the other way round. To ensure conservation of charge, in the first case an electron is produced, while in the second case a positron is produced. The simplest case is the decay of a free neutron; this decay takes place with a half-life of around ten minutes and, in this decay, the neutron changes to a proton, and an electron is emitted.

While this was certainly what was observed, major difficulties with the ideas gradually emerged. An amount of energy, varying continuously between zero and a certain maximum, was missing in the final state of the system, and also spin angular momentum did not balance between initial and final states of the system, as it should have. To explain this difficulty, in 1930 Wolfgang Pauli suggested that another particle was emitted in the decay: the so-called *neutrino*.

Because in some cases there was no observed missing energy, the neutrino had to have zero or extremely small mass, but of course it could take off the missing energy as kinetic energy. It had the same magnitude of spin angular momentum as the electron, proton, and neutron, which also allowed conservation of spin in the decay; in addition, because it interacted only via the weak interaction, it was no surprise that the proposed particle had not yet been detected. In this type of process, it was assumed that a positron would be accompanied by a neutrino, and an electron by an antineutrino, the antiparticle of the neutrino. (The difference between a neutrino and an antineutrino is described in Chapter 2.)

Pauli's proposal solved the problems and fitted the facts. Nevertheless, he and most other physicists were reluctant to take it too seriously. At this time, there were only three particles known: the electron, the proton, and the photon, and physicists were reluctant to accept the existence of a fourth without direct evidence [21]. However, there was a gradual acceptance of the idea, although for a long time it seemed unlikely that the neutrino would ever be detected. By 1956, though, fission reactors produced such a strong flux of neutrinos that Frederick Reines and Clyde Cowan were able to detect the particle for the first time, essentially by reversing the beta decay process by which it was formed [22].

This experiment was clearly a tour de force worthy of a Nobel Prize but, in this case, the not-unusual gap between discovery and award of prize had an unfortunately consequence. Cowan, actually the younger of the pair by a single year, died in 1974, leaving Reines to make the trip to Stockholm alone when the prize was awarded in 1995.

In 1937 another new particle, the muon, was discovered; the discovery was a great surprise, and it was additionally perplexing that this particle was found to have exactly the same properties as the electron, apart from the fact that its mass was about 200 times as high. Then, much later, Pontecorvo [23] in 1959 and Melvin Schwartz [24] in 1960 suggested that there might be two neutrinos, one associated with the electron and the other with the muon, and they also came up with a way of finding out whether the suggestion was correct.

Their proposed method was to produce pions by allowing accelerated protons to collide with a target. These pions would decay into muons and neutrinos; the muons would be blocked with massive shielding, so the experiment would discover what would happen when neutrinos interact with protons. If there were only one type of neutrino, both electrons and muons would be produced, but if there were two types of neutrino, electron neutrinos and muon neutrinos, the neutrinos would be muon neutrinos and so only muons would be produced.

Finding a second neutrino would clearly be an important discovery worthy of a Nobel Prize, and CERN, confident in the knowledge that, from November 1959 until June of 1960 when the Brookhaven 33 MeV AGS (Alternating Gradient Synchrotron) would be completed, the most powerful accelerator in the world was in Geneva, took on the challenge, even although they knew that Brookhaven was already in the chase.

The *History of CERN* [11] talks of 'the difficulty, the sophistication, and the scale of the venture' but, in September 1960, the decision was taken at CERN to dedicate the whole organization to the project. It was given priority over all other experiments, and it became the direct responsibility of the directorate of the laboratory, in particular Gilberto Bernardini, Director of the Synchrocyclotron Division, and Victor Weisskopf, from 1961 until 1965 Director-General of CERN. The experiment would require the work of 40 physicists, two heavy-liquid bubble chambers, a cloud chamber, and a very large amount of electronics.

But failure beckoned. If 1960 was 'the year when all seemed possible', 1961 was 'the year of doubts and anxieties'. In April of that year, with the experiment about to run, it was found that the flux of pions, which was the primary beam of the experiment, was an order of magnitude lower than had been calculated; the experiment as planned was impossible, and therefore it was cancelled. In retrospect, Veltman [20] admitted that, in the rush to be first to the discovery, corners had been cut: a 'magnetic horn' which would have increased the flux of neutrinos was omitted, and the target was merely inserted into the machine itself rather than more technical beam extraction being attempted.

It now had to be acknowledged that the Americans would win the race and the Nobel Prize that would go to the winner. In fact, the muonic neutrino was identified, the Brookhaven paper was published in 1962 [25], and the Nobel Prize, which was awarded to the three main participants, Leon Lederman, Schwartz, and Jack Steinberger, followed in 1988. Later in the decade, Steinberger would carry out important work as a colleague of Bell and came to admire him greatly.

We may take the opportunity to carry the story forward, introducing the term *lepton* (literally, 'light particle', from the Greek) for the electron, the muon, and the two neutrinos. A completely unanticipated discovery of a third charged lepton, to be known as the tau, was made by Martin Perl in the years around 1975. The mass of the new particle was around 3,500 times that of the electron, and it was taken practically for granted that there would be an associated neutrino, the tau neutrino.

Thus, there are three pairs of leptons, and they took their places in the emerging *standard model* of particle physics [26]. In this model, there are three so-called *generations*, so each pair of charged lepton and associated neutrino belongs to a particular generation. There are good reasons for believing that there will be no further generations containing charged leptons of higher mass still. We shall meet other parts of this model later, but it should be added here that, just as the electron has an antiparticle, the positron, each of these leptons has its own antiparticle.

To come back to 1961, for CERN the cancellation was a massive blow. The confidence that, with a state-of-the-art accelerator available, the Europeans could compete on equal terms with Brookhaven or Berkeley had certainly been dented. Different people have suggested different reasons for the failure. One view at CERN [11] was that the European physicists just lacked experience with big machines, and did not even realize the wide range of problems that would be encountered stepping up from small machines.

For Pestre and Krige [27], much of the difference between the Americans and Europeans was that, over quarter of a century—pre-war, wartime, and post-war—there had emerged in America a breed of what might be called engineering physicists, equally at home designing and using accelerators and other machines of big science. In Europe the work was still fragmented between engineers, who were perhaps inclined to overdesign, and physicists, who were content merely to use the machine once it was built.

For Schwartz [28] himself, the problem at CERN was one of 'style'. He felt that there was little drive there to do 'adventurous and risky experiments', and he believed that, even with the problem with the pion flux, the situation could have been saved by shifting a 'bread-and-butter experiment'. Total cancellation, he felt, was a colossal blunder; at Brookhaven, he said, they all 'breathed a sigh of relief' when they heard about it!

Indeed, it has to be admitted that, even for a decade or so after this period, CERN came in behind the American competition. Useful results were produced at CERN, but the dramatic discoveries, those that would lead to Nobel Prizes, were made in the United States, with CERN trailing in behind. Weisskopf liked to say that: 'It is no good in physics to be always late', even if those at CERN would often describe the results they eventually produced as 'far superior' to the American ones [11].

Licking their wounds, the CERN directorate recognized that Brookhaven had 'put practically all their resources to obtain a quick answer to the two-neutrino question', and decided on a 'completely new neutrino programme' with 'a very sophisticated, elaborate, and efficient set-up, providing optimal working conditions for a whole series of experiments to be performed with neutrinos' [11]. In particular a range of techniques were introduced to increase the beam intensity, and a multi-ton spark chamber was constructed to improve the detection rate.

This programme included theory as well as experiment, and Bell and Veltman were centrally involved. Veltman [29] comments that this was a very exciting period, as it drew Bell and himself into intense discussions with the experimental physicists working on the neutrino experiment. He remarks that every theorist should have such an experience 'as it makes you truly aware what reality is'.

Bell's two papers with Veltman describe the possibility of using neutrino beams to produce what they called the *intermediate boson*, written as W, of the weak interaction. To explain this term, it may be said that, by this time, the basic mechanism of each of the four fundamental forces or interactions of nature was broadly understood. The particles interact by *exchange* of a particular type of object, so that, as said in Chapter 2, the situation is rather like tennis, where the players interact by exchange of the tennis ball, rather than boxing, where the interaction is direct.

For gravitation, the exchange particle is called the *graviton*, and that of the electromagnetic interaction is the photon; at this time, it was accepted that the pion was the exchange particle of the strong interaction. The latter belief was not exactly wrong but would be modified over the next few years.

Not much was known of the exchange particle of the weak interaction; in fact, as we shall see, it had been thought that its mass might be low enough that the aborted neutrino experiment might have discovered it. Much more would be found out about the weak interaction in the 1970s, including the fact that there was a Z^0 particle as well as a W^+ and a W^-, and that the masses were far too high for the particles to be found in the simple experiments discussed in this period; that of the W^+ particle and the W^- particle was 80 GeV, while that of the Z^0 particle was 91 GeV. (Although these particles would not be detected for two decades or so, their discovery would be CERN's first major and undisputed triumph.)

The term *boson*, incidentally, means that the exchange particle obeys *Bose–Einstein statistics*, which were invented by Satendra Nath Bose and Einstein; bosons may be created and destroyed without restriction. The other type of particle is the *fermion*, which obeys the so-called Fermi–Dirac statistics, which were invented by Fermi and Dirac. Creation or annihilation of fermions is largely dictated by conservation laws, and typical fermions are electrons, protons, and neutrons.

In their first paper [30], Bell and Veltman considered in some detail, and with a good deal of numerical calculation, part of the formula for cross-section for production of the W particle, specifically, the part which was independent of the particle's mass, which was unknown. A previous calculation had been made by T. D. Lee, Peter Markstein, and C. N. Yang, of whom Lee and Yang had already been awarded the Nobel Prize for their work on non-conservation of parity. However, the detailed models and the attention to detail of Bell and Veltman resulted in the previous values being substantially reduced, in some case by a factor of 2. This result was, of course, extremely important at the time when experiments were being designed and planned.

In the second paper [31], Bell and Veltman answered a question posed by Guy von Dardel, who was the person who had pointed out (actually, in no uncertain terms [11]) the major problem with the cancelled experiment. By this time, it was realized that the neutrino had a helicity, or spin about its direction of travel and, in the formation of the W particle, this property would be passed on to the new particle. When in turn the W particle decayed, this property would then be passed on to the decay products, which would be a pair of pions, or a muon and a neutrino, and it would have a large effect on the energies and distribution of angles of these particles. Again, the authors examined these effects in some detail.

In the experiments, it soon became clear that no such particles would be found [20] and, at a conference at Brookhaven, Veltman reported that, if the W particle (+ or −) did exist, it must have a mass greater than 1.3 GeV as, otherwise, it would have been seen in the experiment.

However, a different story was told at a conference organized at Sienna in September to October 1963. Here the paper was presented by Bell, and he included the names of Veltman and Jørgen Løvseth, the Norwegian theoretician, as joint authors, as they had contributed to the content. Veltman had not even seen the paper, but, as Veltman [20] says: 'He thought he was doing me a favour, but this was not the case!'

The Sienna paper [32] was written on behalf of the neutrino group, experimentalists, and theoreticians. It confirmed the work carried out at Brookhaven, in particular the existence of two neutrinos, and it also included results and ideas from CERN itself, of which by far the most interesting concerned the W.

Bell, Veltman says, broadly presented what the experimentalists told him. They did believe they had seen hints of the W particle in their results. Veltman [20] talks of nightly gatherings at the top of a tower in Sienna, where the CERN people were tormenting themselves. Should they claim the discovery, with the risk of looking foolish if it turned out to be a mistake? Or should they refrain and miss the much-needed triumph of an important discovery?

In the end, they perhaps went for the worst of both worlds. The words used (the 'ominous statement', as Veltman calls it) were: 'We would be very surprised if [the mass of the W] rose as far as say 2 GeV'. This prediction is to be compared with the actual value, which has already been stated, of 80 GeV. Veltman calls this 'the low point in the CERN neutrino experiment'.

At least in retrospect, there were two other problems with the theoretical analysis of the experimental results: the non-discovery both of neutral currents and of scaling. On the latter point, in 2003 Veltman remarked that Donald Perkins [33] (Figure 3.7) 'to this day' blamed the CERN theoreticians for this failure. Veltman calls this section of his book 'Missed Opportunities'.

As a result of theoretical developments in the later 1960s and in particular in the early 1970s, the idea of neutral currents, processes in which a neutrino interacts with another particle while retaining its nature as a neutrino, became very important in elementary particle physics, as we shall see in Chapter 4. Pais

Figure 3.7 Donald Perkins. Courtesy CERN.

[21] calls their discovery 'one of the fundamental scientific contributions of the twentieth century', and in 1973 their discovery was to be an early sign of the genuine coming to age of CERN.

In the 1960s, things seemed very different. Lee and Yang had suggested the possibility of the occurrence of neutral currents in the mildest of terms. Experimentally, it was very difficult to distinguish neutral currents from background events of other types. There is a famous picture [20] of Bernardini at a blackboard on which it is indicated that the ratio of neutral current events to charged current events, in which among the outgoing particles is a muon rather than a muon neutrino, must be less than 5%, whereas today it is known to be around 15%. The Sienna paper [32] is unfortunately fairly dismissive of the possibility of any substantial possibility of neutral current events.

Much more recently, Veltman [20] has reconsidered the analysis, suggesting that errors had been made but also that, in any case, nobody was very interested. He has come to believe that the existence of neutral current events *could* have been established in the experiments of the early 1960s, although admitting that the decision would have been marginal.

Now we turn to scaling. In *History of CERN*, Klaus Winter [34] discussed the relevant information coming from the CERN experiment. In the bombardment of nucleons in the target with neutrinos, it was noticed that there was a marked and highly unexpected increase of inelastic scattering with neutrino energy. (Inelastic scattering may be defined for this problem [20] as cases where extra particles are produced.) At the Sienna conference, this finding was mentioned by Perkins [35] and also by Bell, who said that 'this rise in the inelastic cross-section is one of the most striking features of the experiment.'

Perkins [33] later remarked that various people, of whom he admits that he was one, felt that this finding indicated some pointlike substructure of the nucleon. However, he says that he made the mistake of consulting 'two quite eminent CERN theorists', who advised him to concentrate on the simpler elastic reactions and not to waste his time on complicated processes involving many particles, which probably one would never understand!

Later in the decade, the behaviour was identified as scaling by James Bjorken, and similar experiments using electrons and neutrinos were performed by Jerome Friedman, Henry Kendal, and Richard Taylor, and analysis by Bjorken and Feynman predicted that nucleons were built up out of three quarks. We shall see much more of these ideas later. The three experimenters were to share the Nobel Prize in 1988.

Winter [34] in a sense 'piled on the agony' by mentioning that, in a seminar earlier in 1963, he had calculated that the elastic scattering of neutrinos by electrons would rise in proportion to the energy, precisely because the neutrino and the electron were regarded as being pointlike, just like the quarks-to-be; but, as he said, at Sienna he unfortunately did not make the connection.

In retrospect, one perhaps has to admit that these were not the most glorious days for CERN or for our 'quite eminent CERN theorists'. But, in fairness, they were all, in a sense, novices, entering a highly competitive environment. All would see much better days—CERN with many important discoveries, and our two theorists with work of the highest calibre.

Hidden variables and von Neumann: Bell's first great paper

Since the publication of Bohm's papers, Bell had been thinking fairly consistently about the problems of quantum theory, and he continued to do so on his move to Geneva. Indeed, his thinking on this subject was stimulated further by an encounter with a well-known physicist, Josef-Maria Jauch (Figure 3.8), who was Swiss and, after a period at CERN in the early 1950s, had become Director of Theoretical Physics at the nearby University of Geneva. He was best known for the book he wrote with Fritz Rohrlich in 1955, *The Theory of Photons and Electrons* [36].

By early 1963, Jauch's research was on the foundations of quantum theory, and he came to CERN to give a seminar on what he referred to as 'strengthening von Neumann's theorem', basing the talk on a paper he had written with Constantin Piron [37]. Since Bell was convinced that Bohm's work had discredited von Neumann's theorem, he was naturally keen to hear Jauch's point of view but also convinced that it must be wrong; to Bell, the very concept of strengthening von Neumann was rather like a red flag to a bull!

He had some intense discussion with Jauch, although he actually enjoyed it. Bell was always to appreciate discussions (or arguments) with people like Jauch, who recognized the full power of quantum theory and were thus disinclined to believe that its current status could be questioned. He preferred these discussions to what might be more like mutual congratulatory sessions with those people he was to meet later at conferences on the foundations of quantum physics, people who were only too willing to criticize the theory, and sometimes to suggest savage alterations, just because they had never worked with it and perhaps did not even understand it in any depth.

Bell worked on his response to Jauch and Piron; however, Piron also told him of what seemed to be a greater challenge still, which was in a paper published in 1957 by Andrew Gleason [38]. Gleason, who had been born in 1921, was a pure mathematician of the highest calibre [39], whose work on code-breaking during the war had been greatly admired by Alan Turing. Later at Harvard, he made major contributions to mathematics, including an important contribution to a famous problem known as 'Hilbert's fifth problem', as well as stimulating major developments in mathematical education.

Figure 3.8 Josef-Maria Jauch. Courtesy of Professor Charles Enz and ETH-Swiss Federal University of Technology Zurich.

The paper that Bell was informed about was not explicitly to do with hidden variables but rather on quantum logic; however, Jauch pointed out to Bell that its conclusion could easily be used to produce an argument seemingly ruling out hidden variables. Bell was actually placed in something of quandary. Gleason's paper was only nine pages long but it was dense in rigorous pure mathematics, and Bell had no wish to work his way through the elaborate arguments of the entire paper, merely to reach this conclusion, which was something of a side result.

Fortunately, he was able to produce the particular result concerning hidden variables from a relatively simple argument. The result is sometimes called Bell's 'other theorem' [40], the 'first' theorem being his famous theorem on local realism, to be explained later in this chapter. Since Bell's 'other theorem' was independently discovered by Simon Kochen (Figure 3.9) and Ernst Specker [41] in 1967, it is often called the Kochen–Specker theorem. The theorem, which, of course, was based on particular assumptions which would be analysed in detail by Bell, applied only to systems which included at least three distinct measurable quantities, or, as is said, three observables.

Figure 3.9 Simon Kochen. © Renate Bertlmann; Central Library for Physics, Vienna.

Bell would certainly have recognized that the various theorems were correct mathematically and so, in order to assert the possibility of hidden variables in quantum theory, he would need to demonstrate that they were making use of inappropriate assumptions or axioms. If this task, at any time, seemed difficult to Bell, he must have been reassured by the knowledge that the Bohm theory *did* exist; therefore, the theorems *must* be wrong, so it was just a question of finding the offending axiom.

While he had been thinking about these matters for at least a decade, and certainly they had been very much on his mind since his discussions with Jauch, the opportunity to bring his thoughts finally to fruition and then publication came in 1963, when John and Mary were given leave for a year from CERN and travelled to the United States. At the invitation of Sam Berman [17], they spent most of the time at the Stanford Linear Accelerator Centre (SLAC) in California, one of the most important centres for particle physics in the world, but they also made short visits to Brandeis University in Massachusetts, and the University of Wisconsin at Madison. They actually left Geneva on 23 November 1963, the day after President Kennedy had been assassinated, so they naturally reached a sombre and tense America. They were to return to Europe at the end of 1964 (Figures 3.10 and 3.11).

While they were away, Mary was fully involved in accelerator physics and, of course, John, being supported by the United States Energy Commission, was

Figure 3.10 John Bell; registration photograph at SLAC. © SLAC National Accelerator Laboratory.

morally just as obliged as when he was at CERN to work on particle physics and quantum field theory. Indeed, he published three papers on these topics while he was in the United States. In each of these papers, his address is given as SLAC, although the third paper has a co-author, Chris Goebel, from the University of Wisconsin. These papers will be discussed later in this chapter.

However perhaps the freedom from routine activities allowed him to devote some time to quantum theory. Half a century later, it is clear that the results were spectacular, even revolutionary.

Not only did Bell produce a result—Bell's theorem, or Bell's inequality—of very great importance, but his work removed a self-imposed prohibition on the part of the community of physicists on *discussion* rather than just *use* of the quantum theory, thus opening up an enormous amount of fruitful research and dialogue. Important and interesting as much of Bell's other work on accelerators, particle physics, and other aspects of quantum theory certainly is, it is undoubtedly these papers written in the United States that caused his name to be celebrated in the history of physics.

Figure 3.11 John Bell as a cowboy during his trip to California in 1964; from *Europhysics News*, Volume 22, April 1991, p. 65. Courtesy European Physical Society.

On the first of his two papers on quantum theory [42], Bell's address is given as SLAC, while on the second [43], which will be discussed in the next section of this chapter, his address is given as the University of Wisconsin, and acknowledgement is made to 'colleagues' both there and at Brandeis.

Although Bell had an intense interest in the foundations of quantum theory, he certainly approached publication with great care. On the one hand, he definitely wanted to get his ideas and arguments on the record. The interaction with Jauch had stimulated his thoughts to the point where he was sure he had an important case to make. He would have liked his ideas to be discussed by others and, as we shall see, he was able to propose, at least in outline, important experiments that he hoped would be performed.

On the other hand, he knew that the most common response to the study of the fundamental ideas of quantum theory would be, at best, lack of interest, and more likely that he would be categorized as something of a crank—these issues, it would be said, had been put to bed 30 years before. Both these responses had been seen in the reaction to the work of Bohm, and Bell hoped to do somewhat better.

He had two good reasons for this hope. First, the content of Bohm's papers was extremely ingenious, but the argument itself was long and rather technical. Those physicists who were, in any case, totally convinced of the non-existence of hidden variables in quantum theory (and it is not too much of an exaggeration to say that this included practically everybody except Bell) would certainly not take the time to study them. In contrast, although Bell's arguments were of enormous depth and subtlety, the actual examples he used were simple enough to demand attention, even from some of those who were convinced that he must be wrong.

Second, whereas Bohm had been criticized with the (actually, extremely unfair) argument that the results of his scheme were identical to those of the orthodox interpretation (except for his rather wild suggestion that there might be a difference, yet to be encountered, at short distances), Bell was able, in his second paper, to make a specific suggestion of an area of physics where standard quantum theory might conceivably break down. Indeed, this paper was in effect a call for particular experiments to be performed to test his thesis.

However, even this achievement had a drawback. To some, any suggestion that quantum theory might be wrong even only in very special circumstances was totally anathema. In fact, some people at CERN might have gone as far as to suggest that, in even suggesting such an idea, Bell was undermining CERN! While this was obviously ridiculous, again it showed how careful Bell had to be. It might be said that he was prepared to be brave, but he did not wish to be stupid!

In the first of the pair of papers, which was concerned with demonstrating the possibility of hidden variable interpretations of quantum theory, his way of hoping to attract the attention of those who might be interested, while avoiding that of those who would be scandalized, was to submit the work to a review journal, *Reviews of Modern Physics*. Normally, original research is published in regular journals such as *Physical Review, Journal of Physics*, and so on. Then, every so often, some expert will review the research in a particular area of physics and publish the result in a review journal.

Often, the review is just a gathering together or a summary of all the published papers; it would be useful for a worker in the field or for somebody who hopes to enter the field, but no more than that. However, most reviewers aim to produce more than just a summary; they provide a critical study of the literature and claim to provide fresh ideas and new insights. A review of this type may contain a significant amount of new material and ideas, and it may be heavily used by future workers in the field.

However, it must be said that Bell's 'review' was not even of this nature, as it was restricted to arguments that he knew were new and important. Yet, admittedly, as we shall see, it was written rather in the *language* of a review. He began the article by saying that, since Bohm's work of 1952, 'the realization that von Neumann's proof is of limited relevance has been gaining ground'— actually, as has been said, not much further than to Bell himself! 'However,' he continued, 'it is far from universal'. Indeed not!

He then said that he had found no adequate account of what was wrong with von Neumann's argument and felt that he could 'restate the position' with such simplicity and clarity that all previous discussions will be eclipsed. Again, it might be said that he was being rather disingenuous and inventing a practically non-existent discussion that he could 'review'. Bohm had suggested how von Neumann had erred, a suggestion with which Bell entirely disagreed, but his suggestion had almost certainly been the only one in the last few years.

Bell was unimpressed by Bohm's argument, which he said 'seems to lack clarity, or else accuracy'. It was that hidden variables were required in the measuring apparatus as well as the measured system. While certainly approving Bohm's use of the entire experimental system, Bell pointed out that these additional hidden variables were unnecessary—later in the paper, he presented a scheme that worked without them. Neither, indeed, were such variables sufficient; if von Neumann's assumptions were acceptable, hidden variables in any location would be powerless for the purpose of Bohm and Bell.

Further, Bell spoke of von Neumann and those who produced the other theorems ruling out hidden variables as 'the authors of the demonstrations to be reviewed', while it would seem much more natural to say that he was not 'reviewing' but strongly criticizing the conclusions drawn from their work, if not necessarily the work itself.

It is interesting to speculate why Bell decided to underplay his undoubted achievements in this paper and indeed in other work. Of course, the special nature of publishing on the foundations of quantum theory has already been stressed, but, even apart from that, it did seem that he was, perhaps rather unusually among academics, reluctant to shout of his triumphs. Indeed, he could be rather critical of those workers in the field of elementary particles who, he felt, were rather too keen on self-publicity. It was probably not, as is occasionally suggested, that he was indifferent to recognition, but he preferred it to come from others rather than to push his own merits. A similar suggestion was made in Chapter 2 about his proof of the CPT theorem.

Let us now study the paper in some detail. As just stressed, Bell did have the Bohm argument as a demonstration of the correctness of his general conclusions, but he buttressed this argument with a very straightforward approach of his own, demonstrating a hidden variable model for a spin-½ system. An electron is the most important elementary particle with spin ½. Actually, this means that the value of its spin angular momentum is ½ in the usual units used for this quantity at the atomic level. This unit is always written as $\hbar$, or $h/(2\pi)$, where h is Planck's constant, which we met in Chapter 1.

Anyway, when we measure the spin of a spin-½ system along any particular direction, we always obtain either the value $+½$ or $-½$ (in these units), and the probability of obtaining either result depends on the spin state of the system and the direction along which the spin is measured. As in Chapter 2, we may call these results $(+)$ and $(-)$.

In Bell's first American paper [42], the hidden variable model of such a measurement was broadly algebraical, while in the second he provided a geometrical model for the same situation. The second model is rather easier to follow than the first, so it will be explained here.

In (a slight adaptation of) this model, we consider the case where the direction of the spins is along the z-axis; the hidden variable interpretation of this case is that each spin has the same mathematical spin state corresponding to this direction, but it also has an associated hidden variable represented by a vector. Since the spin is along the z-axis, the directions of all these vectors lie within the upper hemisphere, so they all have positive components along the z-axis, and the directions are uniformly distributed in this hemisphere.

Now, imagine a measurement of spin along the z-axis for this case. We will get the result $(+)$ for a particular electron if the vector has a positive component along the measurement direction, and $(-)$ if it has a negative component. In this first case, of course, all the vectors do have a positive component, so we are certain to get the value $(+)$.

Similarly if we measure the spin along the $-z$-direction, we find that *no* vectors have positive component along this direction, so we will definitely *not* get the result $(+)$.

It is also straightforward to consider a measurement of spin at right angles to the z-axis, or in other words, somewhere in the xy-plane. In this case, half the spins will have components in the direction of the measurement, and half in the opposite direction. We may surmise that, in half the cases, we will get the result $(+)$ and, in the other half, we will obtain $(-)$. All the results so far are in agreement with quantum theory.

However, we now turn to a measurement in any other direction. In this case, some of the vectors will have positive components along the measurement direction, and some negative. It would be nice if the corresponding fractions were exactly the same as the probabilities decreed by quantum theory for obtaining the results $(+)$ and $(-)$. However, things are not as easy as this! Bell was, though, able to obtain this result by a mathematical transformation of angles; this transformation has a geometrical interpretation as a suitable squeezing of the hemisphere. In this way, agreement can be obtained between the hidden variables model and quantum theory, or in other words, between the model and what is obtained in an experiment.

Of course, Bell would admit that the model was totally artificial with no suggestion of physical significance. In particular, a complete theory would have to say how the system would behave during the measurement. But the important point was that Bell had demonstrated in this extremely simple way that hidden variable theories were certainly *not* ruled out in quantum theory. Now he had to show what was wrong with the various theories that argued that they were!

First, he naturally looked at von Neumann's argument. Bell was looking not for a mathematical mistake—von Neumann was certainly a great mathematician—but probably for a dubious assumption.

To explain where he found one, it is necessary to first explain the (rather misleading) term 'expectation value'. In quantum measurement, in the general case a measurement of a particular quantity with the measured system in a particular state may lead to a number of possible results, say X_1, X_2, and X_3, with the corresponding probabilities p_1, p_2, and p_3. Obviously, the quantity $p_1 X_1 + p_2 X_2 + p_3 X_3$ is interesting; it is an appropriately averaged sum of the results. However, it is rather unfortunate that it is called the 'expectation value'. Clearly, except in special cases, not only do we not *expect* to get the result in a measurement—we know we won't get it! But, like it or not, this is the term that is used for this particular quantity.

Now, consider a particular system and the task of measuring two different physical quantities, which we may call O and P. We might, for example, consider a spin system and wish to measure the x- and the y-components of the spin. Now, we may measure a quantity corresponding mathematically to $O + P$. In the spin case, it would correspond to measuring a component of spin midway between the x- and y-axes (with a factor of $\sqrt{2}$ coming in as well).

We may ask the following question: will the result we obtain when we measure $O + P$ be the sum of the results we obtain when we measure O and P individually? We should remember that, in the general case, we will need entirely different experimental arrangements to measure O, P, and $O + P$. This will certainly be the case for the measurements of spin. We should definitely not expect the experimental results to behave in this simple fashion, and quantum theory confirms this expectation. For the spin case, the results for O and P will each be $\pm\frac{1}{2}$, while that for $O + P$ must be $\pm(\sqrt{2})/2$, so there is no way in which the simple result could apply.

Yet, the rather strange fact is that, when we look at the expectation values of the measurements of these quantities, quantum theory tells us that they add up exactly as the rule suggests. The expectation value of $O + P$ is equal to the expectation value of O added to that of P.

Now let us turn to von Neumann's famous argument [44]. Bell pointed out that he assumed that the same relationship—additivity of expectation values—applied to his hypothetical hidden variables just as it did to the variables of conventional quantum theory and, in fact, this assumption was an important part of his proof. Tempting as this assumption may have been, it was actually completely inappropriate, and Bell showed that it applied neither in the Bohm case nor for his own simple model. Without that assumption, the theorem collapsed, and it was clear that it did not imply that there was any prohibition on hidden variables in quantum theory.

Bell would later claim [45] that what von Neumann had called a 'very general and plausible' argument was actually 'absurd'. And with von Neumann's argument there disappeared also (at least in time) what may be called the 'taboo' on discussion of the foundations of quantum theory, a taboo that had inhibited progress for 30 years. Today, many ideas and interpretations of the foundations of the theory are analysed and published, with some exceptionally fruitful consequences.

However, first here we must look at the second of the papers claiming to rule out hidden variables, that of Jauch and Piron [37]. Theirs was actually an argument in 'quantum logic', which may be described as a set of rules for reasoning about logical propositions related to quantum theory. Admittedly, this sounds very abstract, and we may certainly assume that this was Bell's suspicion as well. However rigorous the argument might be in mathematical terms, he must have felt that the translation from mathematics to physics would be extremely fraught.

The argument was based on projection operators. A projection operator projects a particular vector onto its component in a particular direction or onto a particular subspace. For example, in ordinary three-dimensional space, projection operators may project a general vector onto its components along the x-, y-, and z-axes and, if a and b are projection operators projecting onto particular subspaces, then the *intersection* of a and b is defined as the projection operator which projects onto the intersection of the two subspaces.

In their paper, Jauch and Piron discussed the expectation values of projection operators and their intersection. In fact, they assumed that, if the expectation values of both a and b were zero, so was that of their intersection. In response, Bell asked why they considered this to be a natural assumption and suggested that it was because they were making an analogy with propositions in standard logic.

In standard logic, we may say that the value 1 for a proposition means that it true, and the value 0 means that is false, while the intersection of *a* and *b* effectively means just *a* and *b*. So if the expectation values of *a* and *b* are both equal to 1, so is that of the intersection of these projection operators. Similarly, if the expectation values of *a* and *b* are both 0, so is that of the intersection, and with these results it is quite easy to reach Jauch and Piron's conclusion—no hidden variables.

However, Bell pointed out that Jauch and Piron made a mistake analogous to that of von Neumann. The projection operators should not be thought of as logical propositions but as the results of physical measurements and, as has already been stressed, different measurements (except in special cases) require entirely different experimental set-ups. It was thus completely illegitimate to add projectors as Jauch and Piron had done. As has been said, the result of their argument happened to be true for the variables of conventional quantum theory, but there was certainly no justification for extending the argument to the hypothetical hidden variables. Thus Jauch and Piron's argument did not apply to the hidden variable models of Bohm or Bell, and, like that of von Neumann, it failed to prohibit the possibility of hidden variables in quantum theory.

Bell now had to tackle the third of the proofs, that of Gleason–Jauch, as he called it. Addressing their proof was certainly a much bigger challenge than addressing either of the other two arguments had been, since this proof deliberately attempted to avoid the difficulties of von Neumann. As stressed above, Bell's arguments relied on the fact that measuring O and P required different experimental set-ups (and the same for *a* and *b* in the Jauch–Piron case). The Gleason–Jauch argument attempted to avoid this difficulty by restricting itself to cases where O and P could be measured simultaneously. And yet Bell was able to find a flaw in the argument. It will be remembered that the argument only applied to systems where there were at least three distinct observables. At first sight, that would not seem to cause any problem for the argument; it should be easy enough to find a third observable simultaneously measurable with O and P, and we might call this variable Q.

However, Bell trumped this argument by pointing out that, in fact, we have a *choice* of this third variable—say, Q_1 or Q_2. It might certainly be the case that each of Q_1 and Q_2 is simultaneously measurable with O and with P, but it will not be the case that Q_1 and Q_2 are themselves simultaneously measurable, and this observation destroys the Gleason–Jauch argument. To put things a different way, the apparatus required for measuring O, P, and Q_1 would be different from that for required for measuring O, P, and Q_2; thus, the Gleason–Jauch argument fails.

However, it must be admitted that the Gleason–Jauch argument and Bell's response to it eliminate the simplistic view, which could have been obtained from such arguments as EPR, of hidden variables. In this view, the hidden variable has a value before we measure the relevant quantity, and we merely obtain this value when we measure it. However, here it has been shown that the value we obtain for O or P cannot be determined before the measurement, because what we obtain will depend on what else we measure with it; it will depend on 'the complete disposition of the apparatus'.

In his paper, Bell quoted Bohr [46] on this matter, pointing out 'the impossibility of any sharp distinction between the behaviour of atomic objects and the interaction with the measuring instruments which serve to define the conditions under which the phenomena appear'. The technical term for this behaviour of the hidden variables is that they must be *contextual*; any measurement results connected with such a hidden variable depend on the full context of the measurement.

Such a stipulation must seem rather awkward in the abstract—perhaps almost contrived—but Bell would very much stress the point that it seems totally natural in the context of an actual hidden variable theory, in particular that of Bohm. In Bohm's model, measurement is not a straightforward registration of the pre-existing value of the measured quantity; rather, it is an interaction between the measured and measuring systems in which the pre-existing value may certainly change to a measured value.

Indeed, Bell [45] used this point as a moral of the story. He asked why 'serious people' took so 'seriously' axioms which, on closer examination, seem so arbitrary. The reason, he suggested, was the pernicious use of the word 'measurement', which does indeed suggest the ascertaining of some pre-existing property, with the apparatus being entirely passive. In this (misguided) view, measurement results might be expected to obey some simple logic, and Bell suspected that the failure of ordinary logic in this circumstance had led what he considered to be the unfortunate creation of 'quantum logic'. He was to come back to write more strongly about the misuse of the term 'measurement' much later in his life.

Having disposed of these three arguments against hidden variables, at the end of the paper Bell returned to the Bohm theory. Bohm had stated clearly that his theory was non-local and Bell emphasized this point by constructing a model of two particles, each with a magnetic moment, travelling across a space, and being detected by magnetic detectors. He was able to show that the trajectory of each particle depended on the trajectory and wave function of the other one, and so the results of measuring the behaviour of each particle would depend on the disposition of the apparatus measuring the other one, even if the particles and the measurement apparatuses were far apart. Bell remarked that this was the result that Einstein would have liked least: hidden variables or realism restored but at the expense of losing locality.

In the years since 1952, the year when he had seen Bohm's scheme, it is certain that Bell had put considerable effort into the production of alternative models of hidden variables that did not suffer from non-locality, but he had failed comprehensively. It was natural of him to suspect, therefore, that non-locality was a general property of *any* hidden variable scheme, or in other words, that local realism, for which Einstein had argued and which was his solution to the EPR problem, was impossible. At the end of Bell's paper, he suggested the search for an 'impossibility' theorem proving this idea correct. This search was to be the subject of his second great quantum theory paper, which was also written in the United States.

The first paper was brief—six pages—but Bell found space to thank both Mandl, for intensive discussions after the publication of the Bohm papers, and Jauch, for later discussions. It is clear that Jauch in particular either was unimpressed by Bell's argument or just failed to read the paper. In 1973 he published a book titled *Are Quanta Real: A Galilean Dialogue* [47]. This book was hugely enjoyable but certainly included no discussion of any position on quantum theory apart from strict adherence to the Copenhagen interpretation.

When Bell sent the manuscript for his first paper to *Reviews of Modern Physics*, he must have been in some trepidation. The ideas in the paper were obviously of great significance for him, and he also knew that they were important for the future of physics. Yet, he also realized there was a distinct possibility that the referee of the paper would dismiss it out of hand—for many physicists, any discussion of such matters was the mark of an amateur, practically a crank, who failed to appreciate the lessons of modern physics, which had been taught principally by Niels Bohr.

Bell was lucky that the referee chosen by the editor of the journal was Bohm himself. Clearly Bohm would have been exceptionally interested in the content of Bell's paper. It must have been gratifying to him that Bell was taking such an interest in his own work, even though there was some criticism in the paper of his own ideas. Bohm certainly wanted the paper to be published but, as many referees do, he asked for some additional material to be included.

In fact, he asked for additional discussion of the actual measurement process. It will be remembered that his own ideas implied that the detailed behaviour of the measuring apparatus was more significant than Bell had suggested. Bell recognized, of course, that the presence of the measuring system played a central part in his analysis, particularly in the discussion of contextuality, but he did not think that the details of the measurement procedure were necessary for his own discussions.

Bell obviously had to give some ground to Bohm but, as often in these affairs, the referee and the author have different ideas of how much change or how much extra material is required. The referee perhaps thinks a few more weeks or months of effort are called for, while the automatic response of the author is always 'What is the absolute minimum I can get away with?'

Bell in fact added a couple of sentences: 'A complete theory would require an account of the behaviour of the hidden variables during the measurement process itself. With or without hidden variables the analysis of the measurement process presents peculiar difficulties, and we enter into it no more than is strictly necessary for our very limited purpose'. He also gave references to a few papers on the measurement procedure.

Another relevant point in the published paper referred to his belief that all measurements notionally of any physical quantity were ultimately ones of position—a mark on a scale, or a position of a dial. In the paper, there is a footnote saying: 'There are clearly enough measurements to be interesting that can be made in this way. We will not consider whether there are others'.

With those fairly minor additions, Bell sent the paper back to *Reviews of Modern Physics*. Unfortunately, fate was now against him. The revised version of the paper was misfiled by the journal, and when the editor wrote to Bell asking why no revised version had been received, they naturally sent the letter to SLAC, the address from which he had submitted the original version.

Bell was by now back in Geneva and, unfortunately, the SLAC post office was not alert enough to recognize the name of its recent visitor. They returned the communication to the journal as 'not known here'. It was not until much later that Bell had the courage to write to the journal and inquire what had happened. The revised version of the paper was then published but two years had elapsed— at that point, it was 1966!

To add to the confusion, Bell's second quantum paper had been written in 1964 and published in the same year. It was actually answering the question asked at the end of the first paper two years before the question would be posed in print! In fact, Bell was able to add a note reporting publication of the second paper at the end of the first one. Such confusion could certainly not have helped the reception of this pair of highly important papers.

Bell and local causality: Bell's second great paper

In his first paper discussed in the previous section, Bell announced his suspicion that the creation of local hidden variable interpretations of quantum theory was impossible. In the second paper, he was able to demonstrate the correctness of this idea.

While the first paper was extremely important and encouraged, for the first time in 30 years or so, theoretical speculation on the basic ideas of quantum theory, it was this second paper that became one of the classics of twentieth-century physics, exceptionally significant in its own right but also making it clear that the foundations of quantum theory were ripe for experimental as well as theoretical investigation.

Twenty year later, Paul Davies [48] was to ask Bell how long it had taken to reach his result. Bell's reply was that he had had the ideas at the back of his mind in the previous years (presumably since 1952) and had been thinking intensely about the issues in the previous weeks (perhaps since his arrival in the United States or at Brandeis University); however finally getting the central equation into his head and onto paper had taken about a weekend.

One might have thought that Bell at the time might have been rather secretive about working on quantum theory but, at least to an extent, this seems not to have been the case. At various periods in his life, when asked what he was working on, he was inclined to mention hidden variables and, on various CERN documents, he described himself as a 'quantum engineer'.

Even in Brandeis, it must have been fairly common knowledge that he was at least working on something away from the common path. Kamal Datta [49], who was a postgraduate student at Brandeis at the time, said that faculty members pointed Bell out to him in the corridors and remarked that Bell was engaged in work of the most profound significance in quantum theory. Datta also said that Bell was not at all gregarious and that students were advised not to bother him with their own problems because he preferred to work on his own.

In fact, at the end of the second paper, Bell thanked two colleagues based in Stanford, for their discussions. The first was Myron Bander, at that time a postdoctoral fellow at Stanford. In 1966 he became a professor at the University of California at Irvine, where his main work was on elementary particles, although he also maintained an interest in the fundamental aspects of quantum theory. The other was Perring, who had of course worked with Bell at Harwell and was on study leave from Harwell at Stanford. During their period in Stanford, Bell and Perring were to publish jointly an extremely interesting paper in the elementary particle field; this paper will be discussed later in this chapter.

Let us now turn to the actual quantum paper, and it is of the utmost significance that it is titled 'On the Einstein Podolsky Rosen Paradox'. It will be remembered that, up to this time, it was generally felt that the EPR argument had been rather misconceived, that it had been totally refuted by Bohr, and that it certainly deserved no further consideration. Even if any particular scientist happened to have even a limited interest in EPR and the related argument between Einstein and Bohr, it would have still been taken for granted that any discussion was of the 'armchair' variety, and it would have been inconceivable that it could have led to the performance of crucial experiments.

Bell, however, felt that the EPR paper was not only correct but that, in its stress on entanglement, it was extraordinarily significant. It is also important to realize that Bell by no means believed that, in the EPR paper, Einstein had raised matters which were admittedly of some interest but which he had not understood—hence the 'paradox' idea—and had left to others to sort out. Rather, Bell believed not only that the EPR argument was absolutely correct but also that it could be built on in an exceptionally meaningful way.

Before discussing Bell's argument in detail, we will stress his central point, which may be described as the great advance on the ideas of the previous 30 years. In the experiment as discussed, the measurements of spin were (virtually) always along the z-direction for each spin. Clearly, it was trivially easy to explain the quantum mechanical result, that the $(+)$ would be obtained for one spin and $(-)$ for the other, by the hidden variable assumption that these values were already existence before any measurement was made.

The only other case that was very occasionally considered was where the measurement on one spin was along the z-axis, but the measurement on the other spin was along a direction at right angles to the z-axis, somewhere in the xy-plane. In this case, quantum theory tells us that there is no correlation between the results in a sequence of measurements, and, as explained in the previous section, this result can easily be explained by a hidden variable model in which half the spins are 'programmed' to give $(+)$ and the other half $(-)$ for a measurement in this direction.

Bell's great advance was to consider measurements along a general direction but, more particularly, to find out whether the spins could be 'programmed' by a suitable choice of hidden variables to give the results predicted by quantum theory for any such general measurement. His crucial conclusion was that they could not.

It is maybe natural to inquire whether this step was indeed so 'crucial'. The idea itself is relatively straightforward, and the mathematics is not advanced or difficult. However, the fact is that, in the years since the founding of quantum theory, and in particular since the publication of the EPR paper, nobody had realized the significance of such a step—nobody had had the clarity of thought to question whether the arguments of Niels Bohr and his followers could be tested by this relatively simple procedure.

Now let us examine Bell's argument in more detail. As has been said, he started with what was essentially a rerun of the EPR argument, focussing on the correlation of experimental results on the two spins or, as we may say, in the two wings of the experiment. He next introduced what he called a 'hypothesis': 'if the two measurements are made at places remote from one another the orientation of one magnet [which measures the spin] does not influence the result obtained with the other.'

He then suggested that this 'hypothesis' seemed to be 'at least worth considering', basing his remark on Einstein's statement [50] that 'on one supposition we should, in my opinion, absolutely hold fast: the real factual situation of the system S_2 is independent of what is done with the system S_1, which is spatially separate from the former.'

It is interesting to consider why Bell put these ideas in this half-apologetic way as 'a hypothesis worth considering'. He did not actually have to endorse the position, but surely he could have presented it a little more positively! Again, it would seem that he was presenting the argument cautiously— almost as an intellectual exercise rather than a deeply held belief. Clearly, he wished to get his views into the public arena without running too much risk of antagonizing those who might object to the very idea of such matters being raised.

These ideas of Einstein and Bell are, of course, at the very least closely related to the idea of locality as discussed in Chapter 2, but it may be good to be precise. Bell stated that the purpose of hidden variables in the EPR context was to restore 'causality and locality' to the theory. Einstein, of course, believed strongly in determinism, although admittedly, at least from the 1930s it was not as important to him as realism and locality. For Bell, however, realism and locality were extremely important, but determinism much less so, and so the latter does not appear much in his own analysis.

He discussed the requirement mentioned above as one of 'locality or separability' and it is useful to distinguish between the terms as used by most people who work in the area. Einstein is often associated with the idea of 'separability', which is indeed often referred to as 'Einstein separability'. This idea implies that an event at one location cannot have an effect (not just an *immediate* effect) at a separate location.

Now, Einstein obviously understood that a signal could be sent from one location to the other and so, of course, would cause an effect at the second location, and it may be said that his discussion excluded this possibility. With this case excluded, he felt that an event at one location could have no effect at the second location *at any later time whatsoever*, not just for times short enough that light cannot travel from one location to the other.

Indeed, Einstein believed that if this requirement were violated, it would be impossible to carry on the discipline of physics! He was obviously concerned about a situation where such effects took place in a wholesale uncontrollable and unpredictable manner, in which case he was almost certainly right. However, although it would be rather generally accepted by now that physics is subject to non-separability, its restricted occurrence only in connection with entanglement makes physics a little strange but its study by no means impossible in practice.

Locality, on the other hand, would be defined so as to rule out only interactions between the two locations which travel at greater than the speed of light, but allowing any travelling slower than that speed. In addition to specific signals passing information, then, it also allows the possibility that the detailed nature of the heat radiation leaving one body could somehow also pass information to another body.

In quantum theory, there are actually two aspects of locality and it is important to distinguish between them. In the first, the 'cause' may be an action of a human experimenter, such as setting a magnet at a particular angle; the 'effect' is then the direction of a spin at a different point, as shown in a measurement result. Locality would suggest that any information should pass from one to the other at or slower than the speed of light. This aspect of locality was called *parameter dependence* by Shimony [51]. Separability, of course, would suggest that there should be no correlation at all.

Shimony called the second aspect *outcome independence*. This aspect relates to the type of situation we have in EPR, where entanglement entails that the results of the spins in the two wings of the system are correlated. Outcome independence says that the *result* of a measurement in one wing cannot have an immediate effect on a measurement in the other wing. The equivalent statement regarding separability would again rule out *any* effect, not just an immediate one. It is not so obvious that relativity is involved in any breakdown of outcome independence, since it could not be used by an experimenter to send information (The result of the first measurement is not in the control of the experimenter but just the fact that the measurement takes place.)

However, the relationship between locality in Bell's work and relativity is not a simple one and is discussed further by Tim Maudlin [52]. It should also be mentioned that ideas which were similar to those of Shimony were presented by Jon Jarrett [53], although what Shimony called parameter and outcome independence, Jarrett called locality and completeness, respectively. We shall meet Bell's own ideas on the subject in Chapter 5.

We now return to Bell's brief summary of EPR—just one paragraph! In it, he mentioned both of what we have referred to as parameter and outcome independence, but he laid particular stress on the fact that the result of the measurement on one spin makes it possible to predict the result of the same measurement on the second spin. Since, he said, the wave function does not allow such a prediction, 'this predetermination implies the possibility of a more complete specification of the state', or in other words, hidden variables, and indeed deterministic hidden variables, since the value of the hidden variable in each wing of the apparatus must lead unambiguously to the value in the other wing.

The briefness of this account of EPR may have caused considerable confusion over the basic assumptions of this particular paper. More than 15 years later, in a paper [54] in which he discussed the EPR argument, Bell pointed out that what was essential in the argument was 'local causality' or, as he put it, 'no action at a distance'. Determinism played only a limited role, and it was a *result* of the argument, not an *assumption*. Bell remarked how difficult it was to get this point across because 'there is a widespread and erroneous conviction that for Einstein' (Bell included here a footnote which added 'And his followers') 'determinism was always *the* sacred principle': Einstein's saying 'God does not play dice' is so well known even to the general public (perhaps second only to his equation $E = mc^2$) that it is natural to assume quite erroneously that the basic assumption of EPR *must* be determinism.

Bell made it absolutely clear that he considered himself to be one of Einstein's followers in this respect. Bell added that, although his original quantum paper had started with a summary of the EPR argument from locality to deterministic hidden variables, those commenting on the paper had almost universally said that it *began* with deterministic hidden variables.

Let us now return to our examination of that paper. Having sketched EPR and deduced the existence of deterministic hidden variables, Bell came to the novel part of his argument. He was interested in discovering whether, with the use of the hidden variables but respecting locality, he could reproduce the quantum mechanical predictions for the generalization of the EPR situation in which the measurements in the two wings of the apparatus might be along different directions.

He started by showing that hidden variables *can* handle some interesting cases. In the previous section on Bell's first great paper, we discussed his proof that there is no difficulty at all in using hidden variables to produce the quantum mechanical results for a measurement on a *single* spin. In the second paper, he then showed that there is equally no difficulty in explaining the results for an EPR-type system with two spins when the spin measurements are in the same direction or at right angles—as he pointed out, these were the only cases that had been analysed in discussion up to that point. The experimental result being considered is related to the probability of the two measurements being the same—either both $(+)$ or both $(-)$, or different.

It may be mentioned that Bell actually went further than the cases usually discussed. In these cases, the particles move in the z-direction, and the components of spin measured are also both in the z-direction or, in the exceptional case, in the z-direction and at right angles to it. Bell considered the case where the first measurement might be in any direction but the second was in the same direction or at right angles to it. As has been said, he was able to predict again very straightforwardly the quantum mechanical results with an appropriate choice of hidden variables.

And lastly he showed in a very similar argument that there was again no difficulty in reproducing the quantum mechanical predictions for the general case of measurements at general angles in each wing *if* the results in each wing of the experiment were allowed to depend not only on the experimental setting in that wing, but on the setting in the other wing. But, of course, that freedom was *precisely* what he did *not* allow in his actual proof.

This proof was actually comparatively straightforward. Bell defined three different directions: **a**, **b**, and **c**. He then calculated the experimental result for three cases—first, the measurement direction being along **a** in the first wing of the experiment, and **b** in the second, then the directions being **a** and **c**, and finally **b** and **c**. While each of these results depended on the details of the distribution of the hidden variables, Bell was able to obtain a relationship between the results that was independent of the hidden variable distribution.

This relationship was actually an inequality, and the result is often called *Bell's inequality* (sometimes a number of related results are called *Bell's inequalities*) or *Bell's theorem*. He was also able to show that there was no way in which this relationship could be obeyed by the comparable quantum mechanical expression. Indeed, there had to be *significant* differences between the quantum mechanical expression and the results of his own analysis using hidden variables.

This theorem was the all-important result. It had been obtained by what Bell called 'high school mathematics', although, having noted that fact, Bernstein [55] pointed out that its six pages were 'dense with an extremely abstract set of arguments, which even professionals in the field must work hard at to understand.' Kurt Gottfried and David Mermin [56] actually claimed that Bell's analysis was 'extraordinarily elementary' but marvelled at the sophistication and depth of the argument. It had the generality, they said, 'one encounters in the foundations of thermodynamics'. Rather than analysing the state of affairs using the procedures of quantum theory, it uses the experimental *phenomena* to arrive at general conclusions. Henry Stapp [57] was to call the result 'the most profound discovery of science'.

In this paper, Bell pointed out that it required 'little imagination' to imagine an experiment testing his ideas—basically, that experiment would involve (i) a system producing entangled pairs (of electrons in Bell's paper, although in practice usually photons) and (ii) appropriately oriented detectors. In terms of comparison between experiment and theory, we can immediately see the enormous advance made by Bell over the ideas of Bohm. Whereas Bohm could only make the vaguest of suggestions that his approach just possibly *might* give different values to conventional quantum theory for the case of small distances (a case which had not been explored experimentally), Bell was able to point out clearly a specific region where quantum theory might be thought to be suspect. Of course, most physicists might expect or even be fairly sure that quantum theory would be triumphant for this case as well, but some at least, Bell must have hoped, would think that the matter should be settled experimentally.

What about Bell himself? Obviously, he hoped the experiment would be carried out. When John Clauser [58], who was to carry out one of the very first experiments on Bell's theorem, wrote to him in 1969 asking about the importance of such a test, Bell replied that (of course) he would very much like these experiments, in which the crucial concepts would be tested directly, to be performed and to have the results on record.

It will be remembered though that, although Bell thoroughly distrusted the logical basis of quantum theory—he admitted to Bernstein [55] that he *knew* the theory was 'rotten'—rather paradoxically, he was also well aware of the great and wide-ranging practical success of the theory. He told Clauser that it was difficult to doubt the outcome of the experiments—quantum theory would surely triumph. But, he added, there was always 'the slim chance' of a result that would 'shake the world'. One might guess what probability he estimated of arriving at this result—perhaps 5%? Of course, as he said later [54], 'those of us who are inspired by Einstein' would like this result best— quantum theory may be wrong, and 'perhaps Nature is not as queer as quantum mechanics'.

As yet, we have been somewhat evasive as to precisely what experiment is being contemplated; however, this evasiveness is rather appropriate, because it is actually exactly what Clauser was to accuse Bell of—in fact, he accused him of 'bluffing'. Clauser was a tremendous admirer of Bell—he later said that he had been, in fact, 'astounded' by the result of Bell's paper—but he added that, at the time he had noticed that, while Bell was (typically) 'crystal clear' on everything else, he was vague on the experimental status of his prediction.

There are two possible types of experiments: Clauser called them 'first-generation' and 'second-generation' experiments. In first-generation experiments, the directions of the analysers in each wing of the experiment are static; thus, the measurement directions are static. Bell said that, although the results of such experiments would be extremely interesting, it would be quite possible for the settings in each wing to 'reach some mutual rapport by exchange of signals with velocity less than or equal to the velocity of light'. In terms of our previous definitions, it may be said that we are applying separability but not locality.

Then in the second-generation experiments, the directions of the analysers are not static but are determined only during the flight of the entangled particles, so that there may be no exchange of information between the two wings of the apparatus. Clearly, second-generation experiments must be expected to be considerably more complicated and difficult to carry out than first-generation experiments would be.

The difficulty for Bell in 1964, and indeed for Clauser and the others who wished to carry out experiments on Bell's theorem, was to ascertain whether any first-generation experiments had already been carried out. This question was tricky, because there had certainly been experiments at least of a very similar nature to those suggested by Bell's work. Clearly, if they had been *precisely* of this type, their results must have agreed with quantum theory or otherwise there would presumably have been considerable publicity.

Clauser accused Bell of assuming (without quite saying so—the lack of crystal clarity) that first-generation experiments with results agreeing with quantum theory had already been conducted, and so a test of Bell's theorem would require a second-generation test, with all the difficulty of rapidly altering the directions of the analysers. But as Clauser said, a proposal for a second-generation test would be silly if one did not know whether a first-generation test had actually been carried out. Clauser would claim much of the credit for calling Bell's bluff in this matter!

Let us, though, return to 1964. While carrying out the work, Bell gave two seminars on the topic, one at Brandeis and one at Wisconsin; and then, when he had finished his paper, naturally he had to decide to what journal he would submit it. The most likely choice would have been the *Physical Review*, the leading physics journal for papers, while its sister journal, *Physical Review Letters*, was the most important for short papers or 'letters'. Publication in the *Physical Review* would surely have maximized the attention his paper would receive.

Yet, he chose to submit it to a totally new and very unusual journal called *Physics* or in full *Physics Physique Fizika*, because it was willing to receive papers in any of the three languages. The journal was the brainchild of Philip Anderson, who was to win the Nobel Prize for Physics in 1977 for his theoretical work on disordered materials, and Bernd Matthias, who had carried out important work on superconductivity. Their aim [59] was to make the journal one of very high prestige, with papers that were to be chosen from across the whole of the subject but would be important enough that they should be read by all physicists. The journal was subtitled *An International Journal for Selected Articles which Deserve the Special Attention of Physicists in All Fields*. Anderson and Matthias hoped the journal would become the equivalent for physicists of the famous general and literary journal *Harper's*.

Bell was attracted to this journal because the *Physical Review*, like most journals, had page charges, or in other words, authors had to pay an amount for each page of their paper. The aim of the publishers was to keep the prices of the journal down to make it accessible to readers from, for example, developing countries. In most cases, the authors could arrange for the charge to be paid by their university or other employer, and indeed a system has developed in which, when a group of scientists apply for a grant to carry out a particular project, a fairly small amount is included to pay the page charges for the papers that it is expected will be published during the course of the work.

However, Bell was in an embarrassing position. As a guest at Stanford, he did not wish to ask his host to pay the charges, especially for such an unorthodox paper. His solution was to submit to *Physics*, because, in keeping with policy, not only did it impose no page charges, it actually paid authors for their contributions. And, rather amazingly, it seems that Anderson accepted Bell's paper under the misapprehension that it demolished Bohm's theory of hidden variables!

Unfortunately, Bell's solution had a snag. Journals that imposed page charges provided authors with a number of free reprints of the article to be distributed to fellow physicists who they think might or should be interested. Unsurprisingly, *Physics*, having already paid the author, provided no reprints, so Bell was obliged to buy these himself and ended up out of pocket after all.

Such, anyway, is Bell's account of the matter, and it certainly makes sense. However, it is possible to suggest that again there is a measure of self-justification. It could be seen as a method of getting ideas on the table and maybe obtaining the attention and interest of like-minded physicists (if there were any!), without drawing too much flak from those, probably the majority, who would find the ideas unprofessional if not ridiculous.

The paper was published speedily in 1964 and was followed in 1966 by the paper written earlier (discussed in 'Hidden variables and von Neumann: Bell's first great paper'). Bell now had to wait to see if he had made the work presented in this pair of papers visible enough to draw interest.

Incidentally, it emerged that physicists in general were *not* interested in important work carried out in areas other than their own: *Physics* did not become popular in its original guise and it was soon to become a conventional journal restricted to solid state physics; within four years, it had folded altogether.

It is interesting to consider how Bell's papers left the ideas of Einstein and particularly the presentation and conclusions of the EPR paper. Of course, as yet in the 1960s, no experiments had been carried out, and so we do not here address how either EPR or Bell's ideas stood up to nature. Instead, let us study the relation between the three theory papers: the EPR paper, and Bell's two papers.

There is a fairly general belief that Bell's work totally demolished the EPR argument; indeed, it is not uncommon even to hear it implied that Bell set out with that aim in mind. John Gribbin [60] is well known for writing that Bell was 'the man who proved Einstein was wrong', although Robert Romer [61] did achieve some balance by replying that Bell was 'the man who proved Einstein was right'!

A fair assessment of the matter would be that, as was said before, Bell was a 'follower' of Einstein. Both men regarded realism as a central requirement for physical theory, and both were also ardent supporters of locality, although we shall see in Chapter 4 that Bell did have to develop a strategy when it seemed that locality would have to be abandoned. Einstein was also of course a strong supporter of determinism, about which Bell was at best lukewarm. (Admittedly, this fact makes it ironic that Bell had reacted highly positively to Bohm's hidden variable theory, which restored determinism, while Einstein rejected it fairly comprehensively.)

On the clash between Einstein and Bohr over the foundations of quantum theory, Bell was strongly on the side of Einstein. To Bernstein [55] he said that, in his opinion, 'Einstein's intellectual superiority over Bohr was enormous; a vast gulf between the man who saw clearly what was needed, and the obscurantist'.

Of course, Bell accepted the validity of the EPR argument; indeed, it was the cornerstone of his own work. Like Einstein, he would have felt that the combination of locality and realism was the most likely conclusion of the argument, or at least the one that they would both have preferred, although both recognized that there were other possibilities. Since EPR ruled out [62] locality and no realism, the possibilities were (i) locality and realism, (ii) non-locality and realism, and (iii) non-locality and no realism. Although Einstein and Bell, as has been said, both preferred (i), clearly (ii) and (iii) were also possible; in other words, non-locality might be possible.

Einstein, of course, recognized non-locality as a possible solution, although he very much disliked the possibility, as he stated both in the original EPR paper, and even more clearly in a paper he wrote on his own in 1948 [63]. The actual writing of the original EPR paper had been by Podolsky, and Einstein felt that the ideas were somewhat lost inside the rather heavy formalism. The 1948 paper expressed Einstein's ideas in a more straightforward way, and it was reprinted in the collection of letters exchanged by Einstein and Born over a period of 40 years; these letters were published by Born in 1971 [64].

Yet, the unfortunate 'paradox' description hindered acceptance even by those who were not convinced by Bohr's rebuttal [65]. For example, in his excellent early account of quantum information theory, Andrew Steane [66] took the 'paradox' concept to imply that Einstein had argued himself into an explicit and acknowledged contradiction; Steane found that the 'fallacy' in Einstein's work could be removed by an appreciation of what he presumed Einstein had not understood—locality!

It is also quite common for commentators to be unaware of the detailed argument of the EPR paper and to assume that Einstein did little more than to attempt to impose his own wishes—that local realism would be respected—onto the physical universe. As an example of this tendency, in their thorough and wide-ranging text on quantum information theory, Michael Nielsen and Howard Chuang [67] said: 'The attempt to impose on Nature *by fiat* properties which she must obey seems a most peculiar way of studying her laws.'

An assessment of the merits of EPR came from Shimony [68], who was to play an exceptionally important part in taking Bell's work forward. Shimony already had a PhD in philosophy and was beginning a second PhD in physics when his adviser, the well-known theoretical physicist Arthur Wightman, allocated him an exercise—to go through the EPR paper and report back to him with an account of what was wrong with it. Shimony reported back that he didn't think there was anything wrong with it at all!

So, there is no technical fault in the EPR argument; however, of course, the authors did make what turned out to be the wrong choice of local realism for the final result of the analysis, at least according to quantum theory. To this extent, and to this extent only, it might make some sense to say that Bell showed that the EPR analysis was wrong. However, since Bell was actually as big a supporter of local realism as Einstein, it may equally well be said that Bell's theorem proved that Bell was wrong, or at least that Bell's wishes or preconceptions were not respected by quantum theory.

And, of course, it is necessary to stress that we are describing the state of affairs in the years immediately after 1964, so we are discussing the comparison of local realism with quantum theory and not yet with experiment. It was still to be determined whether anybody would carry the appropriate experiments and, if they did so, what the results might be. Possibly local realism might triumph, together with EPR and Bell, and it might be quantum theory that would be found to be wrong.

Bell's original proof of his inequality, and hence of the mismatch between classical notions of physics and quantum theory, was an extremely important achievement. However, as usually happens, the proof is not actually the easiest route to the crucial conclusion. Here, let us look at two alternative arguments.

First, let us look at a simple proof given by Bell in 1981 [54]. In that paper, after presenting an EPR argument to demonstrate the necessity for hidden variables, he compared the predicted results for three situations. In the first situation, the field is at $0°$ to the z-axis in the first wing of the experiment and at $45°$ to it in the second wing of the experiment. In the second situation, the angles are $45°$ and $90°$, respectively, and, in the third situation, they are $0°$ and $90°$. In each case, the probability calculated by quantum theory is for the results to be positive in each wing of the experiment.

We must now remember that, if one of the particles gives a positive result at a particular angle, then the other must give a negative result for the same angle. We may then write our three probabilities as (i) the probability of a single particle giving a positive result at $0°$ but not at $45°$; (ii) the probability of it giving a positive result at $45°$ but not at $90°$; and (iii) the probability of it giving a positive result at $0°$ but not at $90°$. Bell insisted that the sum of the first two probabilities must be at least as large as the third probability because a particle in the third category must either give a positive result at $45°$ and so contribute to the second probability, or fail to give a positive result at $45°$ and so contribute to the first probability. And, of course, no particle can contribute to both the first and second probabilities. This is a simple result obtained from straightforward arguments; and yet it does not agree with quantum theory, which says that each of the first and second probabilities is equal to 0.0732, while the third is equal to 0.2500.

Next, let us examine an argument which is less mathematical and more visual than the previous one, and which Bell explained in an interview with Bernstein shortly before Bell's death [55, 69]. Bell had been asked how he presented his ideas to general audiences, and his remarks, as usual, came in two parts: first, an outline of EPR; and, second, Bell's own work.

For his explanation of EPR, Bell presented as an analogy the case of identical twins separated at birth [70]. He particularly drew attention to one case where twins, separated at the age of four weeks and reunited at the age of 39, discovered that they both suffered from tension headaches, both bit their nails and smoked the same brand of cigarettes, had bought the same model of car in the same colour, had married on the same day, went to the same resort in Florida with their families, and owned dogs with the same name.

The scientific explanation for this behaviour is that the twins have identical genes and that genes are a very important factor in determining behaviour. If the twins are tested in the same way, they will behave in ways which are, at the least, very similar. Genes may thus be regarded as a form of hidden variable, or, as Bell said, a programme telling each twin how to behave in certain circumstances.

What might be called the Copenhagen explanation of this behaviour is that there are no hidden variables—no genes. When one twin purchases a dog and calls him Rover, the other twin feels an irresistible urge to follow suit. Actually, as Bell said, the more dogmatic follower of Copenhagen would say that it is naïve even to ask for an explanation of the correlation.

Bell would say that the explanation of EPR was totally analogous to that for the theory of genes. When tested in the same way, the two entangled particles would behave in totally correlated ways—one (+), the other (−). It would only be sensible to assume that there are correlated hidden variables telling the particles how to behave. Einstein, he said, saw the point; the others had their heads in the sand.

Yet, in the second part of the talk, and basing his argument on hidden variables, Bell admitted that he would have to retrench. One way of putting what we have just said is that, with the two measurements along the same direction, we would get a match in every case: one result being (+), the other one (−). Naturally, if we rotated the direction of one of our measurements by a small angle, we would expect to introduce a small number of mismatches. If we then returned that direction to its original position and then rotated the direction of the other measurement by the same angle in either direction, we would introduce the same number of mismatches as in the first case.

Now, let us imagine that we rotate both directions of measurement in opposite directions. Initially, we might think that, in this case, there should be twice as many mismatches as in each of the previous cases. However, a little thought should tell us that some mismatches might cancel out in the two wings of the experiment, so the final number of mismatches would have to be either less than or equal to twice the number when one direction or the other was changed.

However, this is not what quantum theory tells us will be the case. Above, we looked at effectively the same argument for the specific case where the angle of rotation was 45°. We saw that the prediction made by hidden variables did not agree with that of quantum theory.

It may be said that, in these arguments, like Bell, we have not pointed out the role played by locality, which is, in fact, crucial. In fact, the argument must always be stated in terms of the orientation of the field in *individual* wings of the experiment, as that is all that one photon experiences. For example, in the last experiment discussed, the photon in one wing of the experiment must not 'know' the direction of the field in the other wing, and so must not 'know' that the angle between the fields in the two wings is actually *twice* the angle that the field in its own wing was turned through. The argument is quite subtle and was first spelled out rigorously by Euan Squires, Lucien Hardy, and Harvey Brown in 1994 [71].

Bell's general views on quantum theory in the 1960s

Bell's two great papers of the 1960s—the papers which would make him famous—were written very much to the point; they argued a particular case in an extremely compact fashion. Of course, they could not help at the very least implying some of his views on quantum theory, but these were not stated at all explicitly.

These views did, however, appear in a paper that Bell wrote with Michael Nauenberg in the same period. Nauenberg was a visitor at Stanford at the same time as Bell, although his permanent place of work was Columbia University. In 1966 he was to move to the University of Santa Cruz, where he was to publish widely in the fields of particle physics, condensed matter physics, and non-linear dynamics over the next 30 years. He was also to display an interest in the history of physics, and in particular the work of Hooke, Huygens, and especially Newton.

At this time, Bell was asked to contribute to a book to be published in honour of the highly distinguished theoretical physicist, Victor ('Viki') Weisskopf. Weisskopf was born in Austria but had had been based in the United States since 1937; from 1961 to 1965 he was an extremely successful director of CERN.

Bell had had interesting conversations on the interpretation of quantum theory with Nauenberg, and he asked him to be a joint author of a paper on this topic for this volume. Their paper was submitted in June 1965, when Bell had returned to CERN, and the book was published in the following year.

The joint paper [72] discussed what the authors called 'the moral aspect of quantum mechanics', a term which they used to describe measurement processes in which a second measurement of the same quantity immediately after the first measurement must yield the same result. They used this idea to discuss the details of the 'collapse of wave-function' at a measurement, a concept which was discussed in Chapter 1; in particular, they assumed that this collapse occurs only when the pointer reading produced by the measurement is actually observed. In this paper, they also discussed what would be required for the measuring and measured systems so that measurements will indeed be moral.

They first demonstrated that a macroscopic observation with a pointer is by no means trivial—certainly not merely a subjective adjustment of the wave function to allow for the increase in knowledge—and they showed that, if two additional measurements were subsequently performed, the results from the statistics of possible results of the third measurement would depend on whether or not the pointer position from the second measurement had been observed.

Since Bell and Nauenberg recognized that, in practice, the observation of pointers would not change their state perceptibly, they suggested that forming *axioms* on this level of analysis 'seems distinctly uncomfortable'. After all, instruments are no more than 'large assemblies of atoms'. Perhaps superpositions of different measurement states should remain uncollapsed, just like superpositions of different atomic states.

But the authors, to establish the principle, went further. From the point of view of theorists—which both Bell and Nauenberg, of course, were—measurements performed by experimental physicists—even 'the administration and the editorial staff of the *Physical Review*'—might also perhaps remain uncollapsed.

The experimenters and editors are, of course ('presumably', the authors said) conscious. This fact leaves the authors in something of a dilemma. If consciousness destroys the superposition of different states, the Schrödinger equation is incorrect for such systems. If, however, the superposition remains, and the wave function of the conscious experimenter or editor is left in a superposition of different states, surely quantum mechanics is making an incorrect prediction.

In a footnote, the authors said that they took for granted that conscious experience *is* unique, although jocularly they conceded that there might be *some* people whose state of mind is incoherent. Rather ironically, Bell's former colleague Squires [73] is among those who have subsequently put forward such ideas.

Even if we settle the matter of experimenters and editors by leaving their wave functions uncollapsed, we still have a problem with the reading of the eventual published paper, for surely this act is no more 'elementary' than, say, reading a pointer. Thus, is it not the case that even this final state of the wave function must remain uncollapsed? But then, as the authors said: 'We are entirely lost'! The wave function of the universe may evolve through all time uncollapsed, containing what Bell and Nauenberg call 'all possible worlds'. But the axioms of quantum theory only come into play when something is observed. For the universe, there is nothing else and so no unique way of 'picking out the single unique thread of history'.

Thus, the authors suggested, quantum mechanics is, at the very least 'incomplete', which is Einstein's word from the EPR paper; the authors then referred to Bell's hidden variables.

The authors continued by stating that they looked forward to the development of a new theory that would not rely on describing events in one system in terms of observation by another system.

They added that they felt embarrassed at having brought into their discussion such matters as consciousness and the universe as a whole, as they felt that physics should adopt an objective description of nature before studying these arcane aspects of reality.

They also admitted that, even in considering such ideas, they were very much in a minority of physicists and that this minority view was itself as old as quantum mechanics itself; consequently, any new theory might be a long time coming. In addition, they realized that current interest in the questions was small, as the typical physicist felt that such questions had long been answered and that 'he will fully understand just how if ever he can spare 20 minutes to think about it'.

For themselves, though, even accepting that the idea of consciousness collapsing wave functions or hidden variables supplementing wave functions may, to an extent, save the day, it was much more likely that real progress would involve an astonishing leap of the imagination. In any case, they suggested not only that quantum mechanics would be superseded, but also that, to an unusual extent, its final fate is apparent in its internal structure: 'It carries in itself the seeds of its own destruction.'

As has been said, these were Bell's views in the 1960s, but there was little he would renounce at any later time in his life. Fifty years later, though, Nauenberg [74] had lost his conviction that quantum mechanics was 'rotten' and would feel that some of the replies to Bell did answer these objections to the conventional approaches to studying quantum foundations.

CPT: Ramifications

When Bell wrote his famous paper on CPT invariance, as discussed in Chapter 2, it would have been considered an important theoretical result but not of much practical importance, because it was generally assumed that physical processes were, in fact, conserved individually under charge conjugation (C), parity transformation (P), and time reversal (T). We noted in Chapter 2 that these ideas changed dramatically in 1956–7, when it was discovered that neither C nor P was conserved in the weak interaction. However, physicists still felt reasonably comfortable accepting that the product CP and also T were conserved.

Bell retained an interest in these matters, and indeed Veltman [18] later commented that this interest could be seen as a 'theme in [Bell's] scientific career' and demonstrated his interest in the foundations of his discipline. Chapter 2 discussed Bell's paper on time reversal in beta decay; in addition, Veltman [18] mentioned that, on Bell's initiative, the effects of time reversal invariance were included in one of their joint papers [31] on the possible production of the W particle.

Then, in 1964, this calm was broken. In a crucial experiment conducted at Brookhaven, Jim Cronin and Val Fitch, together with Jim Christenson and René Turlay [75], demonstrated that CP invariance was violated in the decay of the K_2^0 meson. Naturally, since CPT invariance must be maintained, this finding has the rather unexpected result that time reversal invariance is also lost. For this work, Cronin and Fitch were awarded the Nobel Prize for Physics in 1980.

To explain what is involved, a little particle physics [76, 77] is required, broadly as understood in the 1960s, although with some sallies forward in time which will give all that may be needed later in the book. We have already met in this chapter the leptons, the photon, and the graviton. The W particle was anticipated and would be found experimentally in the 1980s.

This leaves two classes of particles: the *baryons* and the *mesons*. It is important to stress that these are particles that interact via the strong interaction, unlike leptons, the photon, and the graviton. The word 'baryon' means 'heavy particle', and the earliest baryons were the proton and the neutron. However, with the advent of accelerators, many more baryons were being found at this time. The new members were of larger mass than the neutron and proton and were unstable and short-lived. Much the same was true for the mesons. The original members of this family were the pions but, again, accelerators demonstrated the existence of other mesons of higher mass.

The discovery that some particles were created in the timescale of the strong interaction (around 10^{-23} s) but lived far longer than that, on the timescale of the weak interaction (about 10^{-10} s), was an observation which appeared to violate a central principle of quantum theory. To attempt to solve this problem,

Figure 3.12 Murray Gell-Mann (second from left), with other Nobel laureates from Caltech: Carl Anderson (far left), Max Delbruck (centre), Richard Feynman (second from right), and George Beadle (far right). Courtesy University of Houston.

as early as 1953 Murray Gell-Mann (Figure 3.12) and Kazuhiko Nishijima [21] invented the new quantum number of strangeness, S, which would be conserved in the strong and electromagnetic interactions but not in the weak interaction. Two particles would be produced together by the strong interaction (known as *associated production*), one with $S = 1$, the other with $S = -1$, so overall strangeness would be conserved. However, once the two particles were separated, each could only decay to a particle with $S = 0$ via the weak interaction. Strangeness would be just the first of a number of quantum numbers of similar type.

As the number of baryons and mesons discovered grew rather embarrassingly large, a great step of classification [21] was taken by Gell-Mann, again, and Yuval Ne'eman in 1961; using the mathematical technique of group theory, baryons and mesons could be collected together into mathematical groups with eight or ten members. Actually, one member was missing from the group containing the most well-known baryons. It was possible to predict quite precisely the properties of this missing particle, and when it was detected with these properties—it would become known as the omega-minus—it was clear that the group theory approach was almost certainly correct.

Incidentally, discovery of the omega-minus was one of three discoveries, the discovery of the muon neutrino and CP violation being the others, that Perkins [78] berates CERN for missing. He says that CERN was recognized for the very high quality and reliability of its engineering, and all these discoveries could and should have been found first at CERN as its technical resources were far greater than those of the Americans, but the Americans had much more experience and know-how than the Europeans.

Once the basic group structure was understood, only a piece of brilliance by Gell-Mann yet again and George Zweig was required [21] to recognize that all the baryons and mesons could be built up out of three basic building blocks that Gell-Mann called *quarks*, the name that has stuck. Mathematically, these objects formed a group of three which was of the same nature as the groups of eight and ten already mentioned, and indeed the larger groups could be constructed from the smaller one. As would be expected, each quark had a corresponding antiquark.

The idea was that there were three types of quark—nowadays they would be called quark 'flavours'—*up, down*, and *strange*. Baryons consisted of three quarks; antibaryons, of course, of three antiquarks; and mesons of a quark and an antiquark. Strangeness worked out easily—the up and down quarks had strangeness 0, while the strange quark had strangeness -1 (the minus sign being a historical accident). Thus, the neutron and proton consist of up and down quarks; in fact, the neutron is udd and the proton is uud. The positively and negatively charged pions consist of a u and an anti-d, and an anti-u and a d, respectively. The neutral pion is not quite so simple; in typical quantum mechanical fashion, it consists of a linear combination of u and anti-u; and d and anti-d.

However, there was great surprise over charge and baryon number. Baryon number B was a quantity which was always conserved—it was equal to 1 for baryons such as the proton and the neutron, -1 for antibaryons, and 0 for mesons. If the quark idea was to make any sense at all, each quark would have to have $B = \frac{1}{3}$, while the charges of the up quark would be $\frac{2}{3}$ in the usual units, and that of the down and strange quarks would be $-\frac{1}{3}$. It is easy to see that the sums add up, but such values were very difficult to believe. Also, if quarks existed, presumably they could be found; however, an exhaustive search turned up nothing!

We now sketch the development of the quark concept up to the present. As we saw in the previous section, the results of neutrino and electron scattering by nucleons were taken as evidence that the nucleon did consist of three quarks. However, the theory of *quantum chromodynamics* (also known as QCD), which was developed to explain these matters, showed that *quark confinement* meant it was impossible to break up a nucleon or meson and interact with an individual quark. On the other hand, *asymptotic freedom* said that, when they are actually in the nucleon, quarks behave very similarly to free particles.

Another important point is that only roughly half of the energy of the nucleon is found in the quarks; the rest is in particles called *gluons*, which, as the name suggests, bind the quarks together; they are the exchange particles of the process. Thus, our view of the strong interaction has changed entirely. In the 1950s, it was an interaction between nucleons, consisting of an exchange of mesons. Today, it is recognized that neither nucleons nor mesons are fundamental and that the strong interaction occurs between quarks and consists of an exchange of gluons.

Two further extremely important discoveries have been of further quark flavours, and quark colours. To u, d, and s were added over many years the charmed quark (c) and the bottom and top (occasionally called beauty and truth) quarks (b and t, respectively). The reason for the period of time of the discovery was the sharply increasing masses of the quarks. Stating the mass of a quark is not straightforward because quarks exist only in combination; however, the trend is obvious in the commonly stated values, which are all expressed in MeV/c^2: u, 3.5; d, 6.1; s, 120; c, 1,350; b, 5,300; and t, 176,000.

Like leptons, quarks may be arranged in three generations: d and u (for conformity between generations, a reversal of the order of masses); s and c; and b and t. There seems little reason for this double repetition, although it may be that three generations are required to allow the process which allows matter to dominate over antimatter, as we find in our universe.

Another rather surprising development is that each flavour of quark is now believed to come in three *colours*, notionally called blue, green, and red. Antiquarks are antiblue, antigreen, or antired. The original reason for this idea was that, even though the famous Pauli exclusion principle says that, in a composite system, no two particles may have all quantum numbers the same, there are baryons where all three quarks are of the same flavour. Giving them different colours solves the problem, because the three quarks will have different colours, although, admittedly, increasing the number of quark types from 6 to 18 seems an expensive way of doing so, particularly when we remember how much Pauli agonized about the suggestion of a single extra fundamental particle, the neutrino.

Yet colour is actually the central element of the theory of quarks and gluons; hence the term *quantum chromodynamics*. Just as one example of its significance, we may say that particles that are actually observed are 'white', such as, for example, a baryon with red, blue, and green quarks, or a meson with, say, blue and antiblue quarks. This approach is certainly a successful way of discussing quark confinement.

Incidentally, there eight types of gluon, a typical one being written as (blue, antired). When a gluon interacts with a quark as part of the exchange process, it changes the flavour of the quark. For example, the (blue, antired) gluon changes a red quark to a blue one. Later in this chapter, we shall see how Bell's work added a very convincing justification to the idea of colour.

We now return to Cronin and Fitch's work, which concerned the particle K^0. This particle is probably the one with the most peculiar properties of all. It is a strange meson; its strangeness S is 1 and, since it is a meson, its baryon number B is 0. Its antiparticle K^0-bar must have $S = -1$, and $B = 0$. This has the remarkable effect that, in the timescale of the strong interaction, where strangeness is conserved, or we may say that strangeness is 'a good quantum number', particle and antiparticle are totally distinct; however, in long timescales, where strangeness ceases to be a good quantum number, the two particles cannot be distinguished in this way. (There would not be a problem with baryons, of course, because baryon and antibaryon would be distinguished by one having $B = 1$, and the other $B = -1$.)

In fact, in this timescale, rather than dealing with K^0 and K^0-bar, in typical quantum mechanical fashion, we need to work with two distinct linear combinations of these; these combinations are called K_1^0 and K_2^0, and the issue with CP invariance concerns these states. Theory says that K_1^0 has CP $= +1$, and K_2^0 has CP $= -1$, while the pion has CP $= -1$. Since CP combines by multiplication, as K_1^0 decayed to two pions and K_2^0 to three, all seemed in order—up till 1964 . The discovery of Christenson and colleagues in 1964 was that, on one occasion in every thousand, K_2^0 decayed to two pions, which would have CP $= +1$ [75].

As Polkinghorne has said [10], the awkwardness of this very small effect was in some ways more of a shock than the original discovery of non-conservation of parity itself: 'It began to be felt,' he said, 'that nothing was sacred'. What he called the 'most interesting and respectable' attempt to avoid the problem was made by Bell and Perring, both at that time at Stanford. (Veltman [18] called their paper 'truly wonderful'.)

Bell and Perring [79] suggested that, before a 'more mundane' explanation of the recent experimental results was found, it was 'amusing to speculate' that they might be a result of the lack of symmetry of the environment, in the form of a massive preponderance of matter over antimatter. They then postulated the existence of a further fundamental and long-range force, which was analogous to electromagnetism, but which couples to strangeness rather than charge and so distinguishes between matter and antimatter. Bell and Perring provided a brief but detailed account of how this force could, in some cases, create what they called the 'wrong' value of CP for each state. Effectively, this force would transform K_2^0 mesons into K_1^0 mesons, which would then decay normally into two pions. They also checked that the effect was much too small to have shown up in the most accurate experimental work on gravitational forces.

Polkinghorne later suggested that, if Bell and Perring had been correct, this suggestion would have been worthy of a Nobel Prize. However, it was not to be! As Veltman [18] observed, the hallmark of a good theory is that it is readily testable. Bell and Perring themselves had pointed out that, if their theory had been correct, the size of the effect should have been proportional to the square of the energy. At CERN this suggestion of a fifth force was tested thoroughly by a team led by Klaus Winter and Marcel Vivargent [80], who repeated the experiment at a high energy—ten times as high as that used at Brookhaven. However, the size of the effect was smaller by a factor of 200 than Bell and Perring would have predicted. Regrettably but clearly, their suggestion had to be rejected.

In *History of CERN* [11], it is reported that the laboratory was, in a sense, tidying up after an American team had made the breakthrough; it talks of the 'bitter consequence' of what was felt to be a massive error in paying little attention to weak interaction physics, with the exception of the neutrino experiment.

There is an interesting historical remark that may be made about this experiment and its interpretation. In his own account, Cronin [81] mentioned that the decision by himself and Fitch to go ahead with the experiment was stimulated by the work of a group led by Robert Adair [82], who demonstrated anomalies in the decay from K_2^0—there were more two-pion decays than had been expected. Adair produced a possible explanation along the same lines as that of Bell and Perring, hypothesizing a fifth force that distinguished between positive and negative strangeness; Cronin called this explanation 'very creative' and said that, had it been confirmed, it would have been a major discovery. Actually, Jeremy Bernstein, Nicola Cabibbo, and T. D. Lee [83] also made a similar suggestion.

In the same area of physics, in 1965 John Bell and Jack Steinberger (Figure 3.13) [84] presented a highly comprehensive account of both the theoretical and the experimental aspects of the weak interactions of kaons at an international meeting at Oxford. On the theoretical side, one important investigation was to assume that CPT is conserved, as would be expected, and to study the results of the loss, necessitated by the non-conservation of CP, of time reversal symmetry. This analysis actually provided the possibility of an experimental check on CPT itself. Many aspects of the loss of CP symmetry were also discussed in some detail in this review.

One very interesting comment was that the conclusion that CP is not an exact symmetry had been arrived at only by very indirect means. The authors said that there was nothing inherently unsymmetric between particles and antiparticles in the experiments of Cronin and Fitch; rather, the conclusion had come from the detailed application of quantum theory and, in particular, the principle of superposition.

This comment had originally been an argument of Laurent and Roos [85], who had suggested that a small non-linear term in the evolution of the wave function could cause a failure of superposition. In the analysis, Bell and Steinberger then suggested a number of experiments that would test the conservation of CP more directly. Otherwise, they said, Cronin and Fitch might

Figure 3.13 Jack Steinberger. © Renate Bertlmann; Central Library for Physics, Vienna.

have demonstrated not the failure of CP but the failure of quantum theory! The latter possibility would certainly have been interesting from Bell's point of view!

The authors examined all the experimental evidence that had been accumulated up to that point with great care and also suggested a variety of experiments to be carried out which would provide valuable information about the kaon system and the weak interaction in general. In his obituary of Bell, John Ellis [33; second item] said that the analysis of Bell and Steinberger had been the basis for many subsequent experiments.

Bell was to interact closely with Steinberger on a number of occasions throughout his life, and Steinberger had enormous respect for him. He [86] wrote: 'Trying to learn the behaviour of neutral kaons in the light of CP violation, I had the pleasure of benefiting from John's penetrating understanding and insight, and his readiness to share this.' (Other remarks by Steinberger are included at the end of Chapter 7.)

Bell, gauge theory, and the weak interaction

In 1967 Bell wrote a paper [87] which Reinhold Bertlmann and Anton Zeilinger [86] later described as one of his two most important contributions to quantum field theory (the other being the so-called anomaly, to be met shortly). In the 1967 paper, Bell suggested that there should be a *gauge theory* describing the weak interaction, and Bertlmann and Zeilinger say that Bell's theory had an immense influence on Bell's friend Veltman, as well as on Veltman's graduate student, 't Hooft, whose work led to a complete understanding of the weak interaction. Veltman and 't Hooft shared the Nobel Prize for Physics in 1999 and, in time, gauge theory became the standard tool used to describe all the interactions of physics. In addition, Shimony, Valentine Telegdi, and Veltman [88] said that, although Bell's 1967 paper was 'unknown, except to a few who learned its lesson', it was 'seminal for subsequent studies of the renormaliza-bility of gauge theories'.

The next few pages will first provide a general understanding of the idea of gauge theory [89, 90], with some examples, and then give an extremely brief account of the long struggle, over several decades, to produce a successful theory of the weak interaction; in fact, this theory actually united the weak and electromagnetic interactions, creating the so-called *electroweak* interaction. Special attention will be paid to Bell's part in this struggle.

First, all the central equations of physics have a certain arbitrariness in that we may change the equation *globally* without changing the physics at all. For example, in classical electromagnetism, the value of the electrostatic potential in the governing equation may be changed by the same amount everywhere, and this change will cause no change in the physics predicted by the equation. Similarly, in quantum theory, if the phase of the wave function is changed globally, the resulting physics will be unchanged. We will call this a *global gauge symmetry*.

However, if we replace the *global* change with a *local* one, varying in space and time, we would very necessarily expect that we would have caused changes in the physics in both these cases. The idea, though, of a gauge theory is that we must add additional terms to the underlying equations in order to cancel out such changes. In electromagnetism, these terms will relate to fields; in particle physics, they will be interpreted as novel particles.

The relevant ideas in electromagnetism were found by James Clerk Maxwell [91] in 1868, in connection with his epoch-making work that put the subject on its firm foundation and introduced the idea of light as an electromagnetic wave. It may be mentioned that history did not follow the neat logical order outlined above; Maxwell did not have to postulate the existence of the mag-netic field, as its existence was already understood, but he did need to recog-nize that, for the time-dependent case, Ampère's law required the addition of an additional term to ensure local conservation of charge.

For quantum mechanics, *local gauge symmetry*, or *local gauge invariance*, is achieved by the addition of electromagnetism in the form of the photon. What happens is that one changed particle emits a photon and, as a result, the phase of its wave function changes. However, when the photon is absorbed by another charged particle, the phase of the wave function of this particle changes also, and the net result is that the appropriate symmetry is maintained. We see here very clearly the idea of the electromagnetic interaction consisting of the *exchange* of photons.

Although the gauge theory of the strong interaction, quantum chromo-dynamics, was not obtained until the 1970s, it is convenient to discuss general aspects of it here. It will be remembered that a nucleon, whether a proton or a neutron, should be 'white' in the sense that it consists of one quark of each of the colours red, blue, and yellow.

If we institute a *global* change in which all red quarks become blue, all blue quarks become yellow, and all yellow quarks become red, since the three col-ours have identical status in the theory, it is clear that there will be no change in any physical predictions.

However, if there is a *local* change in which a single quark changes from red to blue, clearly all is not well unless some other occurrence compensates for this change. In fact, the 'novel' particles that must occur, since quantum chromodynamics is a gauge theory, are just the gluons we met before. As men-tioned briefly before, when this quark changes from red to blue, a (red, anti-blue) gluon is emitted. This gluon travels to the blue quark in the nucleon and, when it is absorbed, it changes the blue quark to a red one, so the nucleon is still white. Again we see very clearly the strong interaction as consisting of the *exchange* of gluons, and it is natural to describe the gluon as a *compensating* field for the initial change of colour in the quark.

There is an interesting distinction between the two cases just described. In the case of quantum electrodynamics, the change of phase may be written as a simple number, and it is clear that, if we take two successive changes, they just add up, so the order in which the changes are made is irrelevant. However, in the case of quantum chromodynamics, the changes involve three colours and so, in general, will be represented by 3×3 arrays of numbers or *matrices*; math-ematically, it is a well-known property of matrices that, except in special cases, if you multiply two matrices together to represent two successive changes in the colour of quarks, the result *does* depend on the order in which the changes were made.

Mathematically, we say that the matrices and hence the colour changes *commute* for quantum electrodynamics but not for quantum chromodynamics, and the two types of system are called *Abelian* and *non-Abelian*, respectively, after the nineteenth-century Norwegian mathematician, Niels Henrik Abel.

This is just mathematics, of course, but it does have an extremely important physical consequence. For Abelian cases, such as quantum electrodynamics, the gauge particle does *not* carry the source of the field. The photon trans-mits the electromagnetic force between two charged particles but is *not* itself charged. However, for non-Abelian cases such as quantum chromodynamics, the gauge particle *does* carry the source of the field. As we have seen, the gluons do possess colour (or, more strictly, colour and anticolour). We shall follow this point through for the electroweak interaction.

Before that, though, let us briefly discuss relativity and gravitation. The theory of special relativity, which does not include gravity, was produced by Einstein in 1905 and deals with *global* transformations of space-time. Then, in 1916, Einstein replaced these *global* transformations with *local* ones by addition of the required and compensating gauge field of gravitation. This addition was the central element of his theory of general relativity, although it must be said that any further consideration of this theory lies well beyond the scope of this book!

It is worth mentioning, though, that the term *gauge theory* is related to this area of physics. It comes from the suggestion of the famous mathematician Hermann Weyl that different points of space-time might be connected to different lengths of measuring rods and so to the different *gauges* of the American railway system.

Now let us turn to the electroweak interaction. Although it may seem awkward, the account must start in 1954 with the ideas of C. N. Yang (three years before his work on violation of parity) and Robert Mills [92]. The awkwardness relates to the fact that this work was actually directed not at the weak interaction but at the *strong* one. The theory was based on quantum electrodynamics but was non-Abelian, the first such theory. This theory was proposed, of course, before the days of quarks, and it treated the proton and the neutron as the two basic states.

In fact, they are actually states of a quantity that has not yet been mentioned—*isospin*. As discussed in Chapter 2, particles like the electron have two spin states which may be called $(+)$ and $(-)$ and which may be referred collectively as a spin doublet. Let us now define the neutron and proton as two states of the nucleon, and let us call those states $(+)$ and $(-)$, respectively. More specifically, let us say that the nucleon is an isospin doublet with isospin or I equal to ½, and a third isospin component, which we write as I_3 and which is equal to $+½$ or $-½$,. The Yang–Mills theory is a theory of isospin.

It should be stressed that isospin is certainly not an actual spin and has no direct physical significance at all. Its name derives from the fact that mathematically it behaves exactly as a real spin. However, it is preferred to call this third component I_3 rather than I_z as, unlike 'z', '3' has no geometrical significance. Actually, many people do not worry about this point and call this component of isospin 'I_z' anyway.

With two states for proton and neutron, matrices are 2×2; so, as has been said, the theory is non-Abelian. The number of gauge particles required, apart from the photon, is 3; the particles were called the W^+ particle, the W^0 particle, and the W^- particle (although it should be remembered that the letter 'W' is usually reserved for particles related to the *weak* interaction). Since the theory was non-Abelian, these particles carried isospin, which meant that two of them were charged.

However, the theory seemed to be doomed because, just as the photon has zero mass in quantum electrodynamics, it seemed that the W particles must also have zero mass. Charged particles with zero mass would certainly have been readily detected if they had existed, so the obvious conclusion was that they did not exist and that the Yang–Mills theory was totally unsatisfactory. In fact, though, it played a highly important part in what came next, as demonstrated by the fact that, at the time of writing of this book, a celebration is being planned for the 60th anniversary of Yang and Mills's work; Yang, at the age of 92, will be able to attend this celebration, although Mills sadly died in 1999.

In fact, in 1961 their model was adapted in an attempt by Sheldon Glashow [93] to produce a theory for the combined electromagnetic and weak interactions. Isospin was actually only applicable to quantities subject to the strong interaction, that is, nucleons and mesons (again, of course, before the coming of the quark), but an analogous quantity, weak isospin, which we may write as I_w, was defined. It would equal ½ for the doublet containing the electron and electronic neutrino, while its third component would be equal to $-$½ for the electron and $+$½ for the electronic neutrino. (As other generations of leptons were discovered, each could be handled identically but separately.)

Glashow's theory was of weak isospin, and he obtained four gauge particles, which, of course, were theoretically massless. However, he allocated them non-zero masses 'by hand'. As before, he obtained a W^+ particle and a W^- particle, but he also obtained two particles with zero charge. By judicious (although, from a formal point of view, totally illegitimate) sleight-of-hand with his masses, he was able to take suitable combinations of his particles with zero charge so as to produce one particle with zero mass, which he decreed to be the photon, and another of high mass, which he called the Z^0 particle.

He also arranged that the electromagnetic interaction, which must be mediated by the photon, conserved parity, but the weak interaction, which was mediated by the W^+ particle, the W^- particle, and the Z^0 particle, did not.

Glashow's work was certainly clever, and it was to turn out to be exceptionally important, as indicated by the fact that he was to share the Nobel Prize for Physics with Abdus Salam and Steven Weinberg for this work in 1979. At the time, things seemed very different, as the theory seemed to face two insoluble difficulties. First, the mass problem had not been solved by Glashow—it had really just been ignored. Theory still said that the masses of the particles should all be zero. The second problem lay with the requirement for renormalization. The need for renormalization in quantum electrodynamics was stressed in Chapter 2, and it would be just as important for electroweak

Figure 3.14 Peter Higgs; from Martinus Veltman, *Facts and Mysteries in Elementary Particle Physics* (World Scientific, Singapore, 1993). © CERN.

theory. However, at the time, all ideas suggested that renormalization would be impossible, especially if somehow the gauge particles became massive.

The first problem was tackled with a good deal of success in 1964. The work is always associated with the name of Peter Higgs (Figure 3.14), a Scottish physicist, but five others, working in two groups, were also centrally involved. The first group consisted of the Belgian François Englert (Figure 3.15), and Robert Brout (Figure 3.16), who was born in the United States but spent most of his working life in Brussels, where he collaborated with Englert. Brout died in 2011, while Englert was to share the Nobel Prize for Physics with Higgs in 2013. A Nobel Prize may be shared between up to three people, but it cannot be awarded to anyone who has died, and it has been surmised that the vacant third place was a tribute to Brout.

However, the second group was ignored by the Nobel Prize committee. This group comprised the British physicist Tom Kibble and two Americans, Gerry Guralnik and Fick Hagen, who were all working in London. Their exclusion from the Nobel Prize may have been because their work was published a little later that that of their competitors.

Figure 3.15 François Englert; from Martinus Veltman, *Facts and Mysteries in Elementary Particle Physics* (World Scientific, Singapore, 1993). © CERN.

Higgs himself was as embarrassed as anybody that only his name was commonly associated with the discovery. He has said that he would prefer the term 'Higgs mechanism' to be replaced by 'the ABEGHHK'tH mechanism', a name which includes the initials of the surnames of those mentioned above, as well as that of Philip Anderson, at the beginning of the list, and 't Hooft, at the end [94]. It has also been said that Higgs would have liked Kibble to have been a third name on the Nobel Prize citation.

Anderson was present because he was a highly renowned solid state physicist who had carried out extremely important work in the fields of magnetism, superconductivity (the fact that the electrical resistance of certain materials drops to zero below a particular temperature), and disordered solids (which do not have a crystal structure); in 1977 he was awarded a share of the Nobel Prize for Physics. Anderson drew the attention of his colleagues, who were working on the physics of elementary particles, to the phenomenon, well known in the theory of solids, of *spontaneous symmetry breaking*.

The most obvious example of this phenomenon occurs in *ferromagnetism*—usually just called magnetism by non-physicists—where, below a certain

Figure 3.16 Robert Brout; from Martinus Veltman, *Facts and Mysteries in Elementary Particle Physics* (World Scientific, Singapore, 1993). © CERN.

temperature called the *Curie temperature* (named after Pierre rather than Marie Curie), a material such as iron may have a permanent magnetism along a particular direction. The formula for the energy of the system does not pick out a unique direction; rather, although there is symmetry between many directions in space, below the Curie temperature this symmetry is spontaneously broken. Anderson suggested that this idea could be useful in the study of weak interactions.

Higgs and the others were able to construct a mechanism [89] whereby, using a local phase transformation as for quantum electrodynamics, a particle analogous to the photon is predicted, as well as two other particles, one with mass but the other one massless and so very much undesirable. However, with the use of a suitable gauge transformation, the massless particle disappears (exactly what is required!), while the photon-like particle acquires a mass. This phenomenon has often been described as the photon-like particle 'eating' the unwanted massless particle and thus becoming 'heavy'.

With the use of Glashow's previous work and this new mechanism, in 1967 Weinberg and Salam independently were able to put together a theory of the electroweak interaction. Four gauge particles are predicted: the photon and the two W particles (the W^+ particle and the W^- particle) were all expected, but the presence of the Z^0 particle was a new development. In fact, the photon and the Z^0 particle do not appear directly from the theory. Rather, as with the work of Glashow, two particles emerge which are essentially linear combinations of the photon and the Z^0 particle.

Actually, four Higgs-type particles are required in the theory: two charged, and two uncharged. The charged particles and one of the uncharged particles disappear in the gauge transformation so as to give mass to the W^+ particle, the W^- particle, and the Z^0 particle. However, the second uncharged particle does not disappear. Rather, it remains as a real particle, to be called the *Higgs* particle or the *Higgs boson*, and its possible existence and the search for the particle was to be a central theme of particle physics for the next half-century. Of course, it was its discovery at CERN in 2012 that led to the Nobel Prizes for Higgs and Englert in the following year.

However, this development led to two previous triumphs for CERN. In the section titled 'John Bell at CERN, and the neutrinos', the non-discovery of neutral currents was discussed. The emergence of the Z^0 particle made the existence of neutral currents practically certain and their discovery in 1973 will be described in Chapter 4. Clearly, detection of the W particles and the Z^0 particle was also highly important. As the predicted masses were around 80 GeV for the W particles and around 90 GeV for the Z^0 particle, their discovery would have to wait for the 1980s. As for the mass of the Higgs particle, nobody had any real idea, which was one of the reasons why it remained so enigmatic over the next decades.

In a sense, though, we are jumping the gun because, although Weinberg's paper [95] would now be looked on as one of the most important of the twentieth century, at the time it was all but ignored. Through 1967, 1968, and 1969, it was not cited in a single published paper, while in 1970 it was cited just once and in 1971 four times. Finally, in 1972 it was cited 64 times and, in 1973, 162 times [89].

Clearly, the logjam had been broken at last—in this case, the logjam was the question of renormalization. Without the possibility of renormalization, the theory of the electroweak interaction could not be a serious contender for any real significance, and virtually everybody was convinced that, particularly with mass, renormalization was impossible. The hard work of proving that renormalization was, in fact, possible was carried out by Veltman, with a crucially important final flourish of 't Hooft, but also, in sporting parlance, an essential assist from Bell.

Veltman's work was long and arduous, taking much of the 1960s. It was also highly technical. Pickering [89] later remarked that, in a meeting of 1973, Veltman said that the type of consideration he was explaining 'cannot be palatable to a general audience', although, actually, the 'general' audience consisted of highly experienced particle physicists who just did not happen to be experts in the topic under discussion!

In outlining Veltman's work, it may be said that he started from the ideas of Gell-Mann and Schwinger—both, of course, physicists of great distinction. In this period, field theory was not popular, because its detailed theoretical models were difficult to work with in any satisfactory way. To avoid these difficulties, Gell-Mann had come up with the idea of 'current algebra' where, instead of working with actual wave functions and thus having to solve extremely complicated coupled differential equations, physicists could identify a number of very general and plausible rules, some of which as *sum rules* were well known in quantum theory, and then use these as the basis of complicated but now manageable calculations.

Considerable success was achieved by quite a number of theoreticians, including Bell who, together with Berman [96], wrote an interesting paper on current algebra in beta decay of pions; however, there were also major difficulties. A particular one was the occurrence of *Schwinger terms* in the analysis, as these led to infinities. Of course infinities themselves, although unwelcome, were not a surprise; they were merely a sign that renormalization was required. But the ones produced from the Schwinger terms were particularly virulent. In the end, Veltman's methods suggested that these terms were actually not of much physical interest—in fact, as we shall shortly see, they were just the result of rather 'fancy operator manipulations' as Veltman called them [19].

Veltman decided to return to field theoretic methods and so was able to get rid of these Schwinger terms and come up with two 'simple equations' with 'an amazing structure' which contained the photon, the field particles of the weak interaction, and a term for the pion field [89]. He was able to discuss these equations with Feynman, who identified them as having a Yang–Mills structure; however, Feynman rather muddied the water by convincing Veltman that the equations related to the strong rather than the weak interaction, and that the gauge particles were some of the recently discovered high mass mesons.

In a conference talk in London at the time, Veltman mentioned that his arguments suggested that pion decay into photons was forbidden. He later reported [19] that Bell picked up this comment and a discussion took place between the two men over the next few weeks. Veltman suggested that this discussion eventually led to the discovery of the *anomaly*, which is the subject of the next section, 'Bell, gauge theory, and the weak interaction'.

We now return to Bell's own ideas. He had already [97] studied Veltman's 'fancy operator manipulations', which were described more technically than that as *canonical transformation equations* [1], in order to show that they had to be used with extreme caution because the quantities they were using were not renormalized. He studied a model (the *Lee model*) which was unrealistic but so simple that it could be solved exactly, and he was thus able to show that, because the model was unrenormalized, several of the results of his calculations did not agree with well-established results coming directly from first principles. Clearly, an alternative was required.

In Bell's 1967 paper [87] and also in a conference paper published at about the same time [98], he was able to make progress by using Veltman's two famous equations. Bell demonstrated that they should be used in the study of the *weak* rather than the strong interaction and that they exhibited a Yang–Mills gauge theory type of structure.

In the 1967 paper, Bell suggested avoiding as much as possible the 'creaking machinery' of standard field theory and instead simply *imposing* gauge invariance. In the abstract of the paper, he stated that the lessons of gauge theory implied the 'irrelevance' of the difficulties connected with the Schwinger terms. He recognized that Veltman, in what Bell called a 'fundamental paper', had taken the essential step of avoiding these problems by studying much of the algebra of the quark model along gauge theory lines; incidentally, Bell thanked Veltman as well as Michael Nauenberg for discussion of the ideas of the paper. According to Bell, the aim of his paper was to spell out the relation of Veltman's equations to gauge theory and to a class of applications of current algebra.

Although Bell introduced his ideas by following Veltman in using the quark model, he stressed that, by concentrating on different aspects of the situation, in particular emphasizing the importance of gauge theory, his own work ended up by covering different classes of theory: weak interactions and electromagnetic interactions.

Using the gauge formalism, Bell reached an equation that he called 'the generalized Veltman equation'. He remarked that it was 'true by definition', which may sound a little like a put-down, but what he meant was that it was an important result that came directly from the basic mathematics of the system. Physics, he said, enters only with assumptions related to the current algebra of the particular interactions being studied and, with typical generosity, he acknowledged that Veltman had mentioned that he was already aware of the possible application of his ideas beyond the strong interaction.

In the conference paper [98], which was actually written before the 1967 paper although it was published afterwards, Bell covered much the same ground, suggesting that there might be some advantage in avoiding the standard 'machinery of time-ordered products and equal-time commutators' and working directly with gauge conditions. This advantage, he said, was the motivation for what followed in the paper; he then presented a systematic study of the method of gauge invariance. In particular, he demonstrated the application of gauge theory to a model studied by a number of physicists, including Lee and Weinberg.

However, Bell concluded the paper by pointing out that the mere introduction of gauge theory was of itself no more significant than the use of the more established methods using currents and commutation relations. It was only, he said, the assumption that the currents enter into weak and electromagnetic interactions in a certain way that led to interesting results. 'In the gauge formalism,' he concluded by saying, 'it is the corresponding body of assumptions that makes physics'.

Bell's considerations were tantalizing but obviously only suggestive. Veltman [19] later described the work as 'a more formal derivation' of his own important equations and said that it had made it clear to him that 'the successes of current algebra must be considered a consequence of gauge invariance'. He spent a month at Rockefeller University in New York working out his future strategy and came to the conclusion that Feynman's suggestion that the equation concerned mesons was a red herring [89]. In fact, he concluded that Bell's advocacy of gauge theory would give him exactly what was required and that the particles involved must relate to the weak and electromagnetic interactions.

As to the question of why nature might choose such a mathematical structure for particle physics, Veltman concluded that it must be so that a renormalizable theory was possible. He was to tell himself the following many times in the future months and years: 'Weak interactions must be some renormalizable Yang–Mills theory. It has to be'. Yet it was to be a struggle!

Veltman wished to consider gauge particles that had non-zero mass (massive particles) because it seemed fruitless to think in terms of ones without mass which did not actually exist. Yet, renormalization was generally thought to be conceivably possible only for massless particles. It took a long time to realize the reason for this fact, that is, that while the spin of massive particles may point in any of three directions in space, the spin of massless ones cannot point along the direction of travel, because massless particles must, by the laws of relativity, travel at the speed of light. In Veltman's work, Feynman diagrams with different number of loops were analysed. Ghost particles, which were one of Feynman's inventions and which did not exist in reality but appeared in closed loops, were required. Many of the infinities of the theory were found to cancel, although not all [19, 89].

By early 1971, Veltman became convinced that introduction of scalar fields into his calculations might allow renormalization. However, in a way his thoughts were pre-empted when a graduate student,'t Hooft (Figure 3.17), commenced work with him at the University of Utrecht, where Veltman was teaching, and soon produced what Veltman considered to be the most elegant proof of renormalization for the massless case.

In conversation with 't Hooft, Veltman remarked that all that was now required was a proof of renormalization for any massive case, however far

Figure 3.17 Gerard 't Hooft. © Renate Bertlmann; Central Library for Physics, Vienna.

removed from reality. 't Hooft replied with the memorable words 'I can do that'—and indeed he could! For Veltman and for later publication, he wrote out his argument, which included spontaneous symmetry breaking and, what neither of them had met up to that point, what is now called the Higgs mechanism.

Although others had to be convinced, it was essentially the end of the story of the *theory* of the weak and electromagnetic interactions, as the importance of the work of Weinberg and Salam would gradually be realized. Experimental proof, of course, would come slowly, as we shall see in later chapters.

Because of the dramatic end of the story with 't Hooft, as compared with the long slog of Veltman, the work of the latter has often been underrated, and indeed there has occasionally been rather disparaging comment about the professor who had to be put right by his graduate student. Fortunately, the Nobel Prize Committee made no such mistake and honoured both. In Veltman's Nobel Prize lecture [99], Bell was paid only a small amount of attention; however, Veltman would be the first to admit that Bell's contribution was crucial.

All that remains to be discussed is the puzzle of the so-called *Bell–Treiman transformation*. Sam Treiman, incidentally, was a well-respected American particle physicist. As explained by Veltman [19], he needed to undertake a transformation similar to a gauge transformation but where the gauge parameter was replaced by a field. Veltman said that this calculation was named the Bell–Treiman transformation, but he laconically added that 'neither Bell nor Treiman was responsible'.

In a paper from 1994, 't Hooft [100] explained that the transformation was named by Veltman but in a footnote remarked that, in doing so, Veltman was: 'using his unique sense of humour. There never existed any references to either Bell or Treiman.' 't Hooft added that 'it would have been more appropriate if the identities obtained were called Veltman-Ward identities', referring to John C. Ward, a British theoretical physicist whose ideas contributed to much of the work described in this section of the book.

However, it is interesting to return to Veltman's 1968 paper [101] in which the term 'Bell–Treiman transformation' was introduced, with a footnote saying: 'The author is indebted to Professors Bell and Treiman for discussions on this point.' Moreover, several other authors use this term in a straightforward way that is clearly devoid of humorous intent.

Bell and particle physics in the 1960s

In this section, some of Bell's less well-known papers on particle physics of the 1960s are discussed briefly. Of course, not all of his papers made major contributions to the discipline but, in their biographical memoirs of Bell, Burke and Percival [102] pointed out that this did not mean that the work in these papers was in any way routine. They added that when, as he felt morally obliged to do, Bell worked on practical matters related to CERN experiments, he was always concerned with matters of principle, and made sure that he developed a full understanding of the problem and its origins. This approach enabled him to make progress in fields that others might have considered closed; however, once the underlying physics had been understood, he often stepped aside and left the field for others to study the routine matters. Burke and Percival said that Bell was one of CERN's brightest stars and, by preference, a loner, although he was happy, or at the very least prepared, to be approached by others for advice and help.

In the first part of the decade, Bell wrote a number of papers on fairly formal matters. In one paper, he discussed the electromagnetic properties—charge and current distributions—of unstable particles [103]. The quantities involved are mathematically complex, and they oscillate with amplitudes increasing exponentially at large distances. The principle aim of Bell's paper was to respond to a suggestion by a Russian physicist, Yakov Borisovich Zel'dovich, that an unstable particle, as distinct from a stable particle, might have a non-zero electric dipole moment. Zel'dovich actually used one of Bell's own favourite techniques of time reversal symmetry in his analysis, but Bell was

Figure 3.18 Rutherford Jubilee Conference at Manchester, 1962; the audience contained most of the well-known nuclear and elementary particle physicists of the time. Back row (left to right): P. C. Gugelot, D. H. Wilkinson, H. H. Barschall, H. E. Gove, J. M. Cassels, C. Rubbia, and G. E. Brown. Middle row (left to right): L. T. B. Goldfarb, S. A. Moskowski, M. Goldhaber, R. J. Blin-Stoyle, S. Devons, H. McManus, H. P. Noyes, R. H. Dalitz, B. J. Cohen, and A. Bohr. Front row (left to right): R. Peierls, J. S. Bell, B. H. Flowers, E. Marsden, D. R. Inglis, and J. B. French. Courtesy University of Manchester.

able to show that Zel'dovich's arguments were misleading and that there can be no dipole moment, provided time reversibility holds.

In another paper [104], Bell tested an idea of the well-known theoretician Walther Heitler. In order to remove a mathematical problem in quantum field theory, Heitler had suggested a form of non-locality that does not obey the usual laws of relativity. His idea was that the mass of a particle, rather than being constant, decreases slightly with velocity (completely independently of any relativistic considerations). However, Bell was able to demonstrate that the results of various experiments show that any effect must be much smaller than Heitler would require for his ideas to be successful.

Another quite formal paper [105] was written with Goebel, who was from the University of Wisconsin, while Bell was in the United States in 1964. Usually radioactive decays are, of course, exponential; however, certain physical situations may lead to wave functions that have mathematically unusual structures (*double poles* in the complex plane). In the paper, Bell and Goebel discussed two situations of this type and showed that non-exponential decay

Figure 3.19 John Bell lecturing at the first Erice School, 1963. © Erice School.

would occur. However, they also showed that experimental results depend on the arrangements for production and detection, and are much more complicated than the sum of exponential terms suggested previously by Marvin Goldberger and Kenneth Watson.

In a pair of short papers, Bell made quite important contributions to the ideas of the two Nobel Prize winners, Yang and Lee. Yang had presented a range of theorems related to superconductivity and superfluids such as liquid helium, and Bell [106] was able to simplify considerably an important part of his argument. Lee had presented a theory of weak bosons in which bosons and their antiparticles are placed together in a simple multiplet. In a paper written by Bell together with Philippe Meyer and Jacques Prentki [107], although the attractions of Lee's scheme were admitted, it was shown to conflict with current experimental results.

Bell was also able to make use of his experience in nuclear physics built up at Harwell [102, 108]. In a paper written with Berman [109], he studied the interaction of high energy neutrinos with nuclei, and the production of pions and strange particles. With Løvseth [110], he examined in some detail the theory of the capture of muons by heavy nuclei, and in a paper published while he was at Stanford, he returned to the nuclear optical model, this time for virtual pions, with an argument [111] that was of considerable importance for the study of neutrino reactions.

In a 1962 visit to the University of Washington Summer Institute for Theoretical Physics, Bell [112] published an interesting paper in an area well away from his usual expertise—superconductivity. Lüders—one is tempted to call him Bell's sparring partner—had shown that a standard statistical mechanics theorem that relates density fluctuations to compressibility was violated in the (John) Bardeen–(Leon) Cooper–(John) Schrieffer (BCS) theory of superconductivity.

The BCS theory was produced in 1957. It was the first comprehensive treatment of superconductivity, and it deservedly gained Bardeen, Cooper, and Schrieffer the Nobel Prize for Physics in 1972. However, it did have some technical issues that needed to be improved. In particular, its original explanation of the Meissner effect, which is the expulsion of the magnetic field from a superconductor, had problems with gauge invariance and lack of continuity of current.

Bell realized that Lüders's difficulties were closely related to the ones associated with the Meissner problem. Bell was able to resolve them in detail, and he also produced an improved version of the standard theorem. While this paper was not of major importance, it has to be said that Bell's ability to move into an area totally new to him and of considerable current interest, and to contribute at the highest level was breathtaking!

In a totally different area of physics, the mathematical technique of group theory had been at the forefront of the development of the theory of particle physics and had been used in, for example, the work of Gell-Mann and Ne'eman and the study of quarks and current algebra. Not surprisingly, many theoreticians studied groups that were different from and larger than these [1]. After his work with Skyrme, Bell became interested in the use of such groups to predict a value of the ratio of neutron and proton magnetic moments quite close to that found by experiment; with Henri Ruegg, he made an interesting contribution to this work [113], although he eventually realized that these groups could not be incorporated in a consistent relativistic theory [114].

Bell also carried out a number of calculations more directly related to experiment. He improved the derivation of a result used for Compton scattering (scattering of X-rays by electrons in atoms) [115], with Rogen Van Royen he generalized a useful result in the radiation of photons [116], and he also improved an important relation for the elastic scattering of particles of arbitrary spin [117].

The last paper to be summarized here was written with Eduardo de Rafael and considered the magnetic moment of the muon [118]. The ratio of magnetic moment to angular momentum for the spin of the electron or the muon contains a numerical factor g, the so-called g-factor, which is equal to 1 for a classical theory. However, Dirac's celebrated relativistic quantum theory says that the g-factor should equal 2, and this prediction is in quite good agreement with experiment. However, for the electron, quantum electrodynamics yields a small correction of around 2×10^{-3}, which makes the value in excellent agreement with experiment. Because of the difference from 1, this is often called the 'anomalous' value.

Bell and de Rafael considered the case of the muon, discussing in particular the so-called *hadronic vacuum polarization*, where a hadron is a term for a baryon *or a* meson. They find that that the contribution to $g - 1$ for the muon is less than 10^{-6}. Jackiw and Shimony [1] later commented that Bell and de Rafael's interest in this quantity appeared to anticipate a considerable current interest, as it is now suggested that, while the electron case seems well understood, there is an unexplained difference between theory and experiment for the muon case, and this may suggest a fundamental failing in the broadly accepted standard model of particle physics.

The 'anomaly': ABJ

Towards the end of the 1960s, Bell wrote, together with Roman Jackiw, his most important paper [119] in the general area of particle physics and quantum field theory. As late as the end of the century, it was Bell's most-cited paper, well in front of any of his quantum papers [120]. Fifteen years later, 'On the Einstein Rosen Podolsky Paradox' is ahead of it by a factor of 3 to 2, but the paper with Jackiw is comfortably in the lead of any of Bell's other papers, quantum or otherwise.

In this paper, Bell and Jackiw discovered and discussed the now-famous *anomaly*, usually called the *ABJ anomaly*, or the *Adler—Bell—Jackiw anomaly*, as another expert in quantum field theory, Stephen Adler, performed analogous work at roughly the same time. The problem may be described in the simplest terms as a failure of well-established classical symmetries at the quantum level. It may also be described [120, 121] as a violation of the *correspondence principle*, the principle initially stated by Niels Bohr towards the beginning of the development of quantum theory; this principle said that, in a suitable limit, usually said to be that of high quantum numbers, the experimental results of quantum theory should become identical to those of classical theory.

Anomalies are 'the key to a deeper understanding of quantum theory' [120]. However, they also have a far wider significance. In the theory of solids, they play a role in the study of *edge states* in the *quantum Hall effect* [120]. And, looking beyond physics to fundamental mathematics, a whole host of related ideas emerged at about the same time—the Atiyah–Singer index theory, zero modes of Dirac operators, Chern–Pontryagin characteristics of gauge fields, and Chern–Simons secondary characteristics [120]. The work at the boundary of physics and mathematics continues to this day, and indeed Bell's obituarists,

Shimony, Telegdi, and Veltman [88], have suggested that the central puzzle associated with the anomaly 'remains largely mysterious to this day'.

To give some idea of the very technical ideas involved, we must return to 1967, when Bell was extremely interested in current algebra, but concerned about two problems where the results of current algebra were in total conflict with those of experiment.

The first was the decay of the eta meson to three pions, a result which occurred experimentally but which Bell's colleague, David Sutherland [122], had shown should not occur at all using the broad theoretical approach of current algebra. Of course, some adjustment of the approximations [123] could produce a small effect, but, in a thorough investigation of the issue, Bell and Sutherland [124] said: 'Theoretically eta decay is *not* enhanced; unfortunately it seems to be very much enhanced experimentally.' They concluded that 'the problem remains unresolved in conventional theory.'

The other problem was the decay of the neutral pion to two photons. In this case as well, current algebra said that the decay should not take place, and the argument seemed particularly clear and rigorous [123, 125]. Again, though, this result is clearly violated experimentally. For these reasons, Bell strongly maintained that the subject of current algebra, in other areas highly success-ful, could not be closed until this problem was resolved.

It was at that time in 1967 that Jackiw (Figure 3.20), having finished doctoral studies at Cornell, arrived at CERN for a year-long research visit. At that time, there was a fair consensus about what the important current problems were in particle physics, and at least broadly what methods should be used to solve them. But Jackiw was keen to tackle a problem that was interesting and chal-lenging outside these constraints, and so he welcomed the willingness of Bell both to be free with his time in discussion, and to have the ability and will to talk creatively about issues that were still open.

In particular, Bell discussed the decay of the pion, as mentioned above, and he stressed that, as long as this problem was open, current algebra had to be regarded as an open topic. Jackiw was quite keen to investigate the puzzle, but it was far from clear what route to take, as both theory and experiment seemed clear-cut, although they differed strongly in their conclusions. He found that fellow theorists were not interested, while some more mathematically ori-ented colleagues suggested that the problem lay in rather casual use of diver-gent quantities. While it was to turn out that they were broadly correct, their suggestion that Jackiw should work through the whole area of study with totally rigorous mathematics did not appeal either.

As it happened, the useful information came from a surprising source. At teatime in the CERN cafeteria, Bell and Jackiw were joined by Steinberger, who was making yet another important appearance in Bell's life. Steinberger was, of course, by now a leading experimental physicist, so he was probably the last

Figure 3.20 Roman Jackiw. © Renate Bertlmann; Central Library for Physics, Vienna.

person from whom they would have expected assistance on this highly technical issue in theoretical physics. But, when told of their dilemma, Steinberger was amazed, telling them he had solved the problem nearly two decades before in 1949. Not only had he obtained a result for the rate of decay that was non-zero, but it had agreed very well with experiment.

At the time, Steinberger [126] had, of course, been working with a worldview containing nucleons, pions, and photons, but his work had been at the cutting edge of quantum field theory, referring to the then very recent work of Feynman, Schwinger, and Tomonaga, and using Feynman's methods, the famous Feynman diagrams. Steinberger was working at the very prestigious Institute for Advanced Study at Princeton and, in the paper, he acknowledged the help of such luminaries as Dyson, Oppenheimer, Pais, Rohrlich, and Yukawa.

Steinberger had used what he called a formal method of subtraction fields; in this method, various fictitious fields were subtracted in turn from the original infinite value, making the value converge. This approach was very much along the lines of that used by the original workers on quantum field theory, although, as Steinberger said, the method was 'not without ambiguities' and should be regarded as an 'algorithm' rather than as having being rigorously deduced from fundamental theory. Nevertheless, as stated, for the decay of the pion to photons, it gave excellent results. In his paper, Steinberger acknowledged the similar work of Hiroshi Fukuda and Yoneji Miyamoto [127], but it is clear that his work was totally independent of theirs.

Other aspects of Steinberger's calculations, however, were more problematic. A calculation of the decay of the pion to the muon via coupling to nucleons had led to inconsistent results. Disturbed by this failure, Steinberger had felt himself no match for Dyson and the other young theoreticians at Princeton and was happy to move to experimental work [128]. As Bertlmann [121] laconically put it, 'Steinberger left theory and became a Nobel Laureate in experimental physics.'

In the CERN cafeteria in 1967, Bell and Jackiw very quickly realized that Steinberger's result could be obtained in what was called the sigma model, a type of field theory invented by Murray Gell-Mann and Maurice Lévy to demonstrate explicitly current algebra and many of the other widely held views on quantum field theory at the time. This sigma model, as analysed by Bell and Jackiw, was broadly equivalent to an updating of Steinberger's ideas but with nucleons replaced by quarks.

Since Steinberger's model and the sigma model agreed with experiment but the recently established ideas did not, it was clear that the later ideas must be in error, and Bell and Jackiw were able to discover where the models broke down. Bell and Jackiw showed [119] that the successful models did not respect what may be called *axial symmetry* or *chiral symmetry*.

Jackiw and Shimony [1] explain chiral symmetry as follows. Fermions without mass may be regarded as consisting of two species which do not mix: in one species, the spin points along the direction of propagation and, in the other, the spin points in the opposite direction. Chiral symmetry is the ability to perform independent gauge transformations on each species. However, in the presence of photons, this symmetry is violated.

The fundamental reason for the symmetry violation is that, as so often in quantum field theory, a divergent integral appears in the mathematical analysis, so a prescription has to be given to remove this infinite quantity. However whatever prescription is given, chiral symmetry is lost. We may say that it is lost when quantization, or the change from classical to quantum physics, occurs. Jackiw [129] called this phenomenon *anomalous breaking of a symmetry*, although he also said that, once the surprise had worn off, it might be better called *quantum mechanical symmetry breaking*.

A slightly technical description of the problem will now be given [129]. It is always stressed that, in terms of Feynman diagrams, the anomaly depends on the occurrence of the famous or even infamous *triangle graph*. In this graph, the bottom two vertices denote so-called electromagnetic vector currents, to which the two photons produced in the decay are coupled. The top vertex is the site of the axial vector current, which controls the coupling to the pion. The lines joining the vertices are the propagators of protons for Steinberger, but of quarks for contemporary theory [1].

Direct evaluation of this diagram yields the non-zero decay amplitude of Steinberger and the sigma model, and thus agreement with experiment.

However, the demand of current algebra and of conservation of axial current is that the evaluation must be entirely different; so, clearly, current algebra in particular, however successful it may be in many applications, is flawed. In their paper, Bell and Jackiw [119] stressed that, although they did point out and correct the limitations of the sigma model, their real purpose was completely different. It was to demonstrate that many of the formal manipulations of current algebra calculations are wholly unreliable. Slightly later, Adler [130] showed that axial symmetry behaves in an analogous way in electromagnetism. Consequently, this phenomenon is always called the Adler–Bell–Jackiw, or just ABJ anomaly.

At first the anomaly was perhaps seen to be just a method of clearing up one corner of current algebra. However, in time, it became understood that the anomaly affected much of particle physics, and was important even more widely that that in the fields of condensed matter theory and gravity; as Jackiw and Shimony [1] said, it was definitely not just 'an obscure pathology of quantum field theory'. Indeed, it turned out to be a central aspect of physical theory. There are many entirely different ways of discussing the relevant phenomena, but all lead to the same fundamental conclusion; consequently, Jackiw and Shimony suggested that we might here be dealing with 'an as yet not understood wrinkle in the mathematical description of physical phenomena'.

Although they provided an extensive list of the implications of the anomaly, here just a few of the most interesting will be mentioned. One of these was pointed out by 't Hooft [131]; it makes the dramatic claim that, in the standard model of particle physics, some fermion processes are affected by the anomaly, and, as a result, protons are predicted not to be stable but to decay, although fortunately with an exceptionally low decay rate.

The standard model may be said to appear to have been designed so as to avoid any evil consequences of the anomaly. It has even been suggested [1] that not only theoretical or mathematical physicists know of the anomaly, but also Nature, who not only knows of it but makes use of it.

For example, to avoid difficulties caused by the anomaly, it is essential that the numbers of types of quarks and leptons are equal [132], in order for the anomalies to cancel. The necessity for this condition to be obeyed played a background role in the development of the standard model. Until 1974, four leptons were known—the electron and the electronic neutrino, in what would become known as the first generation, and the muon and the muonic neutrino in the second. However, only the original three quarks were known—up and down in the first generation, and strange in the second. Then with the famous November revolution of 1974, the charmed quark appeared on the scene—not, of course, as a free particle but combined with its own antiparticle or other antiquarks to give mesons. So things seemed fine—four leptons and four quarks [89].

However experiments suggested the existence of the tau lepton only a year later, and by 1978 its existence was fully confirmed, and the existence of its associated tau neutrino, although not immediately established experimentally, was generally taken for granted. Thus, two further quarks were hoped for, to complete the third generation of each type of particle. The bottom quark was found in 1980, but as accelerator energies increased, the non-appearance of the top quark became of some concern. However, when writing his book in 1984, Andrew Pickering [89] commented that this non-appearance was not regarded as a major setback. He said that 'most theorists decided to award themselves the t quark by default', and indeed it was eventually found when accelerator energies had increased sufficiently in 1995.

This requirement may be regarded as part of the necessity, again if anomalies were to cancel, that the sum of the charges of all particles in the universe should equal 0 [88]. Now, in each generation of leptons, there is a particle with charge -1, and a neutrino of charge 0. In each generation of quarks, there is one flavour of quark with charge $\frac{2}{3}$ and another with charge $-\frac{1}{3}$. All charges are in the units of the electronic charge, of course. Clearly, things do not add up as they should; however, we have not yet included colour—with three colours for each flavour of quark, the sum of charges is $+1$ on the quark side, cancelling out the -1 on the lepton side.

This is an extremely interesting argument. First, of course, it is a very simple demonstration of the need for colour to be included in the standard model. Certainly it was by no means the first argument to this conclusion, but it is perhaps the simplest.

More fundamentally, though, it establishes for the very first time a direct relationship between the most fundamental properties of leptons and of quarks, and thus may have profound implications for further unification in particle physics between the strong and the electroweak interactions, and hence for cosmology.

Indeed, the ABJ anomaly has had the deepest implications across virtually every aspect of theoretical physics and mathematics. Just to give one more example, the development of string theory, which is an all-embracing attempt to explain physics in terms of one-dimensional strings, was hampered by a total lack of specificity of the basic concepts. However, the requirement to cancel all anomalies present in the theory limited, at least to an extent, the range of possible theories and thus led to a revival of research in this area [1]. Incidentally, many assert that string theory must be the future direction for fundamental physics [133, 134], although others [135] are just as convinced that the theory makes no contribution to understanding of physics, since it makes no predictions that may be verified in the foreseeable future.

The general significance of anomalies and their implications across physics and mathematics has led to Jackiw and colleagues presenting two volumes

[136, 137] on the topic and its importance in current algebra while, more recently, Bertlmann [121] has written a very substantial and comprehensive account of anomalies in quantum field theory.

Bell and quantum theory: The first responses

Bell's quantum papers had been written in 1964, although, of course, the first actually written was not published until 1966, and he then had to wait to see if his ideas would attract any interest. It must have occurred to him that there was every possibility that they would not do so. Adherence to the ideas of the Copenhagen interpretation was so widespread that, despite the proof by Bohm that hidden variables could exist, extremely few people had any interest in such ideas at all. In addition, Bohr had convinced the general population of physicists of the irrelevance of Einstein's views so comprehensively that few felt that the EPR paper made any worthwhile point. Moreover, Bell had published his work in journals where they might not demand the attention, not only of those whom they might enflame, but also of those who might be genuinely interested.

In the end, he had to wait until 1969 for any response, and the fact that he did eventually gain an audience was really due to two people who were to play an extremely important part in the remainder of Bell's life and work: Shimony and Clauser [138] (Figures 3.21 and 3.22). The two men were very different in their academic backgrounds and personalities, but their different approaches combined remarkably to further Bell's conceptions in the most expeditious way.

Shimony, who was born in 1928, was not only a thinker of great depth, but also a great humanist and a worker for peace and social justice [139]. In 'Bell and local causality', we met his comment to Arthur Wightman in 1955, when Shimony was beginning a second doctorate in physics at Boston University, that he saw no error in the EPR argument. In fact, in the end, he chose not to continue to work with Wightman at Boston but to study under Wigner. Shimony's work with Wigner was in the field of statistical mechanics; however, during this period, he also absorbed some of Wigner's interest in the foundations of quantum theory, through reading Wigner's papers and also through general discussion.

Wigner (Figure 3.23) was to play an important role in the extremely gradual rise to prominence of such foundational work in the second half of the century. He might be described as one of the grandees of physics—he won the Nobel Prize for his work on the symmetries of physics in 1963, and his sister

Figure 3.21 Abner Shimony. © Renate Bertlmann; Central Library for Physics, Vienna.

Figure 3.22 John Clauser. © Renate Bertlmann; Central Library for Physics, Vienna.

Figure 3.23 Eugene Wigner. Courtesy John C. Polanyi, from I Hargittai, The Martians of Science, Oxford University Press, 2006.

was married to Dirac! Yet, decidedly unlike nearly all such leaders of the field, he was open to discussion about the basic ideas of quantum theory. He had his own main idea [140]—that the collapse of the wave function, discussed in Chapter 1, was caused by the *consciousness* of the observer; however, as we shall see, he also showed generosity in promoting the ideas of others.

Shimony had already developed ideas of his own. His first PhD at Yale had been carried out under the direction of Rudolf Carnap, one of the most important philosophers of the twentieth century and one of the founders of the famous Vienna Circle. The self-appointed task of the circle was to endorse *logical positivism* and so to drive out of physics any vestige of *metaphysics*, which may be defined as any idea which cannot be derived directly from experience and so should be regarded as meaningless. Despite this experience, Shimony emerged proud to call himself a metaphysician!

Shimony did not finish his second PhD until 1962; however, even earlier than that, he had been appointed to the philosophy faculty at MIT, where he lectured on the foundations of quantum theory. However, once he had this second doctorate, he moved to what was his dream position—a joint appointment at Boston University in philosophy and physics, and he remained here for more than 50 years.

At Boston, in 1963, he was to write an interesting and important paper on quantum foundations [141]. Actually, it was directly inspired by another important paper written earlier in the same year by Wigner [142]. Wigner's paper discussed what is often called the *measurement problem* of quantum theory. Over the years, it had been quite common for physicists to 'explain' or 'explain away' the problem of quantum measurement as follows.

Taking the specific case of the z-component of the spin of an electron, we may consider a large number or an *ensemble* of such spins, with the wave function of *each* originally a *superposition* of the states $(+)$ and $(-)$. The idea is that the measurement process changes the *superposition* to a *mixture* in which *each* of the spins is *either* in state $(+)$ *or* in state $(-)$. However, in his paper, using very general arguments, Wigner showed totally convincingly that such a 'process' must violate the most fundamental laws of quantum theory. The measurement problem definitely *was* a problem!

In his follow-up paper, Shimony considered the two most popular conventional approaches to quantum measurement: those of von Neumann and Bohr. He took the von Neumann argument to be similar to that favoured by Wigner—that the collapse of wave function was caused by consciousness—but said that it was unclear how collapse caused by a particular observer would be effective for all other observers.

Shimony regarded Bohr's approach, which was sketched in Chapters 1 and 2, as by all means worthy of attention; however, he felt that its weakness was that, while it discussed quite well the *results* of measurement, it failed to provide an *ontology*, that is, an account of how things actually are. Shimony's response to Bohr was thus similar to his response to Carnap, and, of course, very much along the lines of Bell's own ideas. It seemed to him that a new formalism of quantum theory might be required.

Thus, Shimony was an ideal candidate to react to Bell's work—he was a supporter of the EPR argument, and he was keen to develop an approach to quantum theory that discussed the actual behaviour of the system itself rather than just experimental results. Also he was a philosopher *and* scientist of the keenest intellect and also great integrity—like Bell he would not be satisfied with half-truths and excuses.

The second person to be attracted to Bell's ideas was Clauser. He was quite similar to Shimony in being exceptionally bright, and having a highly critical approach to ideas and to quantum theory in particular, but in other ways he was very different. While Shimony's earliest interests were philosophical, Clauser's were in highly technical experimental physics—electronics, early video games, and so on. He was able to take advantage of the fact that his father was Chairman of the Aeronautics Department at John Hopkins University in Baltimore, and John had free access to the extensive laboratory.

Clauser, who was born in 1942, gained a BA at Caltech in 1964 and then moved to Columbia University in New York, from where he obtained an MA in 1966 and a PhD in 1969. His doctoral work was in astrophysics, and he carried out the third measurement of the cosmic microwave background, which had been discovered by Arno Penzias and Robert Wilson in 1965 and which had gained for them the 1978 Nobel Prize in Physics because it was considered the best evidence for the Big Bang.

Clauser [143] has described himself as 'a simple experimental physicist (who spends much of his time mucking about in cutting oil leaks and chasing down vacuum-system leaks and electronics and software bugs)', but that is merely a part, if quite an important part, of the truth. It is certainly true that he is not, and would have no wish to be, an expert in performing algebraic manipulations with no very clear end in view. Certainly, the last thing that would enthuse him would be Maxwell's equations reduced to a compact single line, which other more avowedly theoretical physicists might regard as mathematically simple and beautiful. Rather, he would like each equation to be expressed fully so as to provide as direct as possible link to the actual physical phenomenon.

But with a physical model to analyse, Clauser is actually excellent at calculating, solving problems, and coming up with novel ideas and arguments. His work on the foundations of physics has clearly demonstrated his ability to criticize coherently the ill-judged ideas of others and to come up with fruitful suggestions of his own.

Given his general approach to abstract theorizing, it is not surprising that, during his studies, Clauser built up a considerable interest in, but also a distinct aversion to, quantum theory. He detested finding no clear and direct connection between the wavelike diagrams of the textbook, and the particle being detected in a given localized region. This emotion seems to have been more than just the unhappiness of a Bell or a Shimony with the Bohr or von Neumann account of measurement. Rather, it was a deep antipathy to the abstractness of quantum theory, that is, the fact that, on the page, quantum theory seemed to say nothing about any actual physical system.

During the mid-1960s, Clauser read assiduously the works of those who claimed to understand and explain Bohr and von Neumann, as well as the work of the main dissenters, Einstein and Bohm. Clearly, he was in an ideal position to appreciate Bell's conceptual analysis of quantum theory and his suggestion of experimental testing.

When Bell wrote his paper at Brandeis in 1964, since it was before the days of regular photocopying, he mimeographed a number of copies and sent them to physicists that he thought might be interested. Among these recipients was Shimony, chosen because of the paper he had published in the previous year. Shimony was at first irritated by the smudging produced by the unpleasant purple ink of his copy, and also by some typing errors, and at first he thought the paper might be 'kooky'—the remarkably common attempt of an amateur scholar to make a major scientific breakthrough without unfortunately any real understanding of the basic issues.

However, as soon as Shimony read the paper, he became convinced that it was certainly not 'kooky' but was indeed an extremely interesting and important contribution to the understanding of the most important theory of physics. Personally, he accepted it as a challenge to take the argument further theoretically and also to attempt to obtain relevant experimental evidence on the question. Indeed, it is no exaggeration to say that the few minutes spent reading Bell's paper were to have a major influence on the rest of his life and work.

However, he felt no need to rush. It must have seemed to him that nobody else would take up such a specialized and seemingly recondite topic. As a newly appointed academic, he had many calls on his time and so made little progress on taking Bell's ideas further until the arrival of Michael Horne.

Horne (Figure 3.24) had graduated from the University of Mississippi in 1965 and, keen to pursue doctoral studies in a major university and to work in a fundamental area of physics, had come to Boston. It was in 1968 that he appeared at Shimony's door to request a problem in the field of statistical mechanics; however, Shimony recognized that Horne had the ability and the drive to contribute to the work on the foundations of quantum theory, and so asked him to read Bell's two papers. Horne himself was entranced by these papers and thus began a collaboration that would last for many years.

Their first task was to ascertain whether any first-generation experiments, in which the directions of the detectors in each frame remained unchanged through the experiment, had yet been carried out. If they had, it would be necessary to continue to the vastly more difficult second-generation experiments, in which the directions of the detectors were fixed while the entangled particles were in motion. Clauser had accused Bell of being a little vague on the matter, and this accusation was unsurprising because it was a tricky issue; two seemingly good candidates for first-generation experiments had previously been reported.

Figure 3.24 Michael Horne. © Renate Bertlmann; Central Library for Physics, Vienna.

The first was an experiment by Wu and her research student Irving Shaknov [144] in 1950. (Wu's future work demonstrating the non-conservation of parity, performed later in the decade, was discussed in Chapter 2.) In 1950 she and Shaknov examined the photons produced by annihilation of an electron and a positron. In the annihilation, since both energy and momentum must be conserved, two photons, rather than a single photon, must be produced; the experiment was performed to check a particular prediction of quantum electrodynamics, that the two photons should be polarized at right angles, one being polarized horizontally, the other vertically. Wu and Shaknov confirmed this result and no further discussion was required; indeed, the paper was less than a single page in length.

However seven years later Bohm and his brilliant young assistant Yakir Aharanov [145] revisited the experiment theoretically. They pointed out that Wu and Shaknov's photons were in an entangled state—the wave function is a superposition of a term in which photon 1 is in the h (horizontal) state of polarization, and photon 2 is in the v (or vertical) state; and another term in which the assignations are reversed. Unbeknown to themselves, Wu and Shaknov had produced the first source of entangled photons.

However Bohm and Aharanov wanted to go much further. They regarded Wu and Shaknov's work as a genuine test of EPR (as the title of their own paper indicated) and wanted to confirm that the experimental results agreed with the theoretical prediction. An alternative had been suggested—that once the photons were a certain distance apart, the wave function might break down into one or other of the simple states rather than a linear combination of both. From their analysis of the experimental results, Bohm and Aharanov were quite happy that quantum theory was obeyed.

But, after Bell's suggestions, Shimony had to answer a deeper question. Did the results of Wu and Shaknov rule out *any* local hidden variable theory? Shimony was actually a great admirer of Aharanov and thus took notice when Aharanov told him that indeed local hidden variables were ruled out, at least for the case of first-generation experiments.

Gradually, though, it became clear that this was not the case. For a start, Wu and Shaknov had only taken measurements when the detectors in the two wings of the experiment were either in the same direction or at right angles, precisely the directions for which Bell had shown that there was no problem for local hidden variables.

However, there was more to it than that, as was pointed out by Horne. Because of the high masses of the electron and positron, the photons were themselves of high energy, about 0.5 MeV each—gamma rays, in fact. Thus, they were detected not by the methods used for detection of optical photons—polaroid or calcite prisms—but by Compton polarimeters. These used the Compton effect, in which a high energy photon interacts with an electron in a solid. The difficulty that Horne pointed out was that, following the interaction between each photon and an electron in the polarimeter, the correlation between the electrons that were actually measured was so small that there was no possibility of distinguishing between the predictions of quantum theory and local realism.

It was obvious to Shimony and Horne that the way forward required an entangled pair of low energy photons, so that simple methods could be used to study the polarization—Clauser was to use a pile of plates. At the end of 1968 and the beginning of 1969, Shimony and Horne travelled to many universities, aiming to pick up information on methods of entangling low energy photons and measuring their polarizations; a breakthrough came when they heard of an experiment carried out in 1967 at the University of California at Berkeley by Carl Kocher and Gene Commins [146].

Kocher and Commins had studied the polarization of a pair of low energy photons which had been produced in a calcium cascade. In such a cascade, an atom in a stable level absorbs the energy of a photon and thus jumps to an excited state; however, this state is itself unstable, and the atom returns to its ground state in two stages, in each stage emitting a photon, in this case, an ultraviolet photon. The essential point is that the two photons are entangled; their wave function consists of a sum of two states, in one of which both polarizations are horizontal while in the other they are both vertical.

This seemed the ideal type of set-up to test Bell's ideas. It must be stressed that the particular experiment of Kocher and Commins certainly did *not* constitute such a test. It had actually been planned merely as a lecture demonstration for students and it was situated on a trolley. Only when the experiment turned out to be more complicated than expected was it decided that it should be Kocher's thesis topic and that the work would be published. The paper did refer to the EPR paper and to Bohr's response to it, and it claimed to be a rather general check of the theory; however, the authors appear to have been unaware of Bell's work, and like Wu and Shaknov, took readings only at 0° and 90°.

Not only was this experiment not a first-generation test of Bell's theorem, but by this time Shimony and Horne had become convinced that no such experiment had been done, that it was essential that one should be done, and that this should be a repeat of the experiment of Kocher and Commins for a wider range of angles than previously used. By coincidence, Clauser came to the same view at roughly the same time; Clauser would call it 'calling Bell's bluff'.

Even better for Shimony and Horne was that they were able to persuade Frank Pipkin, a well-regarded professor at Harvard University, to carry out such an experiment. He was already in possession of exactly the kind of equipment that would be required. Actually, it used a cascade in mercury rather than calcium, but the only important difference from the calcium case was that the two photons would always be emitted with different rather than the same polarization.

The regular purpose of the cascade was to measure the lifetime of the intermediate state that existed between the emission of the two photons. This value could be ascertained by measuring the time delay between the detection of the two photons. However, Shimony and Horne were able to persuade Pipkin and his new PhD student, Richard Holt, to undertake their own project, which would entail including polarizers and the means to rotate them. Both Pipkin and Holt thought that this novel experiment was interesting enough to perform in what they thought would be a short time, after which Holt could get back to his 'real' thesis experiment. It did not quite work out like that!

Shimony intended to submit a paper at the spring 1970 meeting of the American Physical Society in Washington, DC but, unfortunately, he missed the deadline. However, he was still convinced that he would have no competitor in the field, and decided to write up a full paper and submit it to an appropriate journal. However, he then saw the *Bulletin* of the Society with the programme and abstracts for the meeting, and came across an abstract from Clauser.

Clauser himself had not been sent a copy of Bell's paper, but in 1967 he came across a copy in the library and was extremely excited by it, although also sceptical. As an avowed opponent of quantum theory, he was delighted that Bell's work seemed to offer the possibility of showing it to be wrong, but first he wanted to check that Bell was correct. Clauser made every effort to find a local hidden variables model of such an experiment that *did* obey quantum theory, and only when he failed to do so did he move whole-heartedly into the business of checking Bell's inequality.

First, like Shimony and Horne, he had to check whether first-generation experiments had already been performed. He visited Wu in Columbia to see if she and Shaknov had taken data at other angles and had just not published it, but this was not the case. However, he did interest Wu in the whole idea and, with two students, she was to perform a Bell test which would be published in 1965. However, for the reasons pointed out by Horne, the assumptions made in analysing this test were not very convincing.

At the time, of course, Clauser was extremely busy finishing off his thesis and was also looking for a job. Perhaps rather unfortunately from his point of view, he combined publicizing his ideas on quantum theory and the fact that he was also looking for relevant information on the topic, with job-hunting. He visited quite a number of universities; very few of the physicists he met were convinced that Bell's work was of interest, and the one sure result was that Clauser would not be considered for an academic position in that place. Clauser's thesis adviser, Pat Thaddeus, was totally against Clauser's new interest, and Thaddeus's 'recommendation' letters for jobs were actually quite the reverse. Bell himself was to become fully aware of the 'stigma' attached to working in this area and it troubled him deeply.

Clauser, like Shimony and Horne, became aware of the work of Kocher and Commins and, in fact, he met Kocher, who had just become a postdoctoral fellow at MIT. Kocher argued that his experiments had certainly disposed of Bell's questions at the first-generation level, but Clauser realized that this was certainly *not* the case: even apart from the fact that the data had only been taken at $0°$ and $90°$, the polarizers were far too inefficient to provide any conclusive results.

Becoming convinced that no genuine test had been carried out, Clauser decided to write to Bell, Bohm, and de Broglie to ask (i) if they knew of any experimental tests that had been performed, (ii) if they thought that an improved test of the Kocher–Commins type would be useful, and (iii) how important such a test would be. The replies from all three were highly encouraging—'no' for (i), and 'yes' for (ii), while Bell was highly enthusiastic in his reply to (iii): he definitely wanted the experiment to be performed, and there was 'a slim chance of an unexpected result, which would shake the world'! David Wick [147] has said that Clauser's letter was the first response to his 1964 paper that Bell received. Without further ado, Clauser submitted his abstract for the Washington meeting.

Naturally, when he saw Clauser's abstract, Shimony was surprised and even distressed that he and Horne had been 'scooped'. Of course, they could have proceeded to write a paper for submission to a major journal, and some colleagues even advised they should pretend they had not seen Clauser's work. However, Wigner persuaded him to phone Clauser, to tell him that he and Horne had come to the same conclusions that he had, and see if Clauser would be prepared to join forces with the two of them.

Fairly naturally, at first Clauser did not see why he should. He had got ahead of Shimony by his publication in the *Bulletin*, and he felt quite able to earn any glory that might be going on his own, rather than having it diluted by a factor of 3. However, once he learned that Pipkin and Holt had agreed to perform the experiment, his attitude had to change. If he were not careful, he would be left out of the cutting-edge experiment altogether.

Clauser, Horne, Shimony, and Holt met up at the Washington meeting and they got on well together. Indeed, the first three, who had been thinking about the problem for a considerable time, realized that it would not be enough to improve Kocher–Commins; a good deal of further theoretical analysis would be required to draw the necessary conclusions from the experiment. They decided to perform this analysis and write a paper to explain it, and in the end Holt made a substantial contribution as well, so his name was included; the resulting paper, which was to become famous, has always been known as CHSH [148].

And now, just as things seemed settled, came a startling development. Clauser's job-hunting succeeded and he was offered a postdoc position at Berkeley under Charles Townes (Figure 3.25) to study the early universe with the use of radio astronomy. The Kocher–Commins experiment, of course, had been performed at Berkeley, Commins was in a senior position there and, most interestingly of all, it turned out that the original equipment was still in existence.

It was natural for Clauser to ask Commins if he might be allowed to put the equipment back in working order and carry out the experiment, so as to give a conclusive answer to Bell's questions. Unfortunately, it was also natural for Commins to believe, like the vast majority of physicists, that the EPR problem had been answered by Bohr many years ago and that the experiment Clauser was suggesting was a total waste of time. It will be remembered that the Kocher–Commins experiment had originally been a student demonstration, not a genuine search for knowledge.

Of course, even apart from that, Commins must have felt that Clauser had been recruited and would be paid to do radio astronomy, and it was not unnatural to expect him to work on radio astronomy, not a totally different experiment that happened to have taken his fancy!

Figure 3.25 Charles Townes. From Denis Brian, Genius Talk (Plenum, New York, 1995).

But then Townes stepped in. He was an extremely important figure in American physics, the Nobel Prize winner in 1964 for his work in developing the laser and the maser, and a guide at some time to nearly all those working on lasers or astrophysics in the country over several decades. When he heard about Clauser's proposed experiment, he suggested to Commins that, for a period which would turn out to be longer than any of them might have expected, Clauser should be allowed to spend half his time on his own experiment. Commins had no option but to accept this ruling, as well as Townes's equally generous suggestion that Stuart Freedman, a new PhD student, should be assigned to Clauser to help with this experiment.

All this meant that, as soon as Clauser arrived in California, he and Holt would be rivals. Of course, though, until that point they would be still be colleagues in writing CHSH.

Clauser was a very keen sailor and actually lived on his racing yacht, so he decided that the best way of getting his boat to California was to sail it to Galveston in Texas, truck the boat to the west coast and then to sail along the California coast. Although a hurricane was to cut short the last stage,

Shimony kept up communication as Clauser sailed down the east coast, by mailing him sections of the draft of the CHSH paper to be collected at the marina where he would be spending the night. Clauser would reply by phone with arguments and suggestions. By the time Clauser reached Berkeley, the paper was complete, and it was sent to *Physical Review Letters* on 4 August 1969 and published on 13 October.

We have stressed that the great advantage of Bell's ideas over those of, say, EPR and Bohm, was that, as Bell said, 'It requires little imagination to envisage the measurements actually being made.' For all that, as Clauser, the expert experimentalist, was well aware, there was still quite a large gulf to an actual experiment. Clauser commented that Bell, being a theoretician, assumed an ideal apparatus, an idealized configuration, and an ideal preparation of the entangled state, and so his result applied only for an ideal system, not a real one! 'Damn theorists!' he remarked [143]. It was the job of CHSH to adjust the assumptions so that the theory applied to real systems and real experiments.

The first problem that Clauser pointed out was that, in Bell's preliminary argument, which was based on EPR, he assumed that the two photons are travelling in precisely opposite directions. As Clauser realized, the smallest deviation from this ideal situation would mean that the whole proof would collapse. Fortunately, Clauser was able to show that the first stage of Bell's argument could be avoided. In CHSH a small departure from perfect correlation was allowed, and the authors were able to show that the main result is unaffected. This was the beginning of a process, which actually continues to this day, of deciding exactly what is essential for Bell's proof to work.

CHSH also went much further than Bell by producing a mathematical inequality that could be used to test explicitly whether a set of experimental results for an experiment of Kocher–Commins type agreed with local realism. The experiment must be carried out four times; each of two values of the setting of the polarizer in one wing of the experiment is coupled with each of two values of the setting in the other wing.

The most interesting case is when all the directions are in the same plane, and the directions in one wing are 0 and 2θ, and in the other wing are θ and 3θ, where θ is an angle that varies from $0°$ to $90°$. Then, a quantity S is defined to be equal to $3(\cos 2\theta) - \cos 6\theta$, and the CHSH condition for local realism is that S must lie between -2 and $+2$. To be exact, both end points -2 and $+2$ are allowed.

Figure 3.26 shows the quantum mechanical expression for S. First, note that, as Bell had shown, for the only values of θ used in experiments up to this time, $0°$ and $90°$, S lies exactly on the allowed values of -2 and $+2$. Such experiments could never rule out local realism. However, it is clear that, for a considerable range of values above $0°$ and below $90°$, the inequality is violated, *if*, of course, quantum theory is true. The largest deviations from the allowed range occur

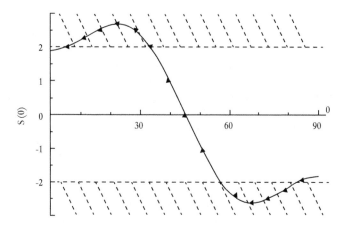

Figure 3.26 The quantity S. According to CHSH, S must not lie in the hatched region. The solid line is the prediction of quantum theory, which lies in the hatched region for large ranges of angle θ. (The triangles are the experimental values obtained by Aspect and his collaborators, as described in Chapter 5.) From A. Aspect and P. Grangier, *Symposium on the Foundations of Modern Physics: 50 Years of the Einstein-Podolsky-Rosen Gedanken-Experiment* (World Scientific, Singapore, 1985).

for $\theta = 22.5°$, and $\theta = 67.5°$, and the corresponding values of S are $2\sqrt{2}$ and $-2\sqrt{2}$, respectively. These angles were discovered by Freedman and are often known as Freedman angles.

But, in a sense, this is putting the horse before the cart. The inequality relation does not refer directly to quantum theory at all. The experimental results may first be used immediately as a check on local realism *without reference to any particular theory at all*. Only after that may they be compared with the predictions of quantum theory or indeed any other theory that might conceivably be of interest.

The new inequality would be called the CHSH inequality. Bell, incidentally, was very happy to recognize that it was superior to his own earlier inequality, which did not go beyond showing that, if local realism were obeyed, quantum theory must be wrong. In fact, this new inequality is often called the CHSH–Bell inequality or just Bell's inequality, although that is very unfair to CHSH!

Another aspect of the experiment where CHSH had to go well beyond Bell was in the study of the experimental results. Bell had, of course, been thinking in terms of entangled electron spins, and it was natural to assume that each member of every entangled pair could be detected. The experiments being analysed by CHSH used photons, and the detection of photons could not be more different from the detection of electrons.

The early experiments used *one-channel analysers*, for which photons with polarization parallel to a given axis are passed and carry on to the detector, while those with polarization perpendicular to this axis are absorbed and so not detected at all. But what is a still worse problem is that detectors of photons were extremely inefficient; in fact, in these experiments, in only about one case in a million are both members of an entangled pair detected. So, even if both photons have parallel polarization, in the great majority of cases, neither will be detected. If a photon is detected in one wing of the system, we don't know whether its partner was polarized perpendicular to the axis and was absorbed, or parallel to the axis and just failed to be detected. The only useful data is when a photon is detected simultaneously in each wing of the apparatus.

In order to solve this problem, CHSH suggested performing every measurement four times: once with polarizers in place in each wing of the experiment, once with each polarizer in turn removed, and once with both removed. The analysis can then be straightforward, because, for example, when the polarizer in the first wing is removed, we see (or technically may see) those pairs of photons for which the polarization is parallel to the axis in the second wing, but it may be parallel or perpendicular to the axis in the first wing.

However, it is important to stress that at the heart of the analysis is what is known as an *auxiliary assumption*, that the probability of a pair of entangled photons both being detected is completely independent of whether polarizers are present in either wing, and also, if they are present, the direction of either axis of polarization. In fact, the probability of detection may be treated as a constant and, since the analysis uses ratios of events in different cases, this constant may be ignored altogether.

We must now focus on the auxiliary assumption. Is it true? One's first reaction would probably be to say that the assumption seems very good, perhaps even obviously true. Why should the probability of detection of a photon depend on whether it has passed through a polarizer, or the axis of the polarizer where appropriate?

However, it must be remembered that the intention of this experiment was to test very deep ideas, ideas that some people feel extremely strongly about (as we shall see in particular in Chapter 5), and they will certainly wish to see every doubt extinguished before they will be prepared to relinquish cherished beliefs.

Also, although it may seem unlikely that photons with slightly different histories should behave differently in a detection process, we must remember that the whole point of hidden variable ideas is that an ensemble of photons, seemingly identical, may consist of different sub-ensembles with different values of a particular hidden variable. Thus, photons of parallel and perpendicular polarization may belong to different sub-ensembles, and so it is after all perhaps not so unlikely that they might behave differently under detection.

Bell had, after all, shown that assumptions such as those of von Neumann may seem natural when described in a rather abstract way, but may turn out to be not just wrong but foolish when considered in the light of particular models. The problem associated with the low efficiency of photon detectors has come to be known as the *detector loophole*.

CHSH took up one more point in Bell's paper that required translating to experimental possibility. As we have seen, Bell had assumed that each pair of entangled particles came out in exactly opposite directions; however, in practice, it is necessary to include photons emerging at some angle to the ideal direction; these photons are gathered with wide lenses which focus the photons onto the polarizer and the detector. Naturally, the presence of the lenses necessitates some adjustment to the mathematical analysis, and it was Holt who showed how to perform this adjustment, thus gaining his position as the fourth author of the paper.

The paper was published in *Physical Review Letters*, which was the leading journal that publishes only research on physics. (The journals *Nature* and *Science* are equally well regarded but publish papers across all scientific disciplines.) In some ways, its publication in this journal was surprising, as work on such topics was, as has been stressed, generally out of favour with the scientific establishment [58]. From the nature of some suggestions that one referee made for minor changes, it was clear to the authors that this referee was Bell, who would naturally have supported publication; however there must presumably have been at least one other positive referee report as well.

It should not be thought that Clauser and Shimony and their associates were the only people interested in Bell's work at the time. There were at least a few others. For a start, there was d'Espagnat (Figure 3.27), a French theoretical elementary particle physicist. As we saw earlier, he was among the very first appointments at CERN; however, in 1959 he returned to Paris, where he spent the remainder of his career as Director of the Laboratory of Theoretical Physics and Elementary Particles at the University of Paris XI at Orsay. However, he spent many long periods at CERN, where he was close to John Bell.

In fact, Bell and d'Espagnat each soon suspected that the other had concerns, ones that were not readily admitted at CERN, about the foundations of quantum theory. D'Espagnat was indeed troubled that virtually no attention was paid to such matters and was delighted to see on Bell's bookshelf a book by de Broglie, being particularly pleased that it was, of course, in French! [149]. There was subsequently much discussion between the two, and d'Espagnat became probably Bell's most constant supporter.

D'Espagnat's paper titled 'The Quantum Theory and Reality' [150] would be published in 1979, and the clarity of its presentation meant that it played an extremely useful role in publicizing Bell's work, as was still necessary even at that rather late date. D'Espagnat's book *Conceptual Foundations of*

Figure 3.27 Bernard d'Espagnat. © Renate Bertlmann; Central Library for Physics, Vienna.

Quantum Theory [151] was an extremely good account of the background ideas, and he also wrote several more philosophical books including *Veiled Reality* [152]. He had amicable disagreements with Bell on philosophical matters but was happy to say [149] that 'Bell was for me one of the best friends I ever had.'

Another scientist who was extremely interested in Bell's work was Franco Selleri (Figure 3.28), an Italian elementary particle physicist who was Professor at the University of Bari from 1968, but for political reasons—he was a Marxist—became frustrated with the current state of physics, and in particular the views of Bohr and von Neumann. He insisted on realism, locality, and *causality*, which for him was less formal than determinism but still demanded a clear link between present and future, and his interests moved towards the foundations of quantum theory. It is obvious that Selleri would be a decided opponent of any experimental results supporting quantum theory against local realism.

His views are explained in his book *Quantum Paradoxes and Physical Reality* [153] and are also discussed by Oliver Freire [154] in his account of 'quantum dissidents'. (Other physicists discussed by Freire include Hans-Dieter Zeh, Bell, Clauser, Shimony, Rosenfeld, d'Espagnat, and Bryce de Witt.) Freire also interviewed Selleri for the Centre for the History of Physics of the American Institute of Physics [155].

Also interested in, and indeed enthusiastic about, Bell's work was Wigner. He produced an independent proof that quantum theory did not respect

Figure 3.28 Franco Selleri. © Renate Bertlmann; Central Library for Physics, Vienna.

local realism [156], using a direct analysis of all the various possibilities; a little later with Freedman, he also refuted a criticism of Bell's argument [157].

In 1969 Selleri was on the board of the Italian Physical Society and he suggested that d'Espagnat should be asked to organize a summer school at Varenna on the foundations of quantum mechanics. Wigner and Bell gave him considerable support, and the meeting (Figure 3.29) was a great success—nearly all those involved in the nascent field of quantum foundations (or perhaps it was just a field coming round from a coma!) gave talks, including Bell, Bohm, Shimony, Jauch, Wigner, and Piron.

A nice touch was Wigner's invitation to Zeh, who, in the late 1960s had invented the theoretical idea of *environmental decoherence*, by which interaction with its surroundings prevents a quantum system from dephasing and losing its quantum nature. However, at that time, Rosenfeld, who had been Bohr's most devoted follower, decided that Zeh's work was in conflict with the Copenhagen synthesis, described it as 'a concentrate of the wickedest nonsense', and, it would seem, took every possible step to wreck Zeh's career.

So it was good that Wigner, who was a member of the editorial board of the new journal *Foundations of Physics*, arranged to publish Zeh's paper in the first issue [158] and, as has been said, invited him to Varenna, where Wigner described Zeh's work as interesting and as one of a small number of possible solutions to the problems of quantum theory. Zeh also reported to Freire [154]

Figure 3.29 The 1970 Varenna Summer School on the Foundations of Quantum Mechanics: *Scuola Internazionale di Fisica 'E. Fermi'*. In the second row (seated), John Bell is sixth from the left, Eugene Wigner is eighth from the left, and Constantin Piron is ninth from the left; Josef-Maria Jauch is first from the right and Bernard d'Espagnat is third from the right. In the front row, Leonard Kasday is fourth from the right, Basil Hiley is fifth from the right, and Hans-Dieter Zeh is second from the right. Emilio Santos is third from the left in the third row. Abner Shimony is second from the right in the fourth row, while Franco Selleri is on his own towards the right in the fifth row. Courtesy, Academic Press.

that Bell was very sympathetic and always asked the right questions: 'This was already a very great thing that there was somebody who was critical against the mainstream and put his finger on the right things. Only very few people did that.'

Bell's talk at the meeting [159] was a little more relaxed in style than his initial publications; clearly, he recognized that his audience was self-selected to be interested. It dealt with much the same ideas as his earlier work, and any differences in presentation will be discussed with that of later presentations in Chapter 4.

A great decade

The 1960s had been a great decade for Bell. Even at the time, it was clear that, during the decade, he had made up ground to become an international authority on elementary particle physics. He had also presented his ideas on quantum theory, and he must have been excited to know that experiments would take place, and he was, of course, very keen to find out what they would tell us. Half a century later, it is obvious just how important these papers had been.

4

The 1970s

Interest Increases

Early successes for Bell and for CERN

Both Bell and CERN received important boosts early in the 1970s (Figure 4.1).
For Bell, it was election as a Fellow of the Royal Society in 1972. Only the very
best scientists from Britain, Ireland, and the British Commonwealth receive
this honour, which is highly competitive. Prospective Fellows have to be proposed by those already in the Fellowship and, in Bell's case, Matthews was
the main proposer, although presumably Edwards and Peierls had also been
involved.

The citation [1] was as follows:

[Bell is] distinguished for his many contributions to Nuclear and Elementary
Particle Physics. Among his contributions to Field Theory, he gave an independent derivation of the important CPT theorem. He has done much work
concerning the theory of many-body systems, advancing particularly our
understanding of the nuclear optical model. He has contributed significantly
to the theory of Weak Interactions, especially concerning processes induced
by high-energy neutrinos, and a unitary relation of central importance for
the analysis of CP violation effects. He is noted for his discussion and resolution of many paradoxes in these fields or other theoretical work.

It is clear that the distinction was definitely *not* for his work on quantum
theory, which might at most just be squeezing into the catch-all grouping
at the very end. Rather, he was being recognized as an eminent theoretical
physicist in the fields of nuclear physics, elementary particle physics, and field
theory.

After his initiation into the Society, Bell wrote to his parents telling of the
events [2]. The Royal Society, he said, has an impressive building in the middle
of London, looking out onto St James's Park near Buckingham Palace. Bell told
them of the new Fellows being given a tour of the building and then lunch
with the president and officers. He sat beside Sir Harrie Massey, a former professor at Belfast whom we met in Chapter 1.

Figure 4.1 John Bell (second row from the front, far left) at the 1972 Trieste conference in honour of Paul Dirac's 70th birthday. Bell liked to joke that this was the nearest he came to Dirac (front row, left) and Heisenberg (front row, right). Courtesy Foto 'Rice'.

Then came what Bell calls the 'initiation rite'. Each new fellow held hands with the President while pronouncing the 'official formula'. The signing of the book was slightly nerve-racking since the signatures went right back to Charles II, and although the signing is done with an old-fashioned quill pen, the authorities are naturally keen that no blot is made. So, there is a special man whose job is to dip the pen into the ink to make sure that enough is taken to sign but not enough to blot.

Bell concluded his letter by saying: 'Now I am FRS and perhaps in some way it will eventually do me some good. Anyway I cannot fret any more about *not* being FRS, and will have to find something else to worry about'.

The important step at CERN was the eventual demonstration of the neutral current, the theory of which was discussed in Chapter 3. What made the success even more pleasant was that it came through a certain facing-out of the American opposition. The work was carried out using Gargamelle [3], a giant bubble chamber, which had been conceived and constructed by a team under a French scientist, André Lagarrigue (Figure 4.3), and was funded largely by the French Atomic Energy Commission. The presence of Gargamelle, which was filled with 4.5 tonnes of refrigerator fluid, meant that it would be very much easier to detect the neutrinos.

Figure 4.2 John Bell at the blackboard; *Europhysics News*. Volume 22, April 1991. Courtesy European Physical Society.

Figure 4.3 André Lagarrigue. Courtesy CERN.

At the very end of 1972, there appeared a single event that looked to have the signature of a neutral current [3, 4], but it was clear that many more such events would be required in order to be able to make a claim of a discovery. However, there was a very delicate balancing act to be followed. Weisskopf, Director-General of CERN, had stressed that CERN must not be 'always late', but being early and wrong was probably even worse. However, by the time 1.4 million images had been scanned, only another two neutral current events had been found.

Then, on 18 July 1973, Lagarrigue received a letter from Carlo Rubbia, who was the leader of the hunt for neutral currents at the National Accelerator Laboratory in Chicago. (A decade later, Rubbia would be working *for* CERN rather than against.) Rubbia reported that his team had around 100 good images of neutral currents and was able to publish. He suggested that each team should acknowledge the findings of the other and that they should share the credit for the discovery.

Lagarrigue, however, was suspicious, and decided that each team should submit its own results for publication, which they agreed to do; the European team was in the lead and in September their paper was published in *Physics Letters*. But what seemed like success soon seemed likely to turn into the complete opposite. The American team had delayed publication of their own results to improve the performance of their equipment. However, when they did so, their neutral current events disappeared; it seemed that their neutral currents had been phantoms [4].

Rubbia travelled the Atlantic to tell Lagarrigue the news, and added that the American team had now written an entirely different paper for publication, dismissing the idea of the neutral current. It seemed that CERN perhaps had to face humiliation. At CERN, there was consternation, and a severe cross-examination of the Gargamelle team by the authorities. However, the Gargamelle team stood their ground and stressed their confidence in their claims.

Gradually it was the Americans who lost confidence. They eventually found some neutral current events on their own film; they were still split—Steinberger, making yet another appearance in this book, bet against neutral currents and ended up by losing a few bottles of fine wine. But, in the end, the Chicago team abandoned their later manuscript and went back to the previous one, which agreed with the paper from the CERN team that neutral current events did exist.

It was certainly a triumph for CERN. Did it deserve a Nobel Prize? *History of CERN* suggests that, had it not been for the untimely death of Lagarrigue in 1975, he might well have received the prize [5]. As it was, the discovery of the neutral current made it fairly sure that the Z_0 existed; a decade later, its discovery would be part of the work that would finally give CERN its first Nobel Prize.

Bell's theorem: The first results

At the beginning of the 1970s, both Clauser and Holt were ready to start work on testing the CHSH inequality [6]. It seems that both initially assumed that the experiment could be performed with a relatively minor adjustment to laboratory systems that were already in existence; however, both found that this was not the case.

Clauser hoped that he and Freedman would be able to get away with repeating the Kocher–Commins experiment, but with, of course, a much greater range of angles than had previously been used, and with minor improvement to the apparatus, including, in particular, a substantial upgrade to the polarizers. However, in practice they found that, in order to obtain reliable results, just about every aspect of the equipment needed to be improved. They were able to get funding to purchase new photomultipliers but, for the other components, Clauser did what he was best at—sorting through the 'junk' left in the wake of earlier experiments and picking out what would be best for the new one. Every aspect of the apparatus, including lamps, lenses, filters, and polarizers, as well as the photomultipliers, had to be chosen so as to maximize performance at the wavelengths of the photons involved in the experiment [7, 8].

Freedman was starting experimental work practically from scratch. Clauser, of course, was vastly experienced in a wide range of experimental techniques, including electronics and computing, which would certainly be essential in the new challenge, but the pair of them still had to learn many new skills, including optics and atomic physics. It was to take them about two years to build and test the apparatus, and then another two months to run the experiment.

In the experiment, calcium atoms emerged in a beam from an oven and were then exposed to radiation from a deuterium arc. The radiation had travelled through a filter that passed only the narrow range of wavelengths that could raise the calcium atoms from their ground state (or lowest energy level) to an excited state. This excited state is the top of a *cascade*, and the atom decays back to the ground state in two stages, emitting a photon each time. The first photon is in the green region of the spectrum, and the second is in the purple, but the most important point is that the photons are entangled. The second photon appears around five billionths of a second after the first one, so the coincidence window has to be long enough to accept both of them.

Then, in each wing of the experiment, one of the photons passes through a primary lens of half-angle about 30° (which had been one of the points where CHSH had to overrule Bell's ideas); a wavelength filter, which selected photons of the appropriate wavelength for that wing of the apparatus; a polarizer, which had to be rotatable and removable; and, after a further focussing lens, the photomultiplier.

Clauser had been well aware of the low fraction of the initial excitations of the calcium atom that would lead to the detection of a photon in each wing of the experiment. The efficiencies of the polarizers were high—up to 96°–97°, but those of the photomultipliers were just the opposite—as low as 0.13° and 0.28°. This low efficiency was despite sterling work put in by Freedman, who spent a lot of time at RCA helping them design and improve these *phototubes*. Also, of course, the comparatively small half-angle of the primary lens meant that only a few photons reached the polarizers in the first place. Consequently, the probability of both photons in an entangled pair being detected was around one in a million. Thus Clauser was determined to construct a system with a large number of initial excitations. Therefore, he made sure that he had a very large system with a large powerful oven, a strong deuterium arc, and large and efficient polarizers.

He also showed his great experimental talent in two other ways. The first was his design of polarizer. He recognized that conventional Polaroid sheets could never be efficient enough and instead used so-called *pile-of-plates* polarizers. These used the fact that, when a beam of light is incident on a plate of glass at the *Brewster angle* (which is named after the nineteenth-century Scottish physicist David Brewster and is about 56°), the only light reflected is polarized parallel to the surface. Consequently, the light that is transmitted is polarized *preferentially* in the perpendicular direction. If the light meets a sequence of such plates, what finally emerges will, to a very good approximation, be polarized *only* in the perpendicular direction, or in other words, the pile of plates just acts as a polarizer. In Clauser and Freedman's experiments, the pile of plates consisted of ten plates each of thickness 0.33 nm.

It should be stressed that, in this first round of experiments, all the polarizers were single-channel ones, that is, they allowed for the detection of only one of the two modes of polarization. Consequently, each experimental run had to be performed not only with both polarizers in place but also with each one present and the other absent, and lastly with both polarizers removed.

Clauser's other stroke of genius was to recognize that their experiment consisted of a huge number of repetitions of the same task: generating photons and measuring their polarizations, for different cases—rotating one of the polarizers or removing or inserting a polarizer in between each repetition. To achieve this effectively, they used a system called the Geneva gear system, which had been used for centuries to transform regular circular motion into a sequence of discrete rotations. It is used, for example, in film projectors, and it works by arranging for a slotted wheel to rotate continuously, engaging and disengaging, as it turns, a peg on a small wheel.

Thus, the experiment (Figure 4.4) worked practically automatically. The mechanism would rotate one of the polarizers, wait for 100 s for an experimental run, and then perform another rotation. Each polarizer could be folded out

Figure 4.4 The Clauser experiment. Courtesy Lawrence Berkeley National Laboratory.

of the way when not required, and all the sequencing was run by an old tele-phone relay which had been dumped and was brought back to life by Clauser. The total run time was 280 hours. All the information from the experiment was sent to computers, and reams of paper tape were produced from which the results of the experiment could be calculated.

In Harvard, Holt was busy constructing his own apparatus which, being a development of the work of Pipkin and his previous students, was much smaller than that of Clauser. Unlike Clauser's experiment, the necessary equipment for which occupied a full laboratory, Holt's experiment could be described as a table-top experiment [6].

Holt was using mercury atoms rather than the calcium atoms used by Clauser and thus had to deal with an important additional complication. Mercury consists of a number of different isotopes, which are versions of the same atomic species but with different numbers of neutrons in their nuclei; so, to obtain the rather rare mercury-198, he had to separate it by distillation into his main glass tube. He used an electron gun to excite the atoms, and the resulting photons were polarized by calcite rather than Clauser's pile of plates. However, the general set-ups of the two experiments were, of course, still the same.

Before we look at the actual results, it is interesting to consider what each of the participants hoped or thought might come out of the experiments. Bell certainly hoped that there might be a disagreement with quantum theory—it would confirm his view of what nature *ought* to be like, and of course it would dramatically increase his standing as a physicist and a thinker—but he was level-headed enough to realize that this result was unlikely. One suspects that Shimony's guess might have been similar. Horne's view was that although, of course, the experiment was extremely interesting and so should be done, quantum theory had been so successful that there was no real possibility that it would fail here. In contrast, Clauser ardently disliked quantum theory and was convinced that the experiments would support local realism; indeed, he took up a bet with Aharanov with odds of 2:1 against quantum theory, threatening to leave physics if the result turned out against him. Holt's rather jocular view was different again. He suggested that, if quantum theory were to be proved wrong, he would certainly be awarded the Nobel Prize, but conservative Harvard would definitely not give him his PhD!

Now let us look at the experimental results [6]. While, of course, one may inspect the results in detail, allowing for the presence and absence of the various polarizers, producing a graph such as Figure 3.26, and determining whether the results stray into the region ruled out by local realism, it is convenient to produce a single-figure criterion, which is called δ. To obtain this value, the results at the two Freedman angles shown in Figure 3.26 are examined. For each of these values, the number of coincidences with both polarizers present is divided by the same quantity with them both absent. The smaller of these numbers is then subtracted from the larger and, for neatness, ¼ is subtracted from the result. Then, the CHSH inequality then tells us that δ must be negative.

Holt produced his results first, and (from most points of view, surprisingly) they showed that local realism was upheld. His value of δ was -0.034 ± 0.013, comfortably negative. Actually, for Holt's particular experimental set-up, the predicted value for quantum theory, 0.016, was not far from the range for local realism; however, this value was still a very long way from Holt's experimental result.

Naturally, Holt and Pipkin were in something of quandary! It is perhaps one thing to make bold statements about rebutting quantum theory in the abstract, but different when the general statement has to be backed up. Publishing the result might indeed give them the Nobel Prize if it turned out to be correct but, if it turned out to be wrong, the trip would be to a scientific Siberia rather than to Stockholm. In the end, they decided to search everywhere for a possible source of error in the experiment, and in the meantime to wait for results from Freedman and Clauser. (Clauser, one may assume, would definitely have published!)

The results from California were just the opposite of those of Holt. Their value of δ was 0.050 ± 0.008, clearly violating local realism, much, of course, to

Clauser's annoyance. For this experiment, the quantum prediction was 0.051 ± 0.007, so the experimental values were extremely close and well within the uncertainties. Indeed, the full experimental curve for all values of the angle θ was extremely close at all points to the quantum theory expectations.

Clauser and Freedman published their results in early 1972—their paper was submitted in February and published in April, and again it is interesting that it was accepted for publication in the prestigious *Physical Review Letters* [9]. The conclusion of the authors was: 'We consider these results to be strong evidence against local hidden variable theories.' Freedman was awarded his PhD in May.

Pipkin and Holt presumably accepted that their result had been wrong. Holt wrote up the work [10] and was awarded his PhD, and in the following year the pair distributed a 'preprint', although with no intention of proceeding to publication; the preprint would tell others what they had done and leave open the possibility that somebody would suggest where there might have been a systematic error. It has been suggested subsequently that the tube along which the mercury vapour travelled was curved, and that possibly reflection at the glass may have rotated the polarization with, unfortunately, a lethal effect on the results.

Clauser, though, was not satisfied with one result supporting local realism and one against, so he decided to repeat Holt's experiment—or at least to do so as nearly as was possible. He was able to borrow suitable apparatus, although he had to use mercury-202 rather than mercury-198, which had been used by Holt. The particular atomic transitions involved for this isotope meant that the work turned out to be far more laborious than either of the previous two experiments. He took readings for 400 hours, but the results were the same as those he had obtained with Freedman—local realism was ruled out and there was excellent agreement with quantum theory [11].

Another physicist had become extremely interested in these experiments. Ed Fry (Figure 4.5) had recently been appointed as an assistant professor at Texas A&M (Agricultural and Mechanical) University in 1969 when he heard a talk on Bell's theorem and CHSH, and immediately wanted to become involved experimentally. In fact, scientific techniques sometimes progress extremely quickly, and the development of narrow-band tunable dye lasers meant that Fry was able to design an experiment that was far more elegant than those of Clauser and Holt. Such a laser could be tuned to just the right wavelength to trigger a cascade in mercury-200, and the use of lasers would make the experiment much easier to perform. Fry duly put in a grant application, only to be turned soundly down. The attitude seemed to be that it was bad enough that money was being wasted at Harvard and Berkeley without wasting more in Texas!

Figure 4.5 Ed Fry. © Renate Bertlmann.

The difference of results between the experiments of Clauser and of Holt led to a change of opinion, and Fry received enough money to carry out his proposal with his research student Randall Thompson. In their experiment, because the laser was tuned to precisely the right wavelength to excite the mercury atoms to the desired energy level rather than to any other level, the numbers of entangled pairs was increased enormously and data could be produced much more speedily. In fact, the entire experiment took only around 80 minutes rather than the hundreds of hours of Clauser and of Holt.

In fact, Fry and Thompson did not need to separate the atoms of mercury-200 from the other isotopes in naturally occurring mercury. Rather, it was the precision of the laser that singled out wavelengths corresponding to atoms of the desired isotope and, in the wings of the experiment, optical filters were used to eliminate all wavelengths but those desired. Fry and Thompson's results confirmed those of Clauser: the value of δ was 0.046 $\pm$ 0.014, clearly positive, and the detailed results agreed very closely with the quantum predictions [12].

During the decade, some other types of experiments were carried out testing Bell's theorem. Wu's interest had been raised by the visit of Clauser in particular, and, with Leonard Kasday and John Ullmann, she carried out an experiment using positronium annihilation. As explained in Chapter 3, this method was been ruled out as a useful technique by Clauser and Horne and, in particular, both Horne and Bell were able to demonstrate that the quantum predictions could be produced by a hidden variable model.

Wu and her collaborators did produce results that ruled out local realism, but their analysis required *auxiliary assumptions* that were far less justifiable than those required by those using cascades [13]. Three other experiments were carried out using this technique; two agreed with Wu, while the third disagreed.

Another type of experiment, carried out by Mohammad Lamehi-Rachti and Wolfgang Mittig [14], was rather interesting because it was nearer Bell's initial conception as it used spins rather than photon polarizations. Lamehi-Rachti and Mittig studied proton–proton scattering, and they also obtained results violating local realism and supporting quantum theory. However, these authors also used auxiliary assumptions that were difficult to justify; so, at the end of the decade, any general analysis was based almost entirely on the experiments using cascades.

We now briefly examine Bell's response to these results. Of course he was disappointed but not really surprised at the outcome, and he was certainly aware of the remaining *loopholes*.

Selleri was much more concerned about the result—he was dedicated to upholding local realism. When he visited Bell at CERN and started to tell him about the various loopholes that rendered Clauser's conclusions unsafe, Bell reached into a drawer of his desk and pulled out a sheaf of calculations that came to exactly the same conclusions.

Bell, though, was measured in his comments on the detector loophole in particular, although there was, of course, no doubt that he was extremely interested in removing the locality loophole, or in other words, moving on to the second generation of experiments. On the detector loophole, he felt it unlikely that imperfect experiments would give results very close to quantum predictions, but the results of improvement would change them to disagree with quantum theory and agree with local realism.

Apart from scientific concerns, he was well aware that his work had encouraged Clauser in particular to go out a limb in support of his ideas. Bell felt that it would be a slap in the face to say that, after all that, the experiment was inconclusive. It definitely did turn out that Clauser was the one to suffer in the long term to what he christened the 'stigma' of working in this area of physics. It is true that CHSH was published in the prestigious *Physical Review Letters*, as were the papers by Freedman and Clauser [9, 11]. Other papers we have mentioned or shall mention were published in the leading journal for regular papers [12, 14, 15] and a leading review journal [16].

So it may indeed be the case that the establishment was coming round to the acceptability of publishing papers in the area, although not so much to the idea of appointing those who wrote such papers [6]. In the mid-1970s, Clauser applied for numerous academic positions but was uniformly rejected. He was to have a highly successful career outside academia, working for a period at the Lawrence Livermore National Laboratory, where he worked on plasma physics, trying to develop a technique for sustainable nuclear fusion, and after that working on projects of his own devising and funded from outside [6, 7].

If there is a sadness about Clauser's career, it is not that he was prevented from carrying out research at the frontiers of knowledge, as such was not the case, but that such a hugely able and visionary scientist was not able to gain employment in the university environment, where his teaching would have been inspirational.

Fortunately, Horne, Fry, Freedman, and Holt were able to obtain suitable appointments in colleges and universities, and all had highly successful careers [6]. Shimony, of course, was well established before becoming involved in the experiments.

Bell's first published comments on the experimental results were presented in 1975 [17] and 1976 [18], included at the end of general discussion. In the first of these papers, he almost seemed to apologize for suggesting tests at all—quantum theory works extremely well, he admitted, even in calculations of the ground state of the helium atom, which has a great deal in common with the kinds of systems he was discussing. He also immediately stressed the importance of the locality loophole, although he admitted that a mechanism taking advantage of this loophole would have to be 'very clever', and so he found the static experiments of the first generation to be 'quite interesting'.

In discussing these experiments, he stressed the very low detection rate of the photons and the geometric inefficiency and also explained the quite drastic auxiliary assumptions that had to be made. Still, as he said, he recognized that even such experiments must be useful, and he merely noted that different experiments came to dramatically different conclusions.

In the second paper, he ran through all the different types of experiment that had been performed—proton–proton scattering and positronium annihilation, as well as the cascade experiments—and again pointed out the low efficiencies of these experiments and the contradictory results obtained. Again, he stressed the centrality, in his view, of the locality loophole.

However, at the beginning of the 1980s, he made some rather decisive comments about the static experiments performed up to that point [19, 20]. While still recognizing that the experiments were far from ideal, as mentioned before, he had to admit that it was unlikely that quantum theory would work so well for inefficient set-ups but fail badly when refinements could be made. Also it had been pointed out [16] that quantum theory predicted *more* correlations between the results in the two wings of the apparatus rather than *fewer*; so, since it seemed likely that experimental errors would lower rather than increase correlation, it was highly likely that imperfections would lead to local realism, rather than quantum theory, being obeyed in error.

He then stressed that the lack of the time factor (the locality loophole) was much more important. In the experiments so far, although it might be difficult to believe in a long range influence between detectors, at least it would not need to travel faster than the speed of light. Because of this he felt that it was of 'capital importance' that Alain Aspect (Figure 4.6) was working on an experiment which aimed to introduce the time factor.

Bell had also referred to Aspect's ideas in his paper from 1976 [18]. Aspect [21] had been born in the southwest of France in 1947 and, after completing a first doctorate in physics, he spent his national service teaching in Cameroon from 1971 to 1974 and thinking about quantum theory. While his previous study had emphasized the mathematics of the theory, he now became absorbed in the physics. In particular, he read the EPR paper and felt that it was not only correct but extremely important. Then, after returning from Africa, he read Bell's paper on local realism; he later admitted that it was 'love at first sight'. He found the argument not only intellectually brilliant but also fascinating emotionally.

He longed to be able to contribute towards experimental checking of Bell's ideas. Finding out that the first-generation experimental work had been performed, and since he was not particularly interested in the detector loophole, he determined to perform a second-generation experiment, in which the polarizer settings were only fixed during the time of travel of the photons. He

Figure 4.6 Alain Aspect. © Renate Bertlmann.

wanted to perform this experiment as the so-called *these d'état* or 'big doctor-ate', which could at the time be a very long task, and he was able to persuade Christian Imbert, a young professor at the *Institut d'Optique* Graduate School at the Université Paris-Sud at Orsay, to act as his adviser; however, Imbert requested him first to travel to see Bell to discuss the experiment.

So, in early 1975, Aspect travelled to CERN and outlined his ideas to Bell. Bell of course, could not have been more delighted that this experiment might be performed; he longed to know the results. Nevertheless his first words were of caution. He was well aware of the 'stigma'—even if he was successful in such a difficult task, Aspect might well receive criticism, even ridicule, rather than acclaim for his labours. Bell asked Aspect: 'Have you a permanent posi-tion?' Aspect was able to assure him that, in the French system, he was indeed technically a permanent employee.

With that problem over, Bell was able to encourage Aspect greatly, and Aspect commenced work. However, he was well aware that it would be a long task. He published a short note [22] in 1975, and then a full experimental pro-posal [23] in the following year. However, obtaining funding and constructing the apparatus were both difficult processes, and Aspect would not conclude his work until the 1980s.

While discussing Clauser's work in the 1970s, it is of interest to discuss a recently published interesting and readable book written by David Kaiser [24, 25] entitled *How the Hippies Saved Physics*. The book reports how, discouraged by decreasing job opportunities caused by government cuts in 1975, a number of past or present graduate students at the Lawrence Berkeley Laboratory formed the so-called Fundamental Fysiks Group (FFG). In rebellion against the rather soulless and even mindless 'shut-up-and-calculate' model of American phys-ics at that point, they returned to an earlier model of physics; in this model, thought about fundamentals and conceptual analysis reigned supreme.

Naturally, they became very much interested in Bell's work, and indeed the more general conceptual issues of quantum theory, in particular, uncertainty, or indeterminism. The rather staid ideas of Bell's non-locality and Wigner's emphasis on consciousness were extended to studies of clairvoyance, extra-sensory perception, psychokinesis (of the Uri Geller spoon-bending type), Transcendental Meditation, mysticism, seances, and Tarot cards, and added in was a deep interest in Eastern religions. And everything was washed down with lashings of LSD!

A typical evening was the attempt to interact with the famous illusionist, Harry Houdini, on the centenary of his birth. One of the group, Nick Herbert, had invented a 'metaphase typewriter', the intention of which was to use quan-tum uncertainty to tap into human consciousness. Unfortunately, Houdini did not show up; however, as Kaiser said, 'after the inevitable paper jams, cele-bratory drinking and psychedelic drug use, a good time was had by all.'

Members of the group were undoubtedly talented, perhaps the most inventive (and interesting) being Herbert. He was actually working in industry during the heyday of the FFG, although, when he turned up for interview 'looking like an insane hippy', the management insisted that he be screened by a psychologist before being appointed. He was to write several best-selling books on quantum theory [26–28]; one of these included what is probably the easiest proof of Bell's theorem—essentially, the one that Bell described, as reported in Chapter 3, and, most interestingly of all, a claim that quantum theory allowed communication faster than light [29].

This claim certainly 'put the cat among the pigeons', since it violated one of the most sacred taboos of physics, as laid out by Einstein in the theory of relativity. Nevertheless, it was difficult to find an error in Herbert's analysis, and an enormous amount of toil and trouble was the result, spelled out in detail by Kaiser [24]. Actually, there was a problem with Herbert's ideas—he had assumed that a quantum state could be copied; the now-famous no-cloning theorem that was subsequently developed demonstrated that such copying was impossible [30–32]. However, this theorem, discovered because of Herbert's work, became a crucial component of quantum cryptography, as we shall see in Chapter 6; thus, Herbert's work had certainly borne fruit, albeit in a far from straightforward way.

Two other associates of the FFG produced books which were very much more sensational than Herbert's work, in particular concentrating on the claimed connections between quantum theory and the sensational ideas bubbling over in the group. These were Fritjof Capra's *The Tao of Physics* [33], which concentrated mainly on Eastern religions, and Gary Zukav's *The Dancing Wu Li Masters* [34], which discussed in particular such matters as human consciousness. Both were runaway bestsellers but rather irritated many of the conventional exponents of quantum theory.

Two of those associated with the group have become very well known. The first is Henry Stapp, who was even at the time a staff member of the laboratory and, while he was extremely interested in the arcane ideas floating around, and remained particularly concerned with the part played by consciousness in the collapse of the wave function [35], he always managed to keep his work broadly acceptable to the powers that be. He is famous for saying [36] that Bell's theorem was 'the most profound discovery of science'.

The other is Clauser himself. There is indeed no doubt that he enjoyed interacting with the members of the FFG. They were probably the only people around sharing an interest in Bell's work or even thinking that it was proper physics. He certainly appreciated their discussions, and later said: 'We definitely had a lot of fun. We asked some fundamental questions, and I think we got some reasonable answers' [24]. He probably influenced the other members of the group a great deal.

But this is not to say, as is sometimes suggested today, that his own important work stemmed from his participation in the group. The relative dates, of course, rule this out—his intense interest in the subject was developed and CHSH was written before his return to California, and years before the beginning of the group. Much as he enjoyed the company of group members, his warm-hearted comment on them was that they were 'a bunch of nuts really' [24, 37]. His serious work on quantum foundations was carried out with Shimony and Horne, who were both, one might say, with respect, East Coast types!

In his book, Kaiser asked why the work of the FFG has disappeared from history. The answer is presumably that those who were making most headway on these topics were serious physicists, who felt that it would be the kiss of death professionally to be identified with such activities as psychokinesis. This description would apply to such as Selleri and d'Espagnat, but particularly, of course, to Bell.

Bell was cautious enough about letting those at CERN know about his own quantum activities. He was always sensitive to the fact that in analysing and even questioning quantum theory, he might, ridiculous as it may seem, be seen as undermining the laboratory. Clauser [8] suggests that nobody at all at CERN knew about Bell's 'double life' until, with Antonino Zichichi and Bernard d'Espagnat, he organized an International 'Ettore Majorana' Conference in Erice titled 'Experimental Quantum Mechanics', in 1976.

He would certainly have been the last person who would want to be considered remotely in the same category as the FFG group. Clauser [8] again reports that, when he and Bell were chosen to be the first recipients of 'The Reality Foundation Prize' of $6000 in 1982, Bell contacted Clauser to check whether it was a 'quack' group that was offering the prize, in which case acceptance would have harmed his reputation as a legitimate scientist. Clauser reassured him that the money was being provided by one of the founders of Federal Express, Charles Brandon, who had a considerable interest in legitimate study of the foundations of quantum mechanics. Despite this reassurance, Bell did not attend the awards ceremony but instead asked d'Espagnat, who did attend, to accept the prize on his behalf.

It does seem [24] that Clauser had perhaps been slightly misled. Brandon had indeed provided the money, but it had been at the request of Nick Herbert and an FFG colleague, Michael Murphy, and the prize ceremony was very much an FFG affair. Herbert wrote to Bell after the event to congratulate him on the award, and he commented that the champagne toasts drunk in Bell's honour had made the participants 'merry' but not undignified.

In fact, Bell's attitude to those interested in applying his ideas was not necessarily negative. To Bernstein [38] he said that he found people who were attracted to a holistic philosophy of the world were often interested in coming to talk to him—and usually, he stressed, they were very nice people. They wished to build a bridge between, as they put it, science and wisdom, and were quite happy to accept that he did not himself see much connection.

He had mixed views on Zukav's book. On the one hand, he felt that by overstressing the importance of quantum foundations, and even of Bell's own work, the book completely misrepresented what the great majority of physicists were actually concerned about. On the other hand, it did bring some of the biggest questions in physics to a public audience, and overall he did not resent the success of Zukav's book. He could not quite say the same about Capra's book, because he felt that the connections Capra saw between physics and Eastern religion were just wrong.

Bell admitted that many of the founders of quantum theory, including Bohr, Pauli and Schrödinger, had had a deep interest in Eastern religion, and Bell himself had considered whether there was indeed any connection. But he approved of Schrödinger, who, in the preface of a brief book he had written on his beliefs, pointed out that these beliefs had nothing to do with quantum theory.

As Bernstein later said, Bell liked to make it quite clear that the non-locality he had demonstrated had no obvious link to such 'activities' as telepathy or psychokinesis. He pointed out that the laws of physics which demonstrate non-locality forbid it to behave in these ways. Yet, he would not seek to offend genuine seekers after knowledge. To an earnest inquiry about the possibility of quantum theory allowing for parapsychology, he remarked [38] that experience as an undergraduate in Ireland in attempting to carry out experiments on electrostatics had convinced him that electrostatics could never have been discovered in such a wet climate.

Perhaps, he added, parapsychological phenomena might be sensitive in an analogous way to as yet unknown features of the experimental background, and they perhaps might occur in certain circumstances. If so, he said, it would be the job of academic physics, which he judged to be in its infancy, and physiology to adapt to explain these new results, not to refuse to admit them. So his response was frank but conciliatory, although it is dubious that he would have behaved as generously to those attempting to communicate with Houdini!

Bell and quantum theory in the 1970s

In the 1970s, Bell wrote about a dozen papers in the area of quantum theory. However, this was by no means a planned programme of publication. Rather, with a single exception [39], they were papers given at conferences and thus usually appeared in unheralded proceedings, unlikely to attract attention from anybody who was not already quite interested.

The only exceptions were a few papers that ended up in the fringe journal *Epistemological Letters*. Clauser [8] described this journal as an '"underground" newspaper' which circulated to a 'quantum subculture'. It was published by the Institut de la Méthode of the Association Ferdinand Gonseth, which was named in honour of the Swiss philosopher, and the journal was practically unique in the 1970s in that it welcomed papers on hidden variables and analogous topics. The plan of the editor was announced as follows: ' "Epistemological Letters" are not a scientific journal in the ordinary sense. They want to create a basis for an open and informal journal allowing confrontation and ripening of ideas before publishing in some adequate journal.'

In this journal, Bell and some of the others interested in ideas related to his work, such as Clauser, Horne, and Shimony, could discuss their ideas and debate with each other, confident that only those who already thought the topic worthy of interest would be reading. This was made particularly easy because the unofficial editor of papers on hidden variables for the journal was none other than Shimony himself.

During the decade, the papers that Bell published were of different types. There were papers which aimed to improve the argument that he had made in the previous decade and make it more rigorous and complete [17, 40–42]. In this process, he entered a debate with Clauser, Horne, and Shimony, and we will discuss their ideas as well [15, 16, 43, 44]. It should be mentioned that these three very much worked as a group during this decade and, even when papers were published under two of their names only [15, 16], this was merely because one of the three had been busy and felt that they had not contributed enough to deserve authorship on that particular paper.

Bell also published some fairly general accounts of his ideas [18, 19, 45, 46]. When he was asked to give a conference paper, he clearly took the opportunity to present an overview of his point of view, always, of course, introducing some new idea or argument. He also published a few papers giving his views on two of the best-known interpretations of quantum theory competing with the Copenhagen interpretation: that of de Broglie and Bohm, and the many-worlds interpretation [46–48]. In addition, he published two papers addressing particular points in his argument [39, 49].

We may start by looking at the papers where Bell took his main ideas forward. The papers to be considered in this chapter are his Varenna talk [40], mentioned but not discussed in Chapter 3, and the paper [17] where he perhaps made his first sustained attempt to improve on his original presentation of his ideas in the 1960s. In this paper, he stated that he had profited from a paper (which we shall call CH) by Clauser and Horne [15], as well as from, of course, CHSH.

However, in describing the 'advances' of the theoretical ideas, it is necessary to warn the reader that there have been and still are very different views on exactly what these ideas implied [50–65]. Indeed, rather than a consensus being reached as time moved on, a recent special issue on Bell's work of *Journal of Physics A* seems to have emphasized the divisions in the community [55–61].

The editors of the issue [55] comment:

> From the papers it can be seen that there are still ongoing discussions about the assumptions and the implications of Bell's work. At first sight one may be surprised that fifty years later not all issues are settled, but in our opinion this only highlights the importance of Bell's work. Only groundbreaking results in physics lead to such long-lasting discussions, extending even into mathematics and philosophy.

Much of the discussion centres on the particular criterion required in Bell's theorem in order to rule out quantum theory: locality, local realism, local causality, or several other possibilities. Another part of it discusses Bell's various statements of the theorem, disagreeing over to what extent they were markedly different as distinct from gradually evolving.

In this book, there is no significant attempt to enter into this debate. Rather, we broadly allow those involved, including Bell, to describe their ideas and achievements in their own language, which for Bell may be 'locality' or 'local causality', for Clauser and Horne 'local realism', and so on.

One point may be of interest though. Bell [38] insisted that he regarded the Copenhagen interpretation as non-local, although he thought it had been carefully presented by Bohr and his supporters so that nobody would ask!

Bell's Varenna paper [40], written in 1970 although not published until 1971, was broadly along the lines of his 1964 paper, although the tone is slightly more relaxed. While the earlier paper had been written with extreme care for the sensibilities of editor, referees, and readers, in Varenna he knew that those in his audience were interested in and presumably broadly sympathetic to his project.

There were a few substantive improvements in the argument. The first was that he included the result of CHSH; so, rather than merely setting quantum theory and locality (or local causality) in opposition, the argument dealt directly with the truth or otherwise of the appropriate criterion, without any necessary reference to quantum theory.

Second, Shimony [66] reports that, as soon as he and Bell met at Varenna—it was the first time that they had met—Bell asked him whether he had been able to prove a particular generalization of Bell's theorem; this particular generalization did not assume that the hidden variables fixed the values of all the quantum mechanical observables in a deterministic fashion. Shimony stressed that Bell asked about this generalization because he did not want to take the credit for it if the lesser-known Shimony had also obtained this result;

Shimony regarded it as a typical example of Bell's consideration and generosity. As it happened, Shimony had to admit that, while he had thought about the generalization, he had not been able to prove it.

So, Bell went on to discuss this generalization in the Varenna paper. In this paper, the hidden variable dictates not each measurement but merely the probability of obtaining any particular result in any particular measurement; of course, however, the locality requirement is maintained. Bell remarked that, although the move to considering probabilities may have been related to the possibility of hidden variables in the measuring instruments themselves or even the result of a fundamental indeterminism, it did not affect the basic mathematics of Bell himself or of CHSH.

Then moving to some generalization of the analysis, he also considered the possibility that either or both instruments fail to register a signal, and he suggested a formal way of handling this difficulty.

At the end of the paper, he commented that his original suggestion of using the entangled spins of two electrons for an experimental check may not have been as promising as the use of two photons—he was aware, of course, of CHSH. He then discussed the demonstration by Day [67] that the two kaons produced in certain decays are entangled, just like the photons produced in the annihilation of positronium. However, he reported the belief of d'Espagnat that the short lifetime of the kaon would be a major obstacle for an experiment based on this effect.

He concluded the paper by remarking that, while, given the general success of quantum mechanics, it was only to be expected that the results of such a crucial experiment would be to confirm the quantum predictions, such a result would still be 'a severe discouragement'.

In 1975 Bell [17] attempted to make his discussion of the assumptions and implications of his theorem more rigorous. He stated that he was particularly conscious of having benefited from CH [15], in which the authors had thought deeply about the essential underlying assumption of Bell's theorem. Clauser [8] later said that both CHSH and Bell had initially assumed that it had to be determinism, an idea based on EPR, and perhaps particularly on the standard view that, for Einstein, the fundamental point was *always* determinism. Everybody knows that 'God does not play dice'!

Yet, as Clauser added, both Bell and CH realized that determinism was not the central issue and indeed that the theorem went through even in circumstances where determinism definitely did not hold; CH came to the conclusion that it should be replaced by an 'objective local theory', which was later to be rechristened a 'local realistic theory'.

CH also introduced what was, in effect, a new loophole. This may be called the *freedom of choice loophole*, and it is an admission that, in the analysis to that date, the experimenters in each wing of the experiment are assumed to have free will. Naturally, if each of the experimenters, rather than being able to choose freely at what angle the polarizers should be set, is somehow programmed to work with certain prearranged values, or cannot behave independently, then the values of these angles may be chosen so that the results of the experiment will agree with quantum theory, or indeed any other theory. The case where the angles are *precisely* stipulated may be called *superdeterminism*.

In later experiments, decisions on angles in the two wings of the experiment are made not by humans but by some mechanism such as a random number generator; in these experiments, closing the freedom of choice loophole requires that the two mechanisms used must be independent. In Chapter 6, we shall meet experiments aimed to ensure that the wings of the experiment are sufficiently far apart that the mechanisms must be independent, although again the absence of some form of superdeterminism must just be an assumption (although it is perhaps just an assumption that is necessary to carry out science in the first place).

Bell's paper [17] was an attempt to be more explicit and general about locality, and it also invented the idea of the *beable*. The theory of beables, he said, was a theory which hardly exists but *ought* to exist! Bell remarked that Bohr had stressed that the results of quantum theory must be expressed in classical terms, but these classical terms are not in the actual quantum equations but 'relegated to the surrounding talk'.

Bell, of course, contrasted the word 'beable' to the word 'observable'. Observables, he said, have very precise mathematical properties in quantum theory; however, physically, it is very difficult to decide which processes are to be regarded as observations and which are to be 'relegated to the limbo between one observation and another'.

Precision might be increased, Bell suggested, by concentrating on beables, which may be described in classical terms; examples are such things as settings for switches and knobs on experimental equipment, and instrument readings. 'Observables' must be constructed out of beables, and the theory of local beables should contain, and indeed provide a precise meaning for, the algebra of local observables. Local beables are defined as those beables that may be assigned to some particular region of space-time.

To introduce the idea of space-time and also the important idea of the backward light cone, it is clear that one's present state may have been influenced by anything that happened one second ago, provided it happened within a distance from oneself less than or equal to the distance light can travel in one second. However if the event happened two seconds ago, the distance may be twice as long, and so on. The regions may be shown as a backward light cone, as in Figure 4.7, which shows space along one coordinate and time along the other, that is, a region of space-time.

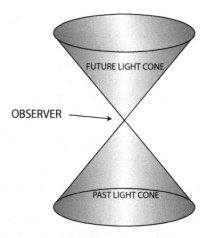

Figure 4.7 Past and future light cones. Any event in the past light cone may directly affect the observer, while the observer may affect anything in the future light cone.

Bell went on to discuss local determinism, which he said was easy to define: a theory is locally deterministic if the state of a particular small region is fully determined by the state in the backward light cone of that region.

However, he wished to work with *local causality* rather than local determinism, with local causality defined as follows. A beable in a particular region *A* may not have an exact value but instead a value given by a probability distribution related to other beables. We also consider a second region *B* completely separated from the first region. Then we may express the probability distribution of the particular beable in region *A* in terms of a complete set of beables in the combined backward light cone of *A* and *B* and in the remainder of its own backward light cone. Then if local causality is obeyed, this probability distribution will not change if we take account also of events in the backward light cone of *B*.

With that definition, Bell was easily able to show that quantum mechanics is not locally causal. Suppose a single radioactive nucleus emits a single alpha particle, and imagine that we are waiting at a particular counter in region *A* to see if it will register a signal. However, if another detector in a separate region *B* does register, then we know, of course, that 'our' detector in region *A* will not do so. The probability of it registering a signal *is* affected by an event in region *B*.

So quantum mechanics as it stands is *not* locally causal; however, Bell said, perhaps it is a fragment of a more complete theory in which there are extra beables, that is, *hidden beables*. Then Bell produced a development of his previous argument of 1964. The argument was rather complicated, involving beables in the backward light cones of A and B, as well as the common light cone, but it ended up with the CHSH inequality.

Bell then went on to develop a nice example of his by now standard demonstration that there are systems in quantum mechanics—in this case, the polarizations of entangled photons—that do not obey this inequality. So, he concluded, quantum mechanics is *not* embedded in a locally causal theory.

Bell then asked, if we are obliged to accept this 'gross non-locality of nature', does this mean that *we* can signal faster than light, in clear violation of relativity? Chapter 3 mentioned the ideas of Jarrett and Shimony, and the distinction between outcome dependence and parameter dependence, that is, that parameter dependence allows signalling whereas only outcome dependence is violated by Bell's type of non-locality.

It should be said that Bell himself was always suspicious of this distinction. However, in the paper being discussed, having queried the distinction between 'us', who can send *and* receive signals, and other beables that can only receive, he tentatively went along with it and was able to show that, assuming this distinction, faster-than-light signalling is not possible. 'In this *human* sense', he says 'relativistic mechanics *is* locally causal' (with his italics in both cases).

At the end of the paper, he did question whether instrument settings should be regarded as free variables, set at the whim of experimenters, as he had assumed, or whether they might be determined in the overlap of the backward light cones of the two events. Essentially, he raised the question of the freedom of choice loophole. However, he answered his own question by saying that, without this freedom, he would not have known how to formulate *any* idea of local causality, 'even the modest human one'.

Bell concluded the paper by acknowledging many discussions of the whole subject with d'Espagnat.

Shimony, Horne, and Clauser disagreed with some aspects of Bell's paper and subsequently in 1976 published a paper [43] in which, in Bell's words, they 'severely criticize[d]' his own paper. (Clauser [8] remarked that Bell's paper used a line of reasoning that had already been rejected in the writing of CH.) Bell replied in 1977 [42], and Shimony [44] completed the debate with a short paper in 1978. All these papers, including Bell's beable paper, were first published in *Epistemological Letters* but, because of their general interest, were reprinted in the mainline philosophical journal *Dialectica* in 1985, with a little editorial comment by Shimony.

In the criticism of the beables paper, there were a few technical points, with which Bell speedily agreed, about the proof; however, the criticism mostly concerned the instrument settings and the experimenters—essentially pointing out the freedom of choice loophole. Bell replied that it was indeed his premise that such settings are in no sense a record of what has gone before. His critics had strongly attacked his assumption that the experimenters have free will, which they called 'a metaphysics which has not been proved and which may well be false'. Bell replied: 'Disgrace, indeed, to be caught in a metaphysical position', although he continued that, in fact, he felt he was just pursuing his profession of theoretical physics, a subject which allows 'free' external variables, allowing a point of leverage for 'free willed experimenters', as well as internal variables conditioned by the theory.

However, he agreed with his critics about his remark about the instrument settings being 'at least not determined in the overlap of the backward light cones'. He said that he recognized that this hypothesis would be quite inadequate for proving the theorem. As he added, he had intended to exploit the freedom of initial conditions, using a standard technique in theoretical physics; however, in the context of his theorem, the initial conditions would have had to have been the creation of the world.

In Shimony's 1978 paper [44], he noted that the positions of Bell and CHS had converged to zero but still remarked that 'the latter can stake claim to have articulated the common position with greater clarity'.

At the end of the decade, Clauser and Shimony [16] published an extensive and comprehensive review article on the whole subject; the conclusions of the work, they claimed, are 'philosophically startling'. Although they still had concerns about the detector loophole, they had far fewer about the locality loophole.

The last paper in this group is a very short note [41], where the editor of *Epistemological Letters* (yet again) had asked Bell to reply to a critical comment of Georges Lochak [68], and Bell took the opportunity to also reply to a criticism made by Luis de la Peña, Ana María Cetto, and Thomas A. Brody [69], and another by Louis de Broglie [70]. (These criticisms had been published in other journals.)

In this note, Bell first tackled de Broglie's criticism, which Bell said, despite the title of de Broglie's paper, did not relate to Bell's theorem at all, but rather to what would normally be regarded as a rather standard equation in quantum theory, an equations which relates to correlations between polarizations of separated spins. As Bell said, it is not clear why de Broglie should have questioned this equation.

Bell said that Lochak had given an interesting account of how the output of a single measuring instrument depends on its setting; but Bell suggested that it would be necessary to move on to two instruments and two settings, in which case the dependence would be non-local. Incidentally, Lochak was a very close disciple of de Broglie, and the presence of both of them in the list of critics is interesting; one wonders whether it might have been a sign of the Paris school becoming interested in Bell or, alternatively, of having already decided that his ideas were mistaken.

Lastly, Bell suggested that de la Peña and his colleagues, who were from Mexico, not Paris, misunderstand the notation in his paper. He pointed out that the measurements involved were not sequential but alternatives.

We now turn to Bell's more general accounts of his ideas. The first [45] was presented in a major international conference held in 1972 to celebrate Dirac's seventieth birthday. The conference included practically every area of modern physics, and the invited speakers included most of its most famous names.

In Bell's short but quite technical paper, titled 'Subject and Object' (actually not his own choice but that of the conference organizers), he began by remarking that the subject–object distinction is at the root of the unease that many people feel about quantum theory. The theory, we are told, is about the results of measurements, but this statement implies that, in addition to the 'system' or object, there is a 'measurer' or subject.

Bell then raised a number of questions. Must this subject include a person? Was there a subject–object distinction in the universe before the beginning of life? At that stage, were there already processes to be regarded as 'measurements' and thus to be represented mathematically by jumps rather than by the Schrödinger equation? He recognized that the pioneers of quantum mechanics were aware of these difficulties and he believed they were quite justified in not waiting for answers. Rather, they developed the theory, which provided marvellous accuracy, and the subject–object division could be disposed of in a manner which was *almost* unambiguous and *almost* self-consistent.

However, Bell speculated that the next development in theoretical physics, while not being complete and final, might at least be unambiguous and self-consistent; it should not be about 'observables' but about 'beables', Bell's first published mention of this term. Beables, he said, must provide an image of 'the everyday classical world'. Observables would be made out of beables, and Bell asked whether some of the observables could be promoted to the status of beables.

He then considered a theory in which the state of a limited space-time region did not change, this state actually playing the role of a deterministic hidden variable. In this theory, 'measurement' of an observable is replaced by 'attribution' of a particular value to a beable; that is, a dynamical intervention from outside the region is replaced by a purely conceptual intervention by a theorist who is remote from the region and who has merely shifted his attention from a wide ensemble of systems to a narrow one.

Bell found it interesting that what he called this minimal programme for restoring objectivity had the result of restoring determinism as well. However, on the whole, he was pessimistic about this scheme being put into operation successfully.

The next paper to be considered [18] was called 'Einstein-Podolsky-Rosen Experiments' and was presented at a 1976 symposium in honour of Gilberto

Bernardini. Bell had been asked to speak on the foundations of quantum mechanics to a 'captive audience' of high energy physicists. How, he asked rhetorically, could he hold their attention with 'philosophy'? He said that he would try to do so by talking about an area where some 'courageous experimenters' had recently been putting such philosophy to experimental test.

He started off by explaining the EPR experiment and suggesting that 'some people' feel that it is reasonable to explain it in terms of predetermined results, but says there are three types of response, summed up as: 'Why worry?' or 'Is this not just like classical physics?' or 'Is it true?'.

He attributed the response 'Why worry?' to what he calls the orthodox approach of Bohr. This approach says that asking such awkward questions is exactly what physicists learned to avoid in order that quantum mechanics could actually be constructed. We should not try to argue beyond the experimental results. We have to learn to consider an experimental arrangement as a whole, not to break it down into separate pieces.

He responded to the question 'Is this not just like classical physics?' by arguing mathematically to the following point, which he also subsequently made to Bernstein (as reported in Chapter 3). Both local theories (as Bell called them) and quantum mechanics of course predict that, when the two polarizers in a Bell-type experiment are aligned, all pairs of polarizations would be anticorrelated but that, when there is a slight move away from alignment, the behaviour of the two theories would be very different. As shown in Figure 4.8, quantum mechanics shows the probability of correlation moving through a gentle minimum, while locality shows a sharp kink!

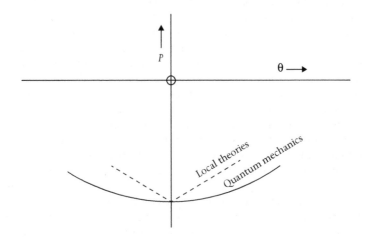

Figure 4.8 Diagram showing the difference between local theories and quantum mechanics.

In answer to the third question, he described the experimental position, as already discussed in this chapter, but finished the paper by defending himself from Jammer [71], who, in an extremely important book published in 1974, had criticized Bell for describing Einstein as an advocate of hidden variables. In reply Bell pointed to many places where he claimed that Einstein clearly did support such ideas.

The next paper to be discussed [19] was an invited talk at the Conference of the European Group for Atomic Spectroscopy, held at Orsay-Paris in 1979. There he presented a rather whimsical scenario in which, in order to check which of the two main French television channels could be responsible for the decline of the birth rate in France, one channel would be shown in Lille, and the other in Lyon. To come to any conclusion, one would have to recognize that factors such as fine weather, which would, to a large extent, occur in both cities on the same days, might cut down both television watching and conception, and so such variables would have to be held fixed. When that is done, it would be surprising, Bell said, if the choice of channel in Lille were found to be a causal factor in the birth rate in Lyon, and vice versa. Yet, he said, that is precisely the situation in quantum mechanics and, indeed, the long-range influence travels faster than light.

He then provided a fairly detailed account of an atomic cascade experiment to test quantum mechanics against the demands of the CHSH inequality. In particular, with the Aspect experiment in the planning stage (although still many years from completion), he insisted that the operations in either wing of the experiment should be arranged to occur sufficiently speedily that information could not pass from one wing to the other.

Having explained how important he thought the Aspect experiment would be, he was still prepared to ask, what possible responses could there be if the results from the Aspect experiment appeared to support quantum mechanics? First, one could mention the inefficiencies of the detectors—the detector loophole. Perhaps if the efficiencies were improved, quantum mechanics would fail and the CHSH inequality would be obeyed, although we have already noted that Bell was very dubious of this possibility being correct.

Second, we could admit that there are influences between the two wings of the apparatus, even though we cannot use these influences ourselves for sending messages, and that these influences travel faster than light.

Third, it may be the case that the two 'independent' settings in either wing of the apparatus are not as independent as one would wish. Whether chosen by independent radioactive devices, separate national lottery machines, or different apparently free-willed physicists, these settings may actually be correlated. In this case, locality survives, but it means that apparently separate parts of the world are actually deeply entangled, and our own free will is entangled with them. This possibility is an elaboration of the freedom of choice loophole.

The fourth attitude, Bell said, would be just to ignore the whole analysis. The message of quantum mechanics, it may be said, presumably by Bohr and his followers, is not to look beyond the actual quantum mechanical predictions. As Bell said, the authorities in Lille and Lyon in the analogous circumstances might just shrug and say: 'Well, that's people.'

The last paper in this group [46] was delivered in 1980 to an Oxford symposium on quantum gravity. It did not deal with Bell's own work but instead concentrated on the tricky question of measurement. Indeed, the second part of the paper dealt with both the de Broglie–Bohm interpretation and the many-worlds interpretations and so will be discussed with other papers with similar content. However, the first part of the paper is extremely interesting.

Bell began with familiar remarks about 'measurement'—that it is surprising to find it appearing in theory *at the most fundamental level*. He asked: did the world have to wait for a single-celled living creature to appear, or to wait even longer for a fully qualified physicist? Are measurement-like processes actually going on all the time?

He then briefly discussed the possibility of having the 'jump' moved from outside the governing Schrödinger equation to an extra term in the governing equation—'a dynamic process in dynamically defined conditions'. This extra term would make the equation non-linear, and here Bell mentioned Wigner's idea that this term relates to consciousness; however, Bell clearly preferred the idea that the term must be insignificant for objects of atomic size, where quantum theory is already extremely successful, but would supress the superposition of macroscopically different states. Bell felt that such superposition would be unphysical, and its suppression would remove the need for the collapse process that he felt was unjustifiable in terms of the standard quantum theory.

He went on to mention a number of those who had attempted to create such theories, but, even though he appreciated the approach of Philip Pearle [72], he had to admit that none had done exactly what he hoped for. In the 1980s he was to be excited by the so-called GRW theory, which will be explained in Chapter 5.

He then discussed some features of quantum theory, starting from what he called 'common ground', and analysed the formation of an alpha particle track in a set of photographic plates. Such tracks are usually called Wilson cloud chamber tracks, after C. T. R. Wilson, whose work on such tracks in cloud chambers in the early twentieth century had been crucial in the development of atomic physics.

The tracks had the form of practically straight lines, with minor kinks where the particle 'collided' or interacted with an atom in the cloud chamber or the photographic plate. With the coming of quantum theory, this was felt by some to cause problems because, if an alpha particle was 'measured' or halted in this way, it was thought that it should, by the Heisenberg principle, have a completely uncertain direction after the measurement and particles would move off at all angles.

However, Mott [73] and then Heisenberg [74] himself cleared up the problem. They showed that the localization is not complete but is limited to a relatively small region, and a simple calculation, using the Heisenberg principle, shows that the direction of travel after the 'collision' will differ only slightly from that before it. If the quantum mechanical measurement procedure is applied at this stage, the 'collapse' of the wave function is to one track among many, all broadly similar, and none differing greatly from the original track.

Such, Bell said, is a measurement—a crude one, but in fact a good model of all applications of quantum mechanics. A more rigorous treatment than this would include the atoms of the photographic plates as well as the alpha particle in the quantum mechanical system. Bell quoted Heisenberg as saying that this procedure is more complicated than the first, but has the advantage that the discontinuous change in probability (or collapse) recedes one step and so seems less in conflict with our intuitive ideas.

Bell then presented the theory of this treatment, the results of which agree at least broadly with those of the first treatment, but made the point that it is by no means final. The first treatment assumed that photographic plates could carry out measurements on alpha particles, the second assumed that we have equipment that can observe atomic excitation in the plates, and we can move on to third and further treatments in which more and more of the system is included in the quantum mechanical 'system'.

In this way, the *Heisenberg split* between what is treated as quantum and what is treated as classical may be moved so as to include more and more in the quantum system, until everyday experience tells us that to include more would make no conceivable further difference to the analysis. Thus we have a satisfactory recipe (or 'common ground') to ensure that the conventional quantum description provides an image of the familiar everyday world. However, Bell concluded this section of his paper by saying that to ask whether such a recipe, however adequate in practice, is a satisfactory general formulation of physical theory would be to leave the 'common ground'!

The problem, Bell said, is that all we have is a recipe that is sufficiently unambiguous for practical purposes only because of practical human limitations. He quoted Henry Stapp [75, 76] asking: 'How can a theory which is *fundamentally* [Stapp's italics] a procedure by which gross macroscopic creatures, such as human beings, calculate predicted probabilities of what they will observe under macroscopically specified conditions ever be claimed to be a complete description of physical reality?' Bell also quoted Rosenfeld [77], who said: 'The human observer, whom we have been at pains to keep out of the picture, seems irresistibly to intrude into it, since after all the macroscopic character of the measuring apparatus is imposed by the macroscopic structure of the sense organs and the brain.'

Bell remarked that these two authors actually felt that the situation was acceptable: Stapp because of pragmatism and because the concept of 'real' truth is generally a mirage, and Rosenfeld because theoretical physics must necessarily be cut to the measure of theoretical physicists. However, Bell replied that: 'In my opinion, these views are too complacent.'

He pointed out that there have been cases in the history of science when the idea of 'real' truth as distinct from a truth that is 'presently good enough for us' has been raised, his example being Copernicus placing the sun rather than the earth at the centre of the solar system. Bell said he could imagine a future where the world becomes more intelligible to human beings, even to theoretical physicists, when they stop imagining that they are at its centre.

He stressed that things are definitely different from how they were in classical theory. Classical theory is not *intrinsically* an approximation, and it does not require an outside observer, the necessity for whom in quantum theory makes the universe as a whole 'an embarrassing concept'. To one who, 'unduly influenced by positivistic philosophy', argues that, even in in classical mechanics, the human observer is implicit, because anything interesting must be experienced, he would reply that, in principle, a human observer causes no problem in classical mechanics, since the observer may be included schematically in the system by a postulation of *psycho-physical parallelism*, or in other words, by correlating the experience of the observer with some functions of the coordinates. Such is not possible in quantum theory, where the observer is resolutely *outside* the system.

Before moving on to his accounts of the de Broglie–Bohm theory and the many-worlds theory, which we shall examine shortly, Bell suggested two principles that should be respected in the formulation of such interpretations of quantum theory. The first is that it must be possible to formulate them for small systems, for, if this is not possible, it is highly likely that some sort of 'laws of large numbers' is being utilized at a fundamental level, and the theory is fundamentally approximate.

The second is that such concepts as 'measurement' or 'observation' or 'experiment' should not appear at a fundamental level. Of course, the theory should allow for 'particular physical set-ups'; these are not very well defined as a class, having a special relationship to not very well defined subsystems, that is, experimenters. But Bell felt that these concepts were too vague to appear at the heart of a potentially exact theory.

We now move on to those papers discussing the two main alternatives to the Copenhagen interpretation of quantum theory during Bell's lifetime: the de Broglie–Bohm interpretation, and the many-worlds interpretation.

Nowadays, there are, of course, far more, as we shall see in Chapter 6, principally because of the conceptual liberation caused by Bell's own work.

It is well known that Bell was a great supporter of the de Broglie–Bohm interpretation, which had played a great part in the development of his own ideas. Since he considered the Copenhagen interpretation to be non-local, the non-locality of the de Broglie–Bohm interpretation was not a particular discouragement to him. We need say no more at this point.

However, we must give a brief fairly general account of the range of rather different ideas often collected together under the title *many worlds*. Let us start from the idea that the idea of collapse at measurement was to reduce the wave function from a sum of terms corresponding to different values of the quantity being measured to a single term corresponding to the answer actually obtained. This idea of collapse has caused problems right the way through this book!

It was Hugh Everett [78] (Figure 4.9) who took the bold step of suggesting that the collapse should not be applied, solving these problems, but only, it might seem, by failing totally to solve the initial problem that collapse was designed to clear up! Squires [79] felt that Everett did not so much invent a new interpretation of quantum theory, as ask for the formalism without collapse to be taken seriously, leaving it as a formalism that still required interpretation. Everett called his ideas the *relative states* approach to quantum theory, as, with observed and observing subsystems, he would choose a state for one and be left with a relative state for the other.

In the late 1960s and the early 1970s, it was de Witt [80] who provided a clear interpretation for Everett's formalism. Each branch of the wave function, together with its corresponding measurement value, *does* exist after the measurement—each in its own world, following a *splitting* of worlds. As such a splitting occurs at each measurement, a tree-like structure emerges. This interpretation may certainly seem bizarre, but it definitely has one major advantage—no observer is required and no classical region, so many cosmologists, including Stephen Hawking, feel quite comfortable supporting the many-worlds interpretation.

Now we turn to John Bell's remarks on these two interpretations, and we shall indeed discover that he found more common ground between the de Broglie–Bohm interpretation and the many-worlds interpretation than most other people.

However, we look first at a 1980 paper [48] which was presented at a quantum chemistry symposium and which discussed only the de Broglie–Bohm interpretation and how the interpretation copes with the so-called *delayed-choice double-slit experiment*. In this intriguing thought experiment, which had been invented by Wheeler [81] (Figure 4.10), a single particle falls on a screen with two slits. Travelling from either of these slits, the associated wave train

Figure 4.9 Hugh Everett. Courtesy Peter Byrne.

is focussed by off-centred lenses and eventually falls onto one of two particle counters. Thus, we may say that the particle has passed through one of the two slits.

However, after the particle has passed the screen, we may insert a photographic plate into the region where the two wave trains intersect. In this case, a single spot will be formed on the plate; if we repeat the experiment many times, an interference pattern will be built up, which is usually said to imply the particle has, at least in some sense, passed through both slits. However, since the plate may be inserted *after* the particle has passed the screen, it seems that we choose at this time whether, at a *previous* time, it went through one slit or two. Incidentally, a favourite saying of Wheeler, who was a supporter of the Copenhagen interpretation, was: 'No phenomenon is a phenomenon until it is an observed phenomenon.'

Figure 4.10 John Wheeler. From Denis Brian, *Genius Talk* (Plenum, New York, 1995).

In the paper, Bell discussed how the de Broglie–Bohm interpretation handled this experiment. Rather than wave *or* particle, we have wave *and* particle. The wave, though, does not in any sense represent a wave of probability for an individual particle. Rather, it relates to a probability density *of particles*. As Bell stressed, the only use of probability in the de Broglie–Bohm interpretation is to take account of variation in initial conditions.

Thus, the wave goes through both slits 'as is the nature of waves', and the particle through just one 'as is the nature of particles'; however, the wave guides the particle towards the places where the wave is large. If the photographic plate is present, the particle contributes a spot to the interference pattern; otherwise, it proceeds to one of the counters. It is even the case that the precise trajectory travelled by the particle has been evaluated using a computer.

There can be no doubt that Bell looked on this 'particularly nice example' as something of a triumph for the de Broglie–Bohm interpretation over the perhaps rather smug Copenhagen approach of Wheeler.

In his papers relating the de Broglie–Bohm interpretation to the many-worlds interpretation [46, 47], Bell admitted that, for a long time, he had found Everett's work completely obscure. Although this obscurity had been somewhat reduced by de Witt's presentations, Bell still felt that his own version of their ideas would probably be accepted by neither of them and indeed suspected that a simultaneous agreement with both might not be possible.

In particular, Bell felt that Everett and de Witt seemed to imagine the division of the wave function into branches in ways that were rather different from the way he imagined it. For de Witt, certain measuring instruments had definite values on different branches, while Bell's impression was that Everett's choice of state for each subsystem seemed arbitrary. 'It is when arbitrary mathematical possibilities are given equal status', he says, 'that it becomes obscure that any physical interpretation has either emerged from, or been imposed on, the mathematics.'

For Bell, Everett's theory is, or at least should be, nothing other than the pilot-wave theory of the de Broglie–Bohm interpretation, without the microscopic trajectories. In other words, macroscopic objects exist, and the values of the associated physical quantities are distributed in the same way that they are in standard quantum theory. But, for Bell, the significant point was that no pairing of values at different times was considered, and he stressed that no continuity between present and past values was required by experience.

Before Bell discussed that last point, he felt it necessary to 'have some mercy' on followers of Everett or de Witt, as they would certainly not have recognized their understanding of Everett's and de Witt's work in Bell's formulation.

First, concerning the many worlds themselves, Bell suggested that this multiplication of universes was extravagant and served no purpose, so it could be dropped without any repercussions. He himself did not see that anything useful was achieved by the assumed existence of other branches of which he was not aware. 'But let he who finds this assumption inspiring make it.'

Second, Bell admitted that the classical variables did not actually appear in de Witt's and Everett's writings but said that they did talk of instrument readings. Actually, though, he suggested that they only did so 'in order to be intelligible to specialists in quantum measurement theory'. He felt that, in principle, they wished to reach a more fundamental level of classical variable than they had succeeded in doing.

However, Bell's main argument concerning this class of theory is that he did not accept the tree-like structure of the past and, in fact, did not wish to associate a particular present with a particular past. Indeed, he felt that the lack of such an assumption did not matter at all, for we have no direct access of the past—we do have 'memories' and 'records', but these are actually present phenomena. In terms of records, the only requirement is that theory should account for the *present* correlations between *present phenomena*; in terms of memories, all we require is that a *present* memory of a correct experiment being performed is associated with a *present* memory of a correct result being obtained.

Bell said that Everett's replacement of the past by memories is a *radical solipsism*, extending to considerations of time the notion of ordinary solipsism (or positivism, Bell said), that is, that everything outside one's head can be replaced by impressions. Solipsism, Bell said, cannot be refuted; however, if such a theory were to be taken seriously, it would be difficult to take anything else seriously. He said that he personally wanted to take more seriously the past of the world (and of himself) than the theory would allow and that he had observed that, when solipsists and positivists have children, they do take out life insurance!

In a similarly ironic mood, he likened Everett's approach to that of creationists, who argue that, when the world was created in 4004 BC, it was a 'going concern': trees had annular rings, Adam and Eve were fully grown with fully grown teeth and hair and with navels (even though Adam and Eve had not been born), and rocks occurred in strata and contained fossils of creatures that had never lived.

Bell made a more technical criticism of the many-worlds interpretation in a conversation with Davies in 1985 [82]. He stated that the very bizarreness of these theories was already enough to make him dislike them, although he admitted that they might have something to say about EPR. However he pointed out that de Witt's approach in particular might seem to be, in some sense, a solution to the difficulties of the Copenhagen interpretation, *if* (but only if) these difficulties were assumed to be the collapse from many possibilities to a single one. *If*, on the other hand, the difficulties related to defining a precise time of 'measurement', as Bell himself would have argued, then the many-worlds interpretation would achieve nothing, for it required a precise time of creation of the many new worlds.

We now turn to two papers where Bell examined particular issues relating to his general ideas. The first was a response [39] to a paper which had been written by Klaus Hepp [83], which many people, although not Hepp himself, thought might provide a 'clean solution' to the measurement problem. However, whereas in his paper Hepp used the rather technical C^* algebra approach, Bell used elementary terms to analyse one of Hepp's models.

The model is as follows. The 'apparatus' is a linear array or lattice of spins, starting at $x = 0$, then at $x = 1, 2$, and so on up to infinity. The 'system' is a single moving spin which travels along the linear array; we may imagine two cases, in the first of which the moving spin is 'up', and in the second of which it is 'down'. The details of the model are arranged so that initially the lattice spins are all up and, when the system spin is up, nothing happens to the apparatus spin; however, when the system spin is down, each apparatus spin in turn is flipped from up to down. Thus in the first of the two cases, all the lattice spins will remain up, while in the second they will all in turn flip to down.

Hepp defined what he called 'macroscopically different pointer positions', which in the example are the two final states of the combined system, which cannot be connected mathematically by what he calls 'local observables'. This means that coherent superpositions of the two final states of the combined system cannot be distinguished from incoherent mixtures, and we may claim a 'rigorous reduction of the wave packet' or collapse.

Bell agreed that this situation would happen as t tended to infinity, but he insisted that infinite time would never come! At any finite time, the two states *could* be connected by local observables, although these observables might admittedly need to be more and more complicated as time moved on; thus, coherence would not be lost, and the claimed reduction would never happen.

Bell stressed that the fundamental point was that wave packet reduction or collapse was incompatible with the linear Schrödinger equation.

The second paper [49], entitled 'How to Teach Special Relativity', was included in both *Speakable and Unspeakable in Quantum Mechanics* and in the quantum mechanics section of QHA. At first sight, one might be surprised at its inclusion; however, for Bell, the non-locality arising from his theorem was very much connected with relativistic arguments. He was never particularly convinced by Shimony's 'peaceful coexistence' between relativity and quantum theory.

Bell's position [38] was that non-locality *as such* was something that one could learn to live with, but what would be much more difficult to accept would be the problems with causality. A standard problem in relativity is to show that, if two events are causally connected, that is, if event A causes event B, then *all* observers will register A as occurring before B. If, on the other hand, A and B are unconnected causally, then in some cases different observers may register the events occurring in different orders, but of course there is no conceptual problem. Technically, if light has time to travel from one event to the other, the events may be causally connected, and the order is fixed; however, if light cannot travel between the events, they are not causally connected, and the order is not fixed.

However, this argument assumes locality. Once one allows non-locality, there is no restriction and it seems that different observers may see causally connected events in different orders, which is certainly troublesome conceptually.

We now turn to Bell's paper. He began by remarking that he had always believed that students might gain a more rounded understanding of relativity if, rather than following Einstein's approach at the outset, they came to his ideas via the work of the turn-of-the-century physicists Joseph Larmor, Hendrik Lorentz, Henri Poincaré, and George Francis FitzGerald.

The problem these scientists faced relates to Maxwell's theory, which demonstrated the existence of electromagnetic waves, including, of course, light. This theory was certainly exceptionally important; however, the problem was that the theory as such only applied in a particular frame, the so-called rest frame of the universe. (It may be helpful to think of physics in a particular frame as being related to a coordinate system fixed in the same frame; thus, physics in the frame of the earth would be described in terms of a set of axes fixed relative to the earth, physics in a frame moving with respect to the earth would be described in terms of a set of axes moving with respect to the earth's axes, etc.)

It was assumed that, if light were a wave then, like any other wave that was known at the time, it would have to be a wave in a particular medium, and this hypothetical medium was called the *ether*. It was taken for granted that the ether was stationary in the rest frame of the universe, and that the speed of light was equal to c, the value calculated by Maxwell, only in this rest frame of the universe; in frames moving with respect to this rest frame, the speed of light would differ from c. To quantify this argument, it was necessary to use the *Galilean transformation* from one frame to another. The equations for this transformation were simple and indeed seemed obvious.

It was natural to inquire what the speed of the earth through this rest frame of the universe might be, and in 1887 Albert Michelson and Edward Morley carried out an experiment to find this out—the famous Michelson–Morley experiment. It was a particular difficult experiment as, unfortunately, one can never measure the one-way speed of light, from say P to Q, although this measurement would give the required information quite easily; rather, one can only measure the two-way speed, that is, the appropriate average from P to Q and then back to P; this experiment had to be carried out along two directions at right angles to each other. It had to be performed with great skill to obtain the required information, but Michelson and Morley should certainly have obtained the answer they were seeking. Yet, they came to the strange conclusion that they could not detect any motion of the earth through the ether.

Over the next 20 years or so, the physicists mentioned above came up with a series of manoeuvres—the FitzGerald–Lorentz contraction, the Lorentz time dilation, and so on, the job of which was to explain (or explain away) this failure to detect the motion of the earth through the ether. The result was a theory that said that there *was* a rest frame of the universe but that the laws of physics were such as to make it impossible to detect—physics in 'moving frames' was the same as physics in the rest frame of the universe. (Technically, we should add that, when we say 'moving', we should really say 'moving at constant speed'; accelerating frames are not considered. The rest frame and the frames moving with constant speed with respect to it are called *inertial frames*, that is, frames in which bodies display their *inertia* by accelerating only when a force is applied to them, as dictated by Newton's first law of motion.)

In 1905 Einstein swept all this away! The terms 'rest frame' and 'moving (with constant speed)' were eliminated, so all inertial frames became of equal status, and the ether was relegated to history. The seemingly obvious Galilean transformation was replaced by the far from obvious Lorentz transformation, the speed of light became a constant, being the same in all inertial frames, and Newton's laws were replaced by new laws of motion, sometimes called Einstein's laws of motion. In all this change, Maxwell's laws remained triumphantly unaltered!

Many, of course, were highly disturbed by this dramatic loss of cherished beliefs, but nobody who thought about the matter seriously, certainly not Bell, could doubt that, conceptually, Einstein's ideas represented an enormous step forward. Bell commented that Einstein's argument 'permits a very concise and elegant formulation of the theory, as often occurs when one big assumption can be made to cover several less big ones.'

However, he still thought it was good for students to be made familiar with the former approach. Thus, he began his article by mentioning a simple problem or puzzle: two spaceships B and C are side by side and gradually accelerate away from another spaceship A. In this case, would a thread suspended between B and C snap? Bell's argument was that, using pre-Einstein relativity, it is easy to see that the thread must 'FitzGerald contract' and so must snap. This answer, he suggested, would be much more complicated to reach if Einstein's approach were followed.

In terms of non-locality, Bell also felt that the original approach had much to offer. For, if there is one rest frame, there is one unique order in which events take place, say, A before B. In other frames, B may *appear* to occur before A, but it is indeed *merely* an *appearance*, just as a 6-foot man may *appear* to be shorter than a 3-foot child if he is much further away than the child. Thus the problems created by non-locality, whether parameter independence or outcome independence, cease to have a genuine existence.

Many years afterwards, Bell discussed these matters with Davies [82] and, in 1989, at a lecture in Trinity College, Dublin, to honour FitzGerald on the hundredth anniversary of the FitzGerald–Lorentz contraction, Bell paid fulsome tribute [84] to the two Irish contributors, FitzGerald and Larmor, to, as Bell would call it, the 'older form of relativity' (Figure 4.11).

Selleri [85] was especially sensitive about locality. Thus, at a meeting to mark the tenth anniversary of Bell's death, he presented a detailed investigation of Bell's ideas, including a study of how the one-way speed of light might be discussed.

(a)

(b)

Figure 4.11 (a) Joseph Larmor, painted by Frank McKelvey in 1945. © Queen's University Belfast. (b) George Francis FitzGerald. Courtesy Trinity College Dublin.

Bell and particle physics in the 1970s

During the 1970s, Bell was busy, of course, on his 'paid' work of nuclear and particle physics. Although none of his papers of the 1970s drew the amount of attention gained by his most important papers of the previous decades, in particular that on the ABJ anomaly, they were still extremely thoughtful and individual attempts at interesting problems, rather than the routine work of many others, to whom Bell usually left the development of his ideas.

His first major collaboration of the decade was with a young colleague, Christopher Llewellyn Smith, and it was in many ways a continuation of the 1959 optical model paper that Bell had written with Squires, as well as of his 1964 paper on the same topic. It centred around *shadow effects*, that is, the idea that, in particular, the magnitude of neutrino reactions on nuclei might be proportional to the surface area rather than to the volume of the nucleus; in other words, it would depend on $A^{2/3}$ rather than A itself, where A was the number of nucleons in the nucleus. This implied that the nucleus might have a *shadow* in the forward direction.

Bell first wrote a paper [86] entitled 'Weak Interactions in the Nuclear Shadow', which extended his 1964 work. Although he was mainly concerned with the neutrino case, he first developed the case of photons in some detail because it was simpler in practice; however, in both cases, he based his work on the role played by pions in the process. These pions would be *virtual*; in other words, although there would not be enough energy to produce them classically, quantum mechanically they might exist for a time short enough that the product of the energy and the time would not be greater than Planck's constant.

He concluded that, in the neutrino case, the $A^{2/3}$ rule might begin to set in at modest energies of a few hundred megavolts, and he argued that, because of the rather low pion mass, the region around the forward direction where the shadow effect would be expected might be substantial.

Then, in 1970 and 1971, Bell and Llewellyn Smith published two substantial papers [87, 88] developing these ideas further. The papers were, of course, theoretically based, but as often with Bell, they were strongly related to experimental work, and detailed estimates of different contributions to scattering and shadowing were made. The general conclusion of these papers was that the region over which some shadowing was predicted was even larger than would be predicted by the $A^{2/3}$ rule, and the amount of shadowing could be as large as 40% of the total effect.

Towards the end of the first paper, the authors made two interesting comments. On a technical note, they noted their frustration at the absence of monoenergetic neutrino beams, which would have enabled experiments that would have distinguished between very different types of effect. The second comment was to 'apologize' for the fact that the theory they had developed was 'naïve in conception and crude in execution' and that it could only be put forward in a 'theoretical vacuum'. Again, this comment makes the point that Bell, the senior collaborator, had the courage to enter an entirely new field.

In the second paper, Bell and Llewellyn Smith were again studying neutrino scattering by nuclei, but in this paper they concentrated on the influence of the shell structure of the nucleus, and the related Pauli exclusion principle. They felt that these must have an important part to play in the study of the nuclear shadow effect.

The exclusion principle had been discovered as early as 1925, and it says that each quantum state of an electron or a nucleon can only contain a single particle. The shell model had been at the heart of atomic theory from the beginning of quantum theory; in that model, as successive electrons were added to an atom, because of the exclusion principle they appeared in shells of steadily increasing energy; in the late 1940s, it had become clear that the same held, at least to some extent, for protons or neutrons being added to nuclei.

Again, the authors performed detailed calculations using some fairly simple shell models. They found that effects due to shell structure did appear but were not large; however, the effects were big enough to make the apparent experimental absence of any effect due to the exclusion principle something of a mystery.

Bell himself concluded this work with another paper [89] which summed up theoretical analysis of the possible shadow effect, and which went along with some preliminary experimental studies. His summary admitted that the work was still necessarily naïve in conception and crude in execution, but he claimed that it did demonstrate how the substantial shadow effect predicted by the process could be diminished by other processes such as quasi-elastic scattering. As he had said before, he looked forward to a time when the various phenomena could be isolated experimentally.

Bell and Llewellyn Smith worked together quite closely in the early- and mid-1970s, and in 1974 they published one other paper, together with Gabriel Karl [90]. This paper contained a brief but rather formal discussion of how many pions might be produced in the annihilation of an electron–positron or nucleon–antinucleon pair. The results could be compared with data produced at CERN, although the authors admitted that the theoretical bounds on numbers of pions they predicted were 'disappointingly weak'.

Llewellyn Smith was to become one of the United Kingdom's leading particle physicists, and he became a Fellow of the Royal Society in 1984 for his work on the theory of quarks, as well as for his work on electromagnetic and strong nuclear forces. However, he was to be most well known as one the most important international administrators of science. He was Director of CERN from 1994 to 1998, during which period he played a crucial role in obtaining approval for the Large Hadron Collider, and from 2003 to 2008 he was Director of the Culham Division of the UK Atomic Energy Authority.

In the latter capacity, he was in charge of the fusion programme and the operation of the Joint European Torus, the most recent European attempt to gain useful energy from nuclear fusion. From 2007 to 2009, he was Chairman of the Council of the International Thermonuclear Experimental Reactor, and for all his many achievements, he was knighted in 2001. He remains a very great admirer of Bell's work.

When gauge theories became popular in the early 1970s, Bell remained fairly aloof, but he did publish one important paper [91] in which he related the behaviour of the theory at high energies to the symmetry structure of the theory. He began by remarking that quantum field theories involving vector mesons might seem to exhibit unphysical behaviour at high energies, but that this behaviour was related to non-renormalizability, and imposition of gauge invariance caused at least partial cancellation of the awkward terms. He suggested that it would be natural to conjecture that this cancellation would be complete in spontaneously broken gauge theories which are fully renormalizable, and in the paper he demonstrated that this was the case. The argument is reasonably substantial and complicated, and it might be suggested that Bell, as he did so often, carried modesty to unnecessary lengths in his comment: 'It seems that nobody had bothered to do this already.'

Another important set of papers by Bell and his collaborators dealt with the spectrum and structure of low mass mesons and baryons [92]. It will be remembered that, while Bell was at Harwell, he had published a paper with Skyrme on the anomalous magnetic moments of protons and neutrons. In the 1960s, the mathematical group SU(6) was particularly in favour, to a large extent because Bram Pais and Mirza Bég had shown that it led to the ratio of neutron and proton magnetic moments being $-\frac{2}{3}$, which was in close agreement with experiment.

However, it soon became clear that the theory had significant faults; in particular it was at best restricted to systems that were non-relativistic and had zero momentum. (See Refs 114 and 115 in Chapter 3.) In that approximation, one cannot rigorously study magnetic moments at all, an observation which suggests that the Pais–Bég result was just a coincidence! However, Bell was not prepared to accept this conclusion.

One way forward seemed to be the so-called *Melosh transformation*. In his PhD thesis in 1973, H. Jay Melosh [93] suggested this transformation as a means of studying the breaking of the relativistic symmetry of SU(6) (although, technically, he and others, including Bell, were actually working with a related group, $SU(6)_W$). The transformation was between what Melosh called the 'constituent quark' and what he called the 'current quark'. The constituent quark is only approximately the subject of the symmetry group and, as a concept, is derived from the idea that baryons and mesons are composed of a few *constituent* quarks, while the current quark is the subject of the full mathematical algebra of operators and currents.

Bell became extremely interested in this idea and studied carefully several further developments of the basic transformation. (Melosh himself incidentally later became famous as a geophysicist and an expert in input cratering.)

In 1974 Bell presented a lecture at Schladming in Austria; this lecture was subsequently written up as an extensive analysis [94] of Melosh's ideas, and a comparison of them with other ideas dating from a quarter of a century earlier. Bell started by describing the tale of relativistic SU(6) as 'a long, and often sad, one' and pointed out that, although Melosh had successfully recovered the value of $-\frac{2}{3}$ for the ratio of magnetic moments, there had been some inconsistency in the argument used to obtain this result.

Later in the same year, Bell published a short paper with Tony Hey [95]. The aim of this paper was to show that a transformation which was at least something like the Melosh transformation was absolutely necessary—specifically, that identification of the current and constituent quarks at the very least excluded the strange result that certain meson systems would have zero magnetic moment.

Bell and Hey interacted a good deal in the middle of the 1970s, and Hey's subsequent career was extremely interesting, in some sense seeming to circle that of Bell. From 1974 Hey spent 30 years at the University of Southampton, at first working in particle physics but then moving to computer science in 1985 and, in particular, specializing in the application of parallel computation to large-scale scientific simulations. He was then Corporate Vice President at Microsoft Research Corporation for a decade from 2005.

He also wrote popular accounts of quantum theory and its applications [96], and, just as Bell's work was making a major contribution to the development of quantum information theory, Hey, together with Robin Allen, edited a volume of Feynman's early work [97] in developing quantum computation, as well as a second volume [98] in which Feynman's achievements are discussed by his collaborators at the time.

We now return to the 1970s and the Melosh transformation. In 1975 and 1976, Bell published two further papers [99, 100] on the topic with Henri Ruegg, a physicist from the University of Geneva and a protégé of Ernst Stueckelberg [101], who had made important contributions to quantum field theory. In these papers, the authors extended Melosh's arguments, which had been developed only for free particles, to the hydrogen atom [99] and positronium [100].

In the first paper, an approximation was made in which c, the speed of light, was assumed to be large, so small quantities could be neglected. A symmetry $SU(2)_W$ emerged, which was analogous to the $SU(6)_W$ we met before. Although $SU(2)_W$, like $SU(6)_W$, leads to zero magnetic moment, the symmetry is broken by small terms and, when a Melosh transformation is applied to eliminate these terms, the correct value of the hydrogen magnetic moment is produced. Although Melosh's ideas are obviously significant, their implications are not exactly clear. Bell and Ruegg pointed out that, in this case, the Melosh transformation was not actually applicable to constituent quarks but was applicable to what they called 'classification' quarks, which are related to classifying states within a symmetry group.

In their second paper [100], positronium-like systems, that is, systems containing a particle and its antiparticle, were studied in an analogous way. In this case, it was possible to express the results approximately in terms of a product of symmetry groups, $SU(2)_A \times SU(2)_B$, where A and B are the two particles in the system. As said, the agreement was not complete, but again it could be considerably improved by the application of a Melosh transformation.

During this work, Bell and Ruegg noticed that, as well as the approximate SU(2) symmetry already discussed, they found a family of states with *exact* SU(2) symmetry; they published this result separately [102]. The publication attracted a great deal of attention; however, unfortunately, it turned out that the central result had been published by G. B. Smith and L. S. Tassie four years before Bell and Ruegg had published their paper [103]. Consequently, Bell and Ruegg published an apology [104] for having missed Smith and Tassie's paper.

Another area of particle physics to which Bell made important contributions towards the end of the decade was the study of models of partons. It will be remembered from Chapter 3 that the quark model that Gell-Mann and Zweig had designed to explain symmetries in the physics of baryons and mesons was, after much discussion, coupled to the parton idea, which had principally been developed Feynman and which explained the scattering of leptons by nuclei, to produce the quark–parton synthesis. In the titles of his papers on this subject, Bell used the word 'parton', because it was that aspect that he was considering.

By 1978, the general quark–parton idea was, of course, well established, but there was still much uncertainty about the detailed dynamics of the partons, and this was what Bell hoped to elucidate. In the first paper, which was published with Hey [105], the authors took an extremely simple model which they called a 'colourless, flavourless one-dimensional box'—in other words, the basic potential was the simplest possible, and the quark–partons had neither colour nor flavour (up, down, strange, etc.), although Bell and Hey were able to adjust their results for a number of colours, and for up and down quarks.

As well as *valence partons*, they considered the distribution of so-called *sea partons*, which resulted from the excitation of quark–antiquark pairs from the 'vacuum' in the box. Their work was definitely an improvement on some previous attempts at the problem, as in these the probability distributions had, unfortunately, assumed negative values. They were able to retain Bjorken scaling, which was mentioned in Chapter 3, even while obtaining some terms which previous workers had felt obliged to drop. However, they still did not achieve conservation of momentum; this failure was obviously a drawback of the theory.

The second of these papers [106] was written with Anne-Christine Davies, now Professor of Mathematical Physics at Cambridge, and Johann Rafelski, now Professor of Physics at the University of Arizona at Tucson and still a regular visitor to CERN. In this paper, the model that had been used in the first paper was extended to three dimensions, in fact to a 'spherical box', and again special attention was paid to the sea partons. Unfortunately, in this case, the results were mathematically divergent, an observation which suggested that Bjorken scaling did not apply to this model, although the authors raised the possibility that this difficulty might be avoided by using a less violent confining potential than they had used.

To demonstrate the breadth of Bell's interests, here a brief discussion will be given of some of his other work during this decade.

In 1970, working with Royce Zia, who has been at Virginia Tech since 1976 and specializes in statistical mechanics far from equilibrium, Bell returned to his interest in CP violation [107]. The work actually stemmed from a visit which Bell made back to Ireland for a spell at the Institute for Advanced Studies at Dublin, and an encounter with Ciaran Ryan there. Ryan had already demonstrated that a formula which was related to CP violation and which had originally been developed by Lee and Yang, would require correction only if certain electromagnetic interactions themselves violated CP significantly. Bell and Zia investigated the problem in detail and showed that, if such violations were found to exist experimentally, the corrections required would provide extremely interesting information on the interactions themselves.

In 1973 Bell investigated the problem of hadrons moving through nuclear matter, where, if one worked non-relativistically, there occurred (in Bell's term) an 'experimentally embarrassing zero' in the forward amplitude. However a number of papers in the 1970s suggested that this zero could be removed by applying the Lorentz contraction of special relativity to the non-relativistic results.

Bell checked this idea [108] by working with a full relativistic treatment and found that the unwelcome zero was still present in the final answer. Clearly, the simple expedient of applying the demands of relativity to the non-relativistic result was a failure. At first sight, there might appear to be some contradiction with Bell's argument [49] that Lorentz–Poincaré relativity and Einsteinian relativity are equivalent theories and so should always agree. Clearly, though, the simple expedient of merely applying the contraction does not equate to carrying out a complete analysis using either the older or the newer approach.

In 1975 Bell wrote a short note with G. V. Dass, who was visiting CERN from Harwell. They recognized that there were relationships between the scattering of neutrinos and antineutrinos from electrons and partons, and indeed, previously some special cases had already been addressed by means of individual techniques. However, Bell and Dass [109] were able to provide a general treatment from simple considerations.

In the following year, Bell was presented with an interesting problem by Frank Krienen, who was an engineer and had been a pioneer at CERN: the challenge was to work out the basic formula for transition radiation for particles with spin. Transition radiation is the passage of charged particles through the boundary between different media, and Bell [110] was able to produce a formula for the mean energy radiated at a vacuum–dielectric interface per unit photon energy per unit sold angle.

In carrying out this calculation, he was also able to obtain a corresponding result for Čerenkov radiation, which is produced when a charged particle passes through a homogeneous dielectric medium at a speed greater than the phase velocity of electromagnetic radiation in the medium. For this case, he produced a value for the mean energy radiated per unit track length per unit photon energy.

Lastly from this group of papers, let us discuss a collaboration with Karl [111]. It will be remembered from Chapter 3 (in 'Bell, and particle physics in the 1960s') that, while it is well known that a stable particle cannot have an electric dipole moment, provided time reversibility is upheld and even if parity is violated, the usual argument does not apply immediately to an unstable particle. However Bell had shown that in fact even an unstable particle cannot have a dipole moment in these circumstances, and when the question re-emerged in the 1970s in connection with the search for parity violation in atomic physics, he regarded its re-emergence merely as an indication that the early literature had 'not been fully digested'.

Bell had shown that an electric field causes no first-order splitting of the energy levels (which are mathematically complex for an unstable particle), so the spin does not precess about the field as long as the particle has not decayed. Thus, Bell and Karl asserted, the dipole moment must be zero.

The counterargument, though, is that the expectation value of the operator for electric dipole moment is not zero, so the electric field does exert a torque (or moment) on the system, and some angular momentum is delivered to the system. But, as Bell and Karl stressed, this angular momentum is delivered *entirely* to the decay products, not to the unstable atom before it has decayed, and they demonstrated, first with a simple model and then with a general argument, that the angle of the mean angular momentum of the decay products was fixed with respect to that of the parent atoms.

In 1979 Bell, together with J. Pasupathy, published two very interesting papers concerning bound states in quantum chromodynamics. However,

Figure 4.12 Mary and John Bell in Vienna 1980. © Renate Bertlmann.

these papers represented the start of a research project which was to involve Reinhold Bertlmann, and which was to continue well into the next decade, so it is convenient here to delay the discussion of Bell's work with Pasupathy until Chapter 5.

The 1970s had been a good decade for Bell (Figure 4.12). In particle physics, he had followed becoming a Fellow of the Royal Society with confirming his position as a leading worker in the area, while in quantum physics, his ideas had been picked up and experimental checks had been made. He was keen, of course, that the Aspect work should come to fruition and must still have hoped that his important ideas would reach a greater audience than they had done so far. By the end of the 1970s, his Einstein–Podolsky–Rosen paper was receiving about 30 citations a year, a number that is healthy enough for an average paper, but it could not be claimed that his work was attracting the attention it undoubtedly deserved. As late as 1979, it took a popular paper by d'Espagnat [112] to ensure that Bell's work became generally known.

5

The 1980s

Final Achievements but Final Tragedies

Summary of the decade

Of course, it is completely impossible to discuss the 1980s in this book without emotional acknowledgement of the sad death of John Bell at the end of the decade. In this chapter, these tragic matters will be kept for the final section and we shall concentrate until then on happier topics and successful endeavours.

In fact, we shall start with CERN, and there is no doubt that this decade established the laboratory as, at the very least, the equal of any in the world. The prime achievement was the discovery in 1983 of the field particles or bosons of the weak interaction: the W^+, the W^-, and the neutral Z particle [1, 2].

Two important and novel aspects of accelerator physics were central to this work, the first being the great advantage of using collider methods over the use of a stationary target. This advantage of collider methods, in which two beams travelling in opposite directions are made to collide, had been realized by the important Norwegian accelerator physicist Rolf Wideröe as early as 1943 [3]. In the especially attractive case where the masses and the speeds of the particles in each beam are the same, then the colliding particles have momenta equal in magnitude but opposite in direction, so the final momentum, like the total initial momentum, is zero, and none of the enormous effort and expense of increasing the energy of the accelerated particles is wasted in useless kinetic energy corresponding to final momentum of the whole system.

Of course, the beams will necessarily be extremely tenuous and collisions few and far between, and so *storage rings* are required, in which many beam pulses accumulate over several hours before the collision is arranged [4]; the development of storage rings has been the second important aspect of accelerator physics.

CERN's second large machine, the Intersecting Storage Rings, was of this type [5]. It was a proton–proton collider, for which construction began in 1966, and its energy was equivalent to that available from a 2000 GeV fixed target accelerator; it was commissioned in 1971 and was in operation until the end of 1983.

Figure 5.1   Carlo Rubbia. Courtesy CERN.

Around 1977 it became clear that the masses of the W particles and the Z particles were expected to be in the ranges 60–80 GeV/c2 and 75–92 GeV/c², respectively, which were inaccessible to any accelerator in operation at the time. However, construction of the 300 GeV Super Proton Synchrotron had been approved in 1971 and, with later improvements, it accelerated protons to 400 GeV from June 1976. Carlo Rubbia (Figure 5.1), together with David Cline and Peter McIntyre, suggested transforming this brand-new machine into a proton–antiproton collider, and it was Rubbia who pushed the idea through and led one of the collaborations that searched for the particles.

Equally essential was the work of Simon van der Meer (Figure 5.2), who invented the technique of *stochastic cooling* to increase the density of the anti-proton beam to that of the proton beam. In this technique, fluctuations in the average beam position are detected electronically, and correction signals are generated and applied at other points of the ring.

By 1983 both the W particle and the Z particle had been detected, and Rubbia and Van der Meer shared the Nobel Prize for Physics in 1984.

The work of CERN has continued very successfully since that time, culminating in the advent of the Large Hadron Collider in 2008, and its use, following an upgrade, in the discovery of the Higgs boson in July 2012.

Figure 5.2 Simon van der Meer. Courtesy CERN.

During the late 1980s, some work which would turn out to be of enormous significance was being carried out at CERN. Tim Berners-Lee [6] was at CERN briefly in 1980 and then on a fellowship programme from 1983. In 1989 he put forward an internal proposal for what would turn out to be the World Wide Web; it was initially used for communication and flow of information within CERN but, by 1991, the basic software was released to the world outside the laboratory. It could justifiably be said that this work was perhaps of more general importance than any of the studies of particle physics for which CERN was supported!

Now let us turn to John and Mary Bell. In the main, their lives continued in the fairly humdrum manner that they liked. They continued to ski, although Bernstein [7] reported that, by the end of the 1980s, they had given up downhill skiing and replaced it with the more sedate cross-country skiing. When Bernstein interviewed them at the end of the decade, John complained that Mary had *retired* and would not help with any mathematics. Mary had spent her whole career in accelerator design, being an expert in mathematics and computing, and subsequently she admitted to Bernstein that the term 'retirement' was an exaggeration.

Perhaps, though, one of the nicest things that happened to them during this decade was that the couple, who would have been on good-enough terms

Figure 5.3 John Bell and Reinhold Bertlmann, choosing teabags in the Bells' flat in 1981. © Renate Bertlmann.

with friends and colleagues at CERN but were probably happiest in each other's company, formed a much closer relationship than usual with a younger Austrian couple who were visiting CERN: Reinhold and Renate Bertlmann, who were around 15 years younger than the Bells (Figures 5.3–5.6).

When he arrived at CERN in 1978, Reinhold (Figure 5.7) was already an established physicist with a strong interest in a topic of great interest to John—the system of quark and antiquark or *quarkonium* [8]. Renate was a well-known artist; she was a lecturer at the Academy of Fine Arts in Vienna from 1970 to 1982, when she went freelance, undertaking drawings, paintings, installations, films, videos, performances, and multimedia events. She often examines 'the erotic and biological conditions of the female existence', and one of her most recent exhibitions has been *The World Goes POP* at the Tate Gallery of Modern Art [9]. Incidentally, Reinhold is also extremely interested in art and enjoys being with artists.

Reinhold felt [10] that quite a deep friendship developed between the older and the younger couple. Initially, John helped Reinhold with some personal matters, and Reinhold felt that the relationship was rather like student and teacher, or even son and father. John certainly helped Reinhold with his physics, and also the communication of physics—Reinhold would admit that, at that time, he was poor at expressing his ideas, and John was always very patient in his explanations, which was not always the case with other questioners.

Figure 5.4 Reinhold Bertlmann and the Bells, in the Bells' flat in Geneva around 1982. © Renate Bertlmann.

Figure 5.5 The Bells and Reinhold Bertlmann, in the Bertlmanns' flat in Geneva in 1982. © Renate Bertlmann.

Figure 5.6 Renate Bertlmann and the Bells, in the Bertlmanns' flat in Geneva in 1982. © Renate Bertlmann.

Figure 5.7 Reinhold Bertlmann lecturing in 2002. © Renate Bertlmann.

When people came to him expressing themselves in a complicated way, John could get quite fierce!

John would never have pushed himself forward in any company, and yet Reinhold would say that he was fascinated by John's personality. John was by no means an extrovert but, as his career developed, he certainly had style! Strangely enough, Reinhold first became aware of this long before he met John personally, at a conference on neutrinos in Vienna in the early 1970s. All the great people were there—Feynman, Lee, Yang, Weisskopf, Steinberger, Rubbia ... When John came out of the hall onto the steps outside, the sun was shining, and with his red hair and red beard, and his still youthful appearance, for Reinhold he was 'like an Apollo'.

Another example of his style was at seminars in the Theory Division at CERN. While at Harwell, he had sat in the front at seminars and made his presence highly noticeable, at CERN he would have a special place at the back right by the slide projector. Nobody dared to enter that place, and from here he asked questions in a low voice, so that the speaker had to be very quiet so as to hear and then respond to the voice from the back.

When Bertlmann came to CERN in 1978 as a fellow from Austria [11], John approached him while they were having tea after a seminar in the Theory Division and introduced himself. John incidentally felt that it was a worthy aim to attempt to welcome new arrivals, who could otherwise be rather left to their own devices, although he did admit that the fact that everybody was extremely busy meant that this nicety often got squeezed out. When Reinhold introduced himself in turn and told John his field of interest, outlining some of the results he had obtained, John became extremely interested, and a fruitful interaction was initiated.

The technical details will appear in 'Pasupathy, Bertlmann, and Rajaraman', but here a few general matters will be discussed. The first [10, 11] concerns an occurrence which must have been extremely embarrassing for Reinhold at the time. Towards the beginning of their work together, John was away from CERN for a day due to illness, and he asked Reinhold to bring the results of some calculations he had been performing to his flat, which was in a big new building in Geneva.

When Reinhold arrived, in naturally rather an excited mood, John gave him a glass of orange juice; however, unfortunately, Reinhold's fingers slipped on the glass, and the juice was deposited over the beautiful carpet. Reinhold says that he nearly fainted and, of course, wanted to clean the carpet. However, John said that he would clean the carpet himself; he announced that he had formed a conjecture and he asked Reinhold to sit down and prove it. Of course, Reinhold was far too traumatized to think about the conjecture—he felt he should have been cleaning the carpet!—but he was highly impressed when, with the carpet returned to good order, John Bell was able to develop a proof of the conjecture himself.

Indeed, Reinhold felt that John's gentle low voice only added to his appearance to create an enormously impressive personality, which totally fascinated the younger man.

Reinhold's initial fellowship ran from 1978 to 1980, but he was to visit CERN on short-term fellowships most years right through the 1980s for between three and six months. Much of the time he worked with John, and they published five papers together between 1980 and 1984. While personal relations between the two were excellent, he would stress that he worked tremendously hard, a lot of the time with John, and then day and night on his own, moving step by step to the solution to the particular problem. Also, although of course in a general way he was in awe of John, he was prepared to disagree or press his own ideas when he felt it was necessary, and he felt that John actually appreciated that.

In both these ways, Reinhold may have contrasted favourably with other junior workers at CERN. As John's renown steadily increased, he became known to some as the 'Oracle of CERN', and junior workers in particular would feel free to visit him and take his time and his thoughts and advice on their own problems and difficulties.

This he would not have minded too much, and indeed he would probably have regarded it as a natural part of his role at CERN, despite the fact that, by the 1980s, he was already effectively doing three jobs—his CERN-type work, his work on quantum theory, and his communication with many of those interested in the latter. (Although he claimed to 'hold the right not to reply to letters to be the most fundamental of human freedoms' [12], he did, in fact, respond to many questions and suggestions.)

However, he did come to feel that in many cases he was rather being used. Sometimes junior colleagues would take advantage of suggestions he made, and publish the work without giving him any acknowledgement, and sometimes he would point out errors in their ideas and they would publish them all the same!

Others would hope to work with him, and when they came to see him, they would be totally sycophantic; they rushed to agree with him even before he had finished the sentence. John much preferred someone like Reinhold, who was willing to disagree with him, even though Reinhold would admit that on these occasions John was very often right!

Reinhold was touched by a remark John made when he was a referee for Reinhold's habilitation (the process which allowed Reinhold to lecture in a university). Reinhold was allowed to see, although not to copy, the reference, and John had written: 'It was a stroke of luck for me to have found Dr Bertlmann interested in themes of duality and the moment method.'

In some ways, the more personal relationship between the Bells and the Bertlmanns may have been the attraction of opposites. There was a standing joke in CERN about the country where members of staff chose to live. Reinhold and Renate were keen on the French—French fun, the casual French lifestyle, the restaurants, and the way of speaking. Although John joked about the cautious Swiss way of life—the banks, the jewellery, the so-so-accurate Swiss watches—in the end, he and Mary probably preferred the regulated way of living in that country.

At one time it was suggested to Reinhold that there was a good chance of him being offered a permanent position at CERN, and he decided to ask his friends for advice as to whether he should accept. Of course, Renate was not an employee of CERN, but she was able to work in Reinhold's office or the library, so she could well have been happy to remain in Geneva, and all Reinhold's friends but one thought it was an excellent idea.

The exception was John, who, although he would certainly have liked the Bertlmanns to have stayed in CERN, questioned whether this was really the right course for them. Reinhold and Renate liked to have contact with non-physicists—with young people, with artists—and he felt that Geneva could eventually be somewhat stultifying for them. He convinced them and fortunately it was at about that time that Reinhold obtained a permanent position in Vienna.

One example of the Bells' regularity, and what might be called their idea of democracy, was in their eating arrangements at CERN [7, 10, 11]. A general rule at CERN was that the important or intellectual members of staff, in particular the members of the Theory Division, took lunch around one o'clock, while their humble colleagues ate rather earlier than that.

As if in protest over this class consciousness, John always stopped work at precisely 11.28, and the Bells entered the restaurant—John called it the canteen—at 11.30. In the early days, there had been no provision for vegetarians, so Mary would arrive with a satchel of fresh vegetables for her and John. Their regular companions would not be high-powered physicists but friends from other areas of the laboratory, including an expert in computing and a colleague in charge of safety, so pleasant normal conversation could ensue over lunch and tea after the meal.

Afternoon teatime was different. John would always stop at four o'clock for tea and, in the canteen he would order 'deux infusions verveines, s'il vous plait'. Then, there would follow relaxed conversation, sometimes about physics, but often about politics or philosophy. When Renate was present, there could be heated discussions about modern art.

Then, in the evening, the Bells would leave rather regularly, and after dinner, they might walk in one of Geneva's parks.

It is well known, of course, that John was a vegetarian, but Reinhold has pointed out that he was also something of an ecologist—well before this was fashionable. He particularly wanted to save trees, and he would put a sticker on letters calling on everybody to do this. He himself was very keen on saving paper to save trees, so he used any scrap paper that he got, he did his calculations on the reams of computer output that were produced in those days, and, when he was forced reluctantly to use a photocopier, he made the copies smaller and smaller so that in the end a magnifying glass was required to read them!

Among the charities that he supported were World Population Control and Replacement of Animals in Medical Research.

Another interesting point made by Reinhold is that something of an informal club developed among those who had worked with John. The very fact that he had accepted them as collaborators gave them something of an aura. For example, when Reinhold first met Christopher Llewellyn Smith, they immediately spoke as though they themselves were old friends and colleagues, and, when Reinhold had a student who would have been well suited for a postdoctoral position in Jackiw's group, Jackiw immediately expressed great interest in the idea and followed the suggestion through.

As the successes and fame of CERN grew, many leaders in politics and religion chose to make a visit to 'this monument of the West'. Among the religious leaders were the Pope and the Dalai Lama, and John was fascinated by his interactions with the latter [7, 13]. It may be said that, while remaining sceptical himself, he was nevertheless always extremely interested in the views of those who claimed to see connections between physics, in particular quantum theory, and Eastern religions.

His views on the books of Zukav and Capra were discussed in Chapter 4, and he was also very interested in the positions of, for example, Joseph Needham [14] and Schrödinger. Needham dedicated most of the latter part of his life to the study of science and civilization in China, but he always maintained that Buddhism was inward-looking and hostile to science. Schrödinger was extremely interested in Eastern religions and described himself as having mystical Eastern beliefs [15], but Bell noted that he took great pains to stress that these beliefs had no connection with quantum theory.

On one occasion, the Dalai Lama came to CERN with around 13 Buddhists and, at lunch, they sat on one side of the table with around the same number of CERN people, who included many of the high-ups in sober suits. Perhaps because of some perceived connection between Buddhism and quantum theory, Bell was invited to join the party, while among the Buddhists was an Englishman, David Skitt, who was the curator of a Tibetan–Buddhist monastery near Geneva and who had a good knowledge of physics.

In the discussion, one particular point of possible disagreement emerged—the Big Bang. With their system favouring eternal recurrences, it was difficult to see how Buddhists could come to terms with a once-for-all start for the universe. However, both sides behaved graciously. Bell, the main spokesman for the scientists on this point, admitted that the Big Bang was perhaps a matter of scientific 'fashion', while the Dalai Lama said that the recurrences were 'perhaps not part of Buddhism to which we are completely committed'.

Bell was interested in what the Buddhists thought about Needham and about quantum theory, and with one exception the answer was—not much! They had not heard of Needham, and were really not very interested in science at all, being much more concerned about strictly spiritual matters. The exception was Skitt, who knew about Needham and thought his views were completely wrong.

Bell regarded Needham as an interesting bridge between West and East, and he would also have liked to have seen any connection between quantum theory and Buddhism explored by genuine Buddhists, rather than Westerners latching onto Buddhism, perhaps quite illegitimately. Subsequently, he became close friends with Skitt, who often visited him in Geneva for discussions, although whether their views ever came close Bell was never sure.

Another interesting occasion was Bell's invited participation at a 1978 symposium on physics and its implications for religion organized at the Maharishi European Research University at Seelisberg near Lake Lucerne [16]. Shimony, who was visiting the University of Geneva at the time, was also invited [17].

The Maharishi Mahesh Yogi, who had been a physics major and was certainly interested in the relationship of the subject to the mind, was present, sitting on a throne in white robes, surrounded by mainly female acolytes who looked sweet but did not say anything, and Bell felt that the atmosphere of adulation was uncomfortable for a scientist. Shimony gave a little speech, and then Larry Domash, Head of the Physics Department, spoke, suggesting that there was an analogy between the state reached in meditation and the ground state of a superconductor. Bell gave his own speech—naturally rather sceptical in nature, although Shimony gained the impression that Bell was more favourably inclined that he was himself, at least to the discipline of Transcendental Meditation, if not to the doctrine.

Throughout the symposium, the Maharishi made various pronouncements which rather shocked Bell. One was that a little relaxation of a believer could create cloud on a sunny day, which could in turn lead to rain. Another was that the Maharishi felt that he had the power of suspending himself above ground, at least to the extent that, when bounding along under gravity, the bounces might be more sustained than Newton would allow.

It was also suggested that, in a town where as many as 1% practised meditation, the crime rate would fall significantly, and that when a thousand or more were organized to 'pray' for safety on the roads, fewer accidents occurred than at other times. It should be mentioned that examination of these claims was felt to be a bad thing; the effects were inhibited by scepticism.

Shimony found the occasion one of the most exotic experiences of his life. Bell admitted that it was an amazing occasion but did not find the day with the Maharishi as intellectually interesting as that with the Dalai Lama had been. But there was one excellent aspect to the affair—the Maharishi set-up was vegetarian, and the vegetables were excellent!

Pasupathy, Bertlmann, and Rajaraman

In the 1980s, Bell's interests, apart from those in quantum theory, returned to a considerable extent to the physics of accelerators, and the only papers he wrote in the area of particle physics were five papers written with Bertlmann—the most he wrote with any collaborator, at least since AERE reports at Harwell—and two with Ramamurti Rajaraman. The last of these papers was published in 1984.

As explained in Chapter 4, the work with Bertlmann was broadly based on Bell's collaboration with Pasupathy at the end of the 1970s, so here we discuss these papers as well. To explain the work with Pasupathy and Bertlmann, we need to remind ourselves of some basic facts about the positron, which was discussed in Chapter 2, and about quarks, which were discussed in Chapter 3.

It is usual to say that, when an electron and a positron meet, they annihilate each other; while this fact is certainly true, as we saw in Chapter 2, the annihilation is not immediate but takes around 10^{-7} or 10^{-10} s, depending on the relative orientation of the spins of the two particles. This time is long enough for the spectrum to be observable; in other words, the system may jump from the ground state to discrete excited states which have higher energy and orbital angular momentum than the ground state, and then drop back to the ground state or another low energy state, emitting a photon of a specific frequency.

The system is called *positronium*, and clearly it is closely analogous to a hydrogen atom. Each system consists of one particle of positive charge, and one of negative charge which is of equal magnitude to the charge of the first. The spectrum of the hydrogen atom was calculated theoretically by Niels Bohr in 1913—in his analysis of the famous *Bohr atom* or *Bohr model of the atom*. In the model, the electron rotates around the proton; or, a little more exactly, both particles rotate around their centre of mass, although of course, for hydrogen, that is extremely close to the proton.

For positronium, again both particles rotate around their centre of mass; however, in this case, this centre of mass is, of course, midway between the two particles, and the result is that the spectra of atomic hydrogen and positronium are the same qualitatively but different quantitatively.

Let us now briefly review some information about the quarks. It will be remembered from Chapter 3 that, when charm was discovered, it was not in a particle which itself had charm, or as is often said, had *naked charm*. Rather, it was in a meson which was a charmed quark combined with its own anti-particle, the charmed antiquark. This was the famous J/ψ particle, sometimes called the *Gypsy particle*, so called because one of its discoverers, Samuel Ting, called it the J particle, while the other, Burton Richter, called it the ψ, or psi, particle. By analogy with positronium, it is also called *charmonium* [18].

Similarly, although perhaps unfortunately, the combination of the bottom quark and antiquark has been given the name *bottomonium* and the symbol Υ (upsilon). A general term for a combination of a quark and its own antiparticle is *quarkonium*, while the combination of the strange quark with the strange anti-quark is given the symbol φ, and the linear combination of the up and down quarks each combined with the corresponding antiquark is called ρ [11, 19].

Lastly, let us discuss the concept of *duality*. Duality may be said to occur when two seemingly opposite phenomena are strongly correlated and may be regarded as *dual* aspects of the same reality. In any physical event, either of the phenomena may be revealed, depending on the precise circumstances. In quantum physics, the best-known example of duality is *wave–particle duality*, according to which light or an entity normally described as a 'particle', such as an electron, may appear as a wave when it travels but as a particle when it interacts. Naturally, the relationship between the two manifestations of the underlying reality must be interesting and subtle and was crucial at the beginning of quantum theory.

As we saw in Chapter 4, around 1978 Bell became very interested in models of partons and he wrote interesting papers with Hey, and with Davis and Rafelski. At the time he had a strong intuition of partons as real particles; thus, he used the term 'partons' rather than 'quarks'. As the significance of quark confinement and asymptotic freedom became apparent, as mentioned in Chapter 3, he wished to consider the bound states, that is, the hadrons, in the context of confinement.

Jum John Sakurai (Figure 5.8), a physicist who had been born in Japan but who had lived in the United States since high school, had discussed these ideas in terms of duality [20]. At high energies, which broadly, following the Heisenberg principle, correspond to short distances, the quarks behave as free particles; however, at low energies, corresponding to large distances, they are confined and generate states of quarkonium, which show up as resonances in a graph of cross-section (probability of interaction) against energy. This phenomenon is sketched in Figure 5.9.

Sakurai was to become very close to Bell. Sadly, there were also to be parallels between their untimely deaths.

Figure 5.8 Jun John Sakurai; from J. J. Sakurai, *Modern Quantum Mechanics* (Addison-Wesley, Reading, Massachusetts, 1985). Courtesy Addison-Wesley.

Sakurai showed that areas under the two curves, one for the resonance case, and the other for the asymptotic high energy case, should be equal and, towards the end of the 1970s, Bell investigated this point in collaboration with J. Pasupathy, an Indian physicist from the Institute for Theoretical Studies in Bangalore, who was visiting CERN [21]. There were two ways of calculating the sums over the resonances: one using numerical computation, a method which had been followed by Sakurai working with Kenzo Ishikawa [22], and the second using approximate formulae for the resonance wave function, which is the method that Bell and Pasupathy used.

Their first paper [21] was short but it discussed the topic in considerable depth. It generalized previous papers to a greatly increased range of *partial waves* (which are wave functions corresponding to a particular value of angular momentum), and it showed the significance of the value of the wave function at the origin, thus playing a central part in the reconciliation of the two points of view. Quite a lot of detailed numerical calculations were performed in the calculation of these values.

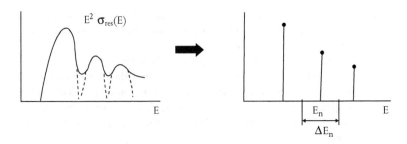

Figure 5.9 Resonances, as presented by Reinhold Bertlmann; from Reinhold Bertlmann and Anton Zeilinger, editors, *Quantum (Un)speakables: From Bell to Quantum Information.* pp. 29–47. © Reinhold Bertlmann.

Bell and Pasupathy were also able to demonstrate the fundamental reason why the two approaches gave the same answer. They showed that both calculations relied on the behaviour of the system for short times, and this short-time behaviour could be shown to be the same for each point of view. The authors did warn that their argument depended on the potential being constant near the origin, although the potential in this case related to the exchange of gluons, and it is commonly thought that the potential would actually diverge near the origin. However, they suggested that the theory may be modified to cope with this difficulty and that the idea of duality should still be correct.

Bell and Pasupathy published one other paper and it was submitted for publication only a month after the first one, but in the meantime Bell had met Bertlmann, who had told him that he was working on models of potential precisely for these quark–antiquark systems, and that he had been able to write a computer program to solve the Schrödinger equation exactly, an impressive feat at that time!

This program was exactly what Bell and Pasupathy needed in order to take their own work forward [23] and, in their second paper, they went into much greater algebraic detail than they had in the first, calculating the magnitude of the wave function at the origin, as they said this function was required to exhibit the appropriate form of bound-state/free-state duality. In particular, the wave functions had oscillatory and non-oscillatory regions, and it was a tricky task to join these successfully.

In their paper, the authors thanked Bertlmann for helpful discussions and for the numerical results which were an essential part of the argument. Incidentally, Bertlmann was very impressed in the way in which Bell handled a slightly embarrassing issue. Bertlmann's numerical values disagreed with some that had been published by Ishikawa and Sakurai [22], and, since Bertlmann's set of values were much more evenly spaced, it was clear that there must have been an error in the earlier values.

Bertlmann expected Bell to highlight this when writing the paper; however instead Bell made the situation totally clear without mentioning the error explicitly. Bertlmann [10] felt that this was an important factor in the relationship between Bell and Sakurai becoming close. Bertlmann himself also developed a good relationship with Sakurai, another example of those close to Bell becoming close to each other.

When Pasupathy had to return to India, it was natural for Bell and Bertlmann to collaborate in this area, and they agreed to study the equality of the cross-sections given by the two models, or in other words, the duality, using non-relativistic potential theory; they expected that this approach might be very successful because, for charmonium and bottomonium in particular, the speed of the quarks inside the meson itself is relatively small, about $0.3c$ and $0.1c$, respectively.

In the first paper that they wrote together [24], which was submitted for publication only two months after Bell's second paper with Pasupathy, they were actually able to extend previous ideas, which demonstrated duality only when the appropriate cross-sections were integrated over *all* resonances; in this paper, it was shown that the relationship was still true if only a *single* resonance was considered [11, 19].

The authors stated that their paper should be regarded as a sequel to Ishikawa and Sakurai's paper [22], although they stressed that, while the latter paper, at least in part, discussed the adequacy of relativistic models, the aim of Bell and Bertlmann was solely to test duality within the non-relativistic area. They considered three forms of potential: Coulombic + linear potential, Coulombic + cubic potential, and logarithmic, and were able to demonstrate generally good agreement between, on the one hand, energy spectra and wave functions obtained by solution of the Schrödinger equation, and on the other, the same quantities predicted by summing over the resonances.

This good agreement may be explained in general terms [11, 19] by recognizing that the potential may be split into two terms: a short-range term corresponding to the Coulomb interaction, and a confining part. Asymptotic freedom means that the confining part becomes significant only for large distances. In the duality relation, there must be a an averaging over a range of energy, but the Heisenberg relation then requires us to focus on small times, for which changes in the wave function cannot spread far enough to be influenced by the confining part of the potential. Essentially, this part of the potential, and so the confinement itself, become irrelevant. Incidentally, it was the presence of the short-range potential that enabled Bell and Bertlmann to demonstrate duality for a single resonance.

While this paper was successful in its own terms, clearly if the concept of duality was to be pushed further, in particular to study the positions of the resonances, two points would have to be addressed: the rather simple nature of the averaging over energy, and the fact that confinement cannot be totally ignored. Bell and Bertlmann would tackle these issues in their next paper.

Before discussing this paper, we shall make two brief remarks about the division of labour in preparation of this first paper. Extensive computation was involved, and Bell by no means expected Bertlmann to carry out this 'donkey work'. Rather, he calculated the numbers roughly himself using a rather old pocket calculator, and then confirmed them on one of the small computers at CERN at the time, not wishing to waste resources! Bertlmann himself used the big machine, and they would then check that their results agreed. Bertlmann would stress that Bell was never an abstract theoretical physicist happy to leave his results as a series of equations. He liked to conclude any investigation with numbers and graphs, and to work with the experimentalists to check the results [10].

The second concerns the writing of the paper [11]. Bertlmann was pleased when Bell asked him to begin this process, and Bertlmann worked all night on this task, bringing his attempt to show Bell the next day. Bell commented that the draft looked nice and he would look at it closely; however, when Bell returned it the next day, Bertlmann was shocked to see that it was completely changed—not a single word was left in the place where he had put it! But fortunately Bertlmann felt that his writing did gradually improve.

Later in 1979, three papers [25]—Bell and Bertlmann [26] refer to them as 'a remarkable series of papers'—were published on the same topic by a Russian group: Mikhail Shifman, Arkady Vainshtein, and Valentin Zakharov, who are always collectively referred to as SVZ. They used a method of moments to calculate the energies of the low energy resonances in quantum chromodynamics, up as far as charmonium. To previous models, SVZ added extra terms corresponding to long-range effects due to the actions of quarks and gluons, or in other words, representing the mechanism of confinement, and their calculations were made with the use of relativistic field theory.

Bell was very surprised by some aspects of the theory and the results, particularly the fact that quite good results were obtained, despite the difficulty of averaging over energy in a suitable way. He knew that, in order to restrict the calculations to short times, which would be expected to work best for the small perturbations used by SVZ, a coarse averaging would be best. However, in order to see properties of individual levels, the averaging must *not* be coarse.

From his point of view, it was surprising that one could 'have the best of both worlds'. He would have expected that, just as properties of individual resonances were becoming dominant, the confining force would itself become of central importance and it would certainly not be small enough to be treated as a small perturbation.

His way of making progress was fairly typical of his methods in theoretical physics. With Bertlmann, he applied the SVZ method to an area where the correct results are known from theory, in order to check how closely the results produced by the novel method would agree. First, they decided to make a non-relativistic theory, and they hoped that, in the non-relativistic limit, they might be able to construct a model with a potential.

Bell stressed that the term involving the gluons had a long-range fluctuation but, because of the confinement, it was not possible to create a potential like the Coulomb potential. However, if you look at the moments, they cut off the large fluctuations, so a potential *could* emerge in that way. This is exactly what the authors were able to do.

Because the method of moments achieved this 'best of both worlds', the authors called it *Magic Moments*, and they named the paper, which was easily the longest of the five that Bell and Bertlmann wrote together, after it. They applied the method to a variety of potentials, including potentials of the kind used for non-relativistic descriptions of charmonium, particularly in attempting to locate the positions of the resonances. It might be most accurate to sum up the results by saying that they are surprisingly good, particularly because of the quandary as to why the method works at all, although not exceptionally good. In particular, they suggested that the size of the effect due to gluons might have been substantially underestimated (a possibility that SVZ were aware of).

At this point, Bell and Bertlmann recognized that, although the types of potential they had considered were often used to mimic the asymptotic freedom together with confinement of the quarks in charmonium, they were unrelated to the specific quantum chromodynamics calculations of SVZ. Therefore, they turned to the problem of constructing a potential for this case in their next paper [27], which they were able to submit by the end of 1980, although by this time Bertlmann was back in Vienna. As before, a problem was that the mathematical terms used by SVZ to describe the gluons were singular, that is, they diverged at the origin. However, the solution to this problem, again as before, was to work with the magic moments.

Two new terms were introduced in this paper. The first was used to describe the mathematical terms involving the gluons that caused confinement—the *gluon condensate*, while the other one was for the potential that Bell and Bertlmann aimed to construct—the *'equivalent' (non-relativistic) potential*.

Having calculated the moments, the authors related the answers to a superposition of power-law potentials and thus identified the appropriate 'equivalent' potential, which turned out to be the sum of a Coulomb term, which was, of course, negative and proportional to the reciprocal of r, and a quartic term, that is, a term in r^4, with the strength of the latter term being proportional to the mass (and hence varying with the flavour) of the quark, and also broadly to the magnitude of the gluon condensate. Unfortunately this model bore little similarity to phenomenological models that others had constructed for charmonium systems [19].

Using the experimental parameters from the SVZ approach, it was found that the mass predicted by the non-relativistic moment method is around 0.1 GeV above that predicted by the Schrödinger equation, although that is an error of only around 3% of the actual mass. However, what is more striking is that, if one decides to adjust the gluon condensate parameter to obtain the correct energy, this parameter has to be increased by a large factor of somewhere between 2 and 5 for the different models.

Bell and Bertlmann argued that this result confirms their previous conjecture [26] that SVZ had substantially underestimated this term. However, they did have to agree that this was indeed merely a conjecture, because they could not disagree with SVZ, who, in a private communication, had stressed that potential theory is *not* field theory!

At the time of the next joint paper [28] in 1983, Bertlmann was a visiting professor at the University of Aix-Marseille, and the Bells had also been invited to visit the university. In this paper. they took their approach to obtaining a potential corresponding to the gluon condensate a little further, particularly studying the case where the Coulomb effect dominates, and the gluon condensate effect may be regarded as a fairly small perturbation; these conditions would be appropriate for heavy quarkonium systems [11]. Mikhail Voloshin and Heinrich Leutwyler had independently also studied this case.

Although the attempt to use their work to produce a form of potential was not particularly satisfactory, Bertlmann and Bell were able to argue for this case to a potential proportional to r^3, and thence, by adding terms in the inverse of the mass, to obtain rather a complicated form of potential, with terms in r^3, r^2, r, and a constant term, which Bertlmann [19] calls 'the second equivalent potential of Bell and Bertlmann' and which, he says, agrees with the results of Leutwyler and Voloshin to within six parts in a thousand for all quantum states.

Overall, though, this study was not able to generate any significant coming together of the SVZ approach and the analysis of heavy quark–antiquark systems, nor, particularly in the fact that all the results depended on mass, was it able to form a bridge to general potential models.

There is one amusing point [10] about this paper, which is the order in which the names of the authors are given—Bertlmann and Bell, and the related footnote. At this time at CERN, there had been a (perhaps not-unusual) complaint by a junior scientist that this person had done all the hard work, and yet came second in the listing of authors. Bell decided that this paper should be 'Bertlmann and Bell' but under protests, added the footnote —'The order of the authors is random'. However, the editor of the journal changed the word to 'arbitrary', which itself annoyed Bertlmann greatly!

Bell enjoyed these little jokes. In a paper he wrote with Veltman (discussed in 'John Bell at CERN, and the neutrinos' in Chapter 3), there is a note: 'One of the authors wishes to thank the other author for helpful discussions.' This is, of course, highly irregular when the authors are expected naturally to discuss the whole content of the paper, but the editor in this case must have thought that the line of least resistance was just to print the note!

Let us now move on to the last paper published by Bell and Bertlmann [29]; this paper was submitted towards the end of 1983 and had been inspired by a fairly critical response by SVZ and two co-workers to their previous work. As Bell and Bertlmann said, their work had been 'welcomed only with some reserve by those who inspired it'.

While they agreed with the Russian group that it was appropriate to retain an attitude of reserve in a general way to the topic, they naturally disagreed with the specific criticism of their own work. Broadly, this criticism was that the methods of Bell and Bertlmann became unreliable as the perturbing term, that is, the effect of the gluon condensate, became large; in other words, when the relevant formulae were used outside their range of validity.

Bell and Bertlmann did not so much dispute this, as question the claim that the work of SVZ did not have the same limitation. They themselves suggested that the Russian work presented the mathematics in such a form as to show the work of Bell and Bertlmann in an unfavourable light, without admitting that their own mathematics could be rewritten in exactly the same way. Nonetheless, SVZ was clearly an important and influential collaboration; their three papers [25] have been cited nearly 7,000, 3,000, and 1,000 times, respectively, and the work of Bell and Bertlmann had certainly provided useful and broadly positive analysis of their work.

Let us now turn to Bell's other interaction in particle physics during the first part of the 1980s. This interaction was with Ramamurti Rajaraman (Figure 5.10), who at the time was working at the Indian Institute of Science in Bangalore, although he had been based in Delhi for a considerable period before he moved to Bangalore and he returned there again afterwards. He had also spent two periods as a member of the Institute for Advanced Study at Princeton.

The pair met during a visit Bell paid to Bangalore in 1982. As Rajaraman [30] reported, Bell politely asked him what he was working on. Rajaraman chose rather to turn the conversation around by asking Bell himself about the reports on fractionally charged states that were emerging. Thinking that he meant the idea of quarks, Bell rather warily suggested that surely these were well established; however, Rajaraman was able to convince Bell that the states he was being asked about were ones theoretically shown to have fractional numbers

Figure 5.10 Ramamurti Rajaraman. Courtesy Jawaharlal University.

of electrons. Bell turned and entered Rajaraman's office, and Rajaraman could scarcely believe his luck; he had Bell trapped in his office, and they were about to start a joint study of the problem that he had suggested to Bell.

They would work together for much of Bell's fortnight in Bangalore, and Rajaraman would also pay visits to CERN to carry forward the collaboration.

The phenomenon had been discovered in theoretical studies of solitons in quantum field theory by Roman Jackiw and Claudio Rebbi [31] in 1976. (In Chapter 2 we met a soliton as a solitary wave that maintains its shape as it travels.) The argument [30] may be sketched as follows.

For a system with a potential given by a symmetric double well, there are two types of solution. The first consists of pairs of broadly wavelike solutions: one with positive energy, and one with negative energy. Using quantum field theory to evaluate the total charge, in, of course, units of the electronic charge, meets no problems for this case. There is an occurrence of ½ for each solution, but it neatly combines with that of its partner to give an integer charge.

However, there is also a single solution with zero energy, which is not wavelike but decreases steadily as one moves out from the origin in either direction.

Because this solution has no partner, when the charge is calculated there is a single ½ in the expression, and so the result is inevitably half-integral.

If this claim, related to the rather abstract mathematics of quantum field theory, was remarkable, perhaps even more so was the suggestion of Wu-Pei Su, John Schrieffer, and Alan Heeger [32], who had worked quite independently of Jackiw and Rebbi, that exactly the same phenomenon may occur in the highly practical example of *polyacetylene*, a long-chain molecule of formula $(CH)_n$, where n is very large. As Rajaraman [30] said: 'A molecule of polyacetylene has after all some finite number of electrons and it is hard to imagine how, even if it twists itself into some topological soliton state, it could have a fractional number of electrons.'

It is actually quite easy to show that the two systems, so different physically, may be represented by the same basic equations, so it is no surprise that they show the same surprising behaviour. Indeed, Rajaraman and Bell, in their study of the problem [33, 34], showed that the equation for polyacetylene was equivalent to a basic equation of field theory. It is interesting to speculate whether Bell remembered, or even conceivably took advantage of, the undergraduate project he had carried out with Ewald 35 years before.

In their analysis, the authors did not assume that the system was of infinite length. Rather they took it to be of length $2L$, with, of course, the possibility of allowing L to tend to infinity at the end of the calculation. In such a technique, it is necessary to impose a *boundary condition* for the behaviour of the system at $+L$ and $-L$. In this case, as is indeed fairly standard in problems of this type, *periodic boundary conditions* were imposed: the behaviour at $-L$ and at $+L$ must be the same.

The change from the analysis of Jackiw and Rebbi was striking. Instead of the solution with zero energy being partnerless, it now has a partner. Whereas the original solution is localized around $x = 0$, the new solution is localized around $x = +L$, and $x = -L$. It is straightforward to see that this solution would not have shown up if L had been effectively put equal to infinity at the start of the calculation. Nevertheless, the beautiful point is that the factors of ½ from this solution and the previous one cancel to remove the initial problem.

As Rajaraman [30] said: 'While integral charge has been restored, if we were to stop there we would have lost all the important physics unearthed by Jackiw and Rebbi.' After all, while one of the localized modes is right at the edges of the system, the other is in the middle and can move about as an excitation.

To examine this point, an operator may be defined that takes account of the charge that lies only between $-\ell$ and $+\ell$, where ℓ is less than L. Provided $L - \ell$ is greater than the width of the mode 'at the edges', this operator will indeed measure a half-integral value, and some quite sophisticated analysis may be performed, but the fact remains that the *total* charge of the system is always integral.

Rajaraman [30] also said that, like so many others, he developed great respect for Bell, 'not just for his great originality and prowess as a theoretical physicist, but also for the precision of thought and language that he demanded of himself and of those who were fortunate enough to work with him. John Bell was more than an intellectual force in physics—he was a moral presence.'

Accelerator work in the 1980s

Apart from his work with Bertlmann and Rajaraman, John Bell's interests in the 1980s, apart, of course, from his quantum 'hobby', moved back to his field of study of the first part of the 1950s: accelerators. On this topic, during this decade he wrote six papers with Mary, and three others with Jon Leinaas, Richard Hughes being a co-author on one of the latter papers.

One may perhaps surmise that a plausible reason for this movement of field was that it would have been pleasant to work with Mary. However, Bell provided a more technical reason than that in his interviews with Bernstein [13] at the end of the 1980s.

Up to this time, accelerators had been designed by physicists who needed to be highly adept at handling classical physics (as we saw for Bell in Chapter 2) but were required to know very little quantum theory. It was true that for circular electron–positron storage rings, such as the *Large Electron–Positron Collider* (LEP), the final beam size was determined by quantum fluctuations—not just zero-point oscillations but also the recoil from synchrotron radiation, as this recoil causes photons to jitter and then to splay out like bullets. This process opposes the effect of radiation damping, which narrows the beam, and the two processes together control its size. But this quantum effect was fairly easily allowed for as, in this type of accelerator, space charge was not very high, and high luminosity was achieved by repeated collisions. Mary, incidentally, had been involved with designing and optimizing one stage of the LEP: the electron–positron accumulator, where electrons and positrons were dumped, ready for the next stage of the accelerator.

For the hypothetical linear accelerators being considered at the time of the interview with Bernstein, that is, the late 1980s, such a simple picture was not good enough. To get enough luminosity, there had to be extremely concentrated bunches of electrons and positrons where the beams crossed, and thus the space charge would be so high that intense radiation would be emitted, the photons taking away an appreciable fraction of the energy. A thorough, rigorous quantum treatment was definitely required.

Incidentally a new word was coined for this type of radiation. The word *bremsstrahlung*, German for 'braking radiation', had long been used for radiation produced by the acceleration or, more particularly, the deceleration of a charged particle passing near to an atomic nucleus; in the late 1970s, the word *beamstrahlung* was invented for radiation and hence loss of energy from one beam in a storage ring or a linear collider caused by interaction with the other beam.

The first person to apply quantum theory to this problem was the Nobel Prize winner Schwinger, and since then it had been investigated by a number of Russian scientists and had also been studied at SLAC. Unfortunately, the only case that had been worked out thoroughly was that where the magnetic field is uniform. Where it was non-uniform, it was possible to use the results for the uniform case at any position, integrate over the region where it varies, and then check whether the integrals converge, but it was far from obvious that this approach was satisfactory. This was one of the problems that John and Mary Bell were investigating.

Mary [35] has been a little more explicit about her joint papers with John than he had been. Three of the papers had actually come from a few years before John's interview with Bernstein. The first two papers [36, 37], which were published in 1981 and 1982, discussed electron cooling, and were in connection with the Initial Cooling Experiment (ICE) at CERN. In preparation for the Antiproton Accumulator (AA) machine, the task of which was to generate and store high energy antiprotons, both electron and stochastic cooling were studied, although in the end stochastic cooling was utilized as it would be less expensive than electron cooling. Electron cooling *was* used, though, for the Low Energy Antiproton Ring (LEAR) [38].

The study of radiation damping [39], published in 1983, was in connection with the LEP, while, as Bell had told Bernstein, the three papers on quantum bremsstrahlung [40–42], which were published in 1988 and 1989, were to provide information and data for linear colliders.

Their first paper [36] was described by the authors as a revision of the theory of electron cooling in storage rings. The technique had been invented by the Russian physicist Gersh Budker in 1966, the idea being that a beam of approximately monoenergetic electrons is made to merge with the particle beam travelling at the same speed. Consequently, energy is lost from the beam to the electrons, thus cooling the beam; in other words, the oscillations of the beam are reduced [43].

Although quite a number of papers on the theory had been published over the previous five or so years, there did seem to be much confusion, and unfortunately there were cases where misprints in one paper had been propagated as errors into others. The Bells therefore aimed to give a clear and, they hoped, final account. They even challenged the reader: 'The arguments for [the final equations] are set out in detail, so that you can see if we go wrong.'

Actually, they agreed with the published literature for the Maxwellian (thermal) distribution of electron velocities, but there was considerable disagreement for the results for the 'flattened' distribution [44], in which the longitudinal component of velocity, that along the direction of the beam, is suppressed, as the beam is nearly monoenergetic. While previous literature suggested the rate of damping would be reduced by a factor of 4 in moving to the flattened case, the Bell's figure of improvement was only 2.4.

This is actually for the case with no magnetic field. When, as is usual, a magnetic field is applied, technically the drag force becomes a divergent integral. This integral must be cut off at some value of velocity, but the result will depend on particular choices of parameters. The authors stated that only simple cases had been considered up to that point, and indeed suggested that computer simulation would surely be required, mentioning that Mary had already made a start on this process.

In the next paper [37], which was submitted just a couple of months after the previous one, the Bells investigated another question of electron cooling: analysing the capture of cooling electrons by cool protons. When this happens, the proton becomes a neutral hydrogen atom, and the observation of these atoms can give useful information about the electron and proton beams. The process is most likely to occur in the later stages of cooling, when the proton is nearly at rest in the electron gas, or in other words, when the proton is already cool.

As the authors stressed, the theory of electron capture by protons had, of course, been worked out in detail, leading to complicated expressions requiring evaluation by computer. In this paper, they preferred to look for simplified approximate formulae for the cross-section (essentially the probability) of capture, as these formulae would be adequate for the low energies involved in the system they were investigating, and could be easily integrated over Maxwellian distributions and flattened distributions.

First the authors used a rather simple expression given in the book by Lyman Spitzer [44]. Secondly they used a Maxwellian distribution and then a flattened distribution, hoping that the former would be comparable with Spitzer's results. Indeed, they found agreement within less than 2%; however, unfortunately, these results were much further than that—in fact, by a difference of 12%—from the values that had been obtained from the formal and complicated theory, and which were more accurate than the results from the simplified approach. However, a relatively small adjustment of the theory, suggested in a paper by Michael Seaton [45], enabled the still simple calculations to agree with the more complicated theory to within something like half a percentage point.

While all the results mentioned so far relate to the Maxwellian case, it was extremely important that the same calculations for the flattened case, which was definitely more relevant for the electron cooling technique, increased the capture rate of protons by a factor close to 2.

The next joint paper by the Bells dealt with classical radiation damping in storage rings [39]. As the authors say, this topic had been dealt with, mostly over the previous decade, in a way which was totally satisfactory but still ad hoc. Thus they felt that it would be worthwhile to show how the results could be obtained in a straightforward manner from a general formula. This work brought up many of the ideas from the 1950s, when John in particular was working on the development of strong focussing. Among the literature cited are many papers of the 1950s, including two Harwell reports by John, one of them being his crucial account of strong focussing (discussed in 'Accelerators' in Chapter 2), two reports by Mary, and papers by John's competitors at the time, Courant, Snyder, and Lüders.

The paper uses the powerful but rather abstract methods of Hamiltonian mechanics and Lagrange invariants, following the most rigorous approaches [46, 47]. Various types of oscillation are analysed—vertical and horizontal betatron oscillations, and synchrotron oscillation—and a general and formally quite simple expression is produced for the amount of damping predicted.

The authors concluded by suggesting that it would be of interest to ascertain how the particular results already present in the literature would fit into the more general framework of their own paper. They also remarked that all the work performed so far, including their own, used certain approximations: extreme relativity and weak radio frequency. If it were necessary to improve on these approximations, they suggested that their method might be 'less painful' than those previously published.

The next paper [40], published in 1988, dealt with quantum beamstrahlung, and was written specifically in connection with linear accelerators. It concerned a recent report, directly associated with the beamstrahlung problem, that had been written by two SLAC scientists, Richard Blankenbecler and Sidney Drell, and dealt with the energy loss of a relativistic electron passing though the field of a uniform cylinder of space charge.

The Bells recognized that this result should be very closely connected with an expression produced in a series of papers by Russian physicists between 1957 and 1971. Not wishing to insult any of the physicists by omitting their names, the Bells called this the Sokolov–Ternov–Matseev–Nikishov–Ritus–Baier–Katkov formula for synchrotron radiation. When integrated, it should have yielded the SLAC result; however, detailed calculation by the Bells showed that agreement was not complete, and they were able to demonstrate that the SLAC workers had omitted a term relating to helicity flip. They pointed out that this term could change the overall results by up to about 8%. While this was quite a small effect, it should be realized that, when enormous sums of money are being spent on constructing accelerators, it is important that the underlying theory is not just almost right but absolutely right!

Their last two papers [41, 42] were specifically on the lines outlined by Bell to Bernstein [13]. Accepting that the theory of beamstrahlung was generally understood for uniform deflecting fields, these papers studied what happened when the uniformity was broken, for the case where the beam ends, that, is the end effect [41], and then where the field was merely 'almost uniform' [42].

The authors started by pointing out that, when the process is classical, the important characteristic length is given by the radius of curvature of the track divided by what they call the FitzGerald factor of special relativity; this factor is close to unity when the speed of the beam is low but increases steadily as the speed itself increases. In this classical case, which occurs if the energy of the photon emitted is much less than the initial particle energy, end effects and non-uniformity would be relatively unimportant, provided the bunch of particles is uniform and long compared to this characteristic length.

However, for the extreme non-classical case, or what they call the *ultra-quantum limit*, for which the photon takes off a substantial fraction of the initial energy, the characteristic length is very much larger than in the classical case, and thus the changes in the nature of the system may take place over a longer distance without affecting significantly the system's behaviour.

The ratio of these two distances, or in other words, the ultra-quantum limit, divided by that for the classical case is an important parameter for the end effect. The mean energy loss by beamstrahlung does indeed increase as this ratio increases—we may call it the *end effect*, but it does so only as the *logarithm* of this ratio, which increases much more slowly than the ratio itself does; and so the general conclusion of this paper is that end effects will be relatively unimportant, except in very special cases.

Let us turn now to inhomogenities in the field [42], for which topic the Bells did extremely detailed calculations, again in the ultra-quantum limit. They found, as confirmed in Bell's interview with Bernstein, that for typical cases with small inhomogenities, the treatment where local uniformity is assumed and the results are integrated over the beam will give satisfactory results.

It might be said that the work on accelerators that John carried out with Mary in the 1980s was highly functional, although, of course, it was exceptionally important for the construction of accelerators. In addition, it was carried out with immense skill and care.

The other topic in this area that Bell tackled, with Leinaas (Figure 5.11) and, more briefly, Hughes, could scarcely have been more different—it might be said to be in a highly exotic field of physics. Jackiw and Shimony [48] described the work as 'one of the most remarkable of Bell's achievements'.

Bell and his colleagues [49, 50] studied the Unruh effect [51], which is itself similar to the Hawking effect [52]. Both effects rely on the fact that 'vacuum'

Figure 5.11 Jon Leinaas (far left) and John Bell (far right) with two others at CERN in 1987; *Europhysics News*. Volume 22, April 1991, p. 78. Courtesy European Physical Society.

should not be interpreted as 'empty space'. Rather, space is filled with quantized fields, and vacuum is merely the lowest possible energy state of these fields.

The Hawking effect is related to the problem of defining a vacuum state in the gravitational field of a black hole. In 1974 Stephen Hawking showed that the vacuum modes are unstable in the presence of this field and spontaneously emit energy quanta—*Hawking radiation*.

Two years later, William Unruh, following previous work by Stephen Fulling and Paul Davies, showed that a detector that is accelerated through a vacuum will observe radiation at a particular temperature; in other words, it will experience the vacuum as a warm (although, as shall see, not very warm) gas. It is interesting that this temperature is directly analogous to that of the Hawking radiation. In fact, to move from the theory of the Unruh case to that of the Hawking case, the only change is that the acceleration through the vacuum is replaced by the gravitational acceleration on the surface of the black hole.

However, the effect is certainly small. If the acceleration is that due to gravity at the surface of the earth, the Unruh temperature is of order 10^{-20} K, while correspondingly, to obtain an Unruh temperature of 1 K, an acceleration of order 10^{20} m s^{-2} would be required.

Leinaas joined CERN on a fellowship for a year in 1981 and soon got to know and to work with Bell. They discussed the Unruh effect but soon came to the conclusion that any straightforward application was out of the question, as any regular thermometer would presumably be destroyed by the required acceleration. It occurred to them, though, that maybe in some way elementary particles might be used as detectors.

At this point in their discussion, Bell remembered a paper that he had recently received from John Jackson [53], which discussed the spin polarization of electrons circulating in a storage ring. It occurred to Bell and Leinaas that this polarization might be considered as constituting a thermometer, the degree of occupation of the upper of the two levels being a measure of the temperature.

In their first paper [54], they calculated what they called a 'natural unit' of the acceleration of an electron: the electron's mass multiplied by the cube of the speed of light, expressed in units of Planck's constant. They obtained a value of order 10^{29} m s^{-1}, a reassuringly high value, that is, many orders of magnitude greater than the 'required' 10^{20} m s^{-1} above.

In terms of using its polarization as a thermometer, they noted that, in a field of 10^4 gauss (today, we would say 1 tesla), the splitting of the 'spin-up' and 'spin-down' levels corresponded to a temperature of around 1.3 K.

The authors then calculated that the value of the electric field required for sufficient splitting would be around 1400 MV m^{-1}. This value is far greater than that for the field in a linear accelerator—for example, they reported that the field in the SLAC linac was about 7 MV m^{-1}—but the electric fields generated by motion through a magnetic field are much larger than that. In fact, for motion with a FitzGerald factor of 10 across a 1 tesla field, they report that the electric field would be around 3000 MV m^{-1}.

Proceeding to consider 'ultra-relativistic' electrons, with a FitzGerald factor of around 10^5 in a storage ring with a bending radius of about 3 km, they were able to say that the acceleration would be of order 10^{23} m s^{-2}, corresponding to a temperature of around 1200 K. These values were appropriate not only for the then-planned LEP but also for the existing SPEAR at Stanford.

This array of numbers convinced Bell and Leinaas that it might be possible to see thermal effects due to acceleration in particle detectors and, in fact, they went even further by suggesting that the effect might already have been observed. They pointed out that there was already both theoretical and experimental evidence that, although circulating electrons became polarized, they did not do so completely, and asked whether this depolarization, or in other words, the failure of the polarization to be complete with all the spins in the lower energy-level, could be regarded as a demonstration of the Unruh effect.

Before discussing the situation for circulating electrons in detail, the authors analysed the case of linear acceleration; they admitted that this case was less significant practically than the circular case but they felt that it would be highly instructive. The relativistic behaviour of the combined electric and magnetic field in accelerating frames is certainly an extremely complicated problem, although with his experience as far back as his early days at Harwell, there was probably nobody in the world who could handle it as assuredly as Bell.

Values of fields corresponding to those of the Stanford linac were used; these values were certainly small compared to those for the case of circulating electrons. However, the authors said that these values presented no immediate problem, although two other factors did.

The first problem was that the equivalent temperature was around 0.01 K, which would be very difficult to see on the background of other depolarizing effects, and in particular, at the actual temperature of around 300 K. Even more important, though, was the fact that the timescale of the process, that is, the rate of transfer of spins from the upper to the lower state, would be enormous. In the accelerating frame, it might be around 10^{18} s, which is itself longer than the age of the universe, while in the laboratory frame it would be much larger than that, something like $\exp(10^{28})$ s. Overall, though, one could say that the analysis was straightforward for the linear case, even though the theoretical values made any experiment inconceivable.

Fortunately, because of the higher possible acceleration in the circular case, the study of the case of circulating electrons gave results which were far more manageable than those from the linear case from the experimental point of view. These results, though, were considerably more complicated than a straightforward application of Unruh's formula would suggest, and it was not possible to define a unique Unruh temperature. Nevertheless it was possible to produce a clear mathematical expression for the polarization, as will be discussed shortly.

First, though, it may be said that the time taken for building up the polarization was also easy to calculate. The formula was the same as for the linear case, but because the acceleration was so much larger for the circular case than for the linear case, the time was very much shorter: perhaps two hours for LEP, but much shorter still—around ten minutes, for SPEAR. These values were also in agreement with experiments that had been carried out at SPEAR.

However, the experimental results for the polarization in the rotating case presented distinct difficulties. The authors said that it was clear that the underlying mechanism was the same as for the linear case, but the numerical results showed substantial disagreements with those found from experiment, which themselves actually agreed with a rather more naïve theory. It was suggested that the discrepancy might be at least partially explained by the fact that the particles followed a circular orbit rather than being accelerated linearly; it might also be due to the complication of handling spin in a rotating frame.

At the end of his year at CERN, after a year in Oslo, Leinaas went to Stavangar in Norway, moving to Oslo permanently in 1989. During the 1980s, he visited CERN frequently to work with Bell, and he was delighted that Bell himself came to the 1987 meeting of the Norwegian Physical Society, after which the two had a pleasant walk in the hills outside Oslo.

In 1984, while he was visiting CERN, Bell and Leinaas considered the Unruh effect again, together with another visitor to CERN, Hughes [55], who was originally from England. In this paper [56], they pointed out that a serious treatment of the Unruh problem would involve taking account of the fluctuation of the electron motion about its mean, and thus would involve dealing with an extended quantum system, in other words, with an extended model thermometer.

The argument involved aspects of relativistic quantum theory related to CPT symmetry—a subject, of course, dear to John's heart. Hughes [57] had, in fact, already written a substantial account of these ideas. As it turned out, the three authors did not need to use the CPT theorem itself; however, they did use the central idea, and also, 'in a primitive way', some of the mathematical ideas at the basis of the proof.

There were some interesting comments in this paper, referring the reader to a paper by Geoffrey Sewell [58] for axiomatic rigour, and commenting that the treatment Bell and colleagues had given in this paper 'may also have some value to readers who find difficult the rigorous mathematical treatment of Sewell'. Bell was certainly capable of working as rigorously as anyone else, but it seems that his idea was to work as rigorously as necessary—but no more!

One of the main points of the paper was to demonstrate that there was an ambiguity in the definition of temperature in this case, resulting directly from an ambiguity in that of acceleration. Fundamentally, the authors pointed out that 'standard clocks attached to different points in the "uniformly" accelerated object do not keep time with one another.'

Finally, the arguments were applied to the electron thermometer, and it was concluded to a reasonable order of approximation, although not to higher orders, that the results of the paper of Bell and Leinaas were correct.

The third paper on this topic appeared in 1987 under the names of Bell and Leinaas only. Hughes was in general contact with Bell from 1978 until Bell's death, but his fellowship at CERN was only from 1982 to 1984. He now works at Los Alamos National Laboratory, where he has become an expert on quantum information theory.

In the 1987 paper, Bell and Leinaas [59] considered in some detail the effects on the spin of fluctuations in the position of the particle around the classical orbit, and they found that the fluctuations do cause important effects. The authors concluded that the spin excitations correspond to a thermal spectrum that is higher by a factor of around 1.4 than that provided by the formula used for linear acceleration. They found also that the calculations of polarization were still problematic, and the measured value could not be used directly as a demonstration of the (circular) Unruh effect.

One last point mentioned by Hughes is interesting. He pointed out that one of the paradoxical aspects of the Unruh effect is that the accelerated observer views a quantum mechanically pure state, which is the quantum field vacuum, to be a thermal mixed state. The resolution of this paradox is that there are correlations in the vacuum state over spacelike intervals; these correlations could, in theory, maintain the pureness of the state, but the observer is unable to detect them when they extend over his event horizon.

In his account of Bell's work [55], Hughes mentioned a lecture titled 'Gravity' [60], which Bell gave to the International School on Physics with Low Energy Antiprotons, held at Erice in October 1986, in light of the upgrade of the antiproton source at CERN. It is an indication of Bell's reputation as a theorist of very broad knowledge that he was asked to talk on this topic, and that he gave a wide-ranging and extremely detailed account, including a historical journey from Newton and Einstein to the most recent theories, as well as a discussion of 'old and new experiments in gravity'. In fact, Bell did not write his lecture up for the proceedings, so the published version of the lecture consists of verbatim lecture notes which had not been edited by Bell.

In his summary of Bell's paper, Hughes concentrated on one of the most interesting experiments that Bell described: the use of the gravitational red shift to test gravitational interactions. He pointed out the error, commonly held by particle physicists, that the gravitational red shift is a result of 'the weight of light'. Bell showed that this cannot be correct, by using a thought experiment in which atomic clocks are placed at the top and bottom of an (existing) 22 m high tower at Harvard, and then one waits for 10^{10} years, during which a 10-minute discrepancy in the times shown by the clocks builds up.

Rather than sending down photons and blaming the discrepancy on their weight, Bell suggested that a man at the top could shout down to one at the bottom, 'What's the time?' In case, as he said, shouting was not considered to be bon ton at Harvard, carrier pigeons could be sent down, and the resultant paper for the *Physical Review* could be called 'Apparent Weight of Carrier Pigeons [rather than photons]'. As Hughes pointed out, the true explanation is that the discrepancy is due to the coupling of gravity to the energy content of the clocks.

Results and thoughts on quantum theory in the 1980s

As the 1980s started, it is certain that Bell could hardly wait for the results of Aspect's experiments, to find out whether, if the subsystems in each wing

Figure 5.12 John Bell (right) and Alain Aspect (left), in Paris about 1985. Courtesy European Physical Society.

of the apparatus were *not* given time to react to the polarizer setting in the other wing, the results might just possibly violate the quantum prediction and agree with Bell's inequalities (Figures 5.12 and 5.13). As well as switching the polarizer settings, Aspect [61] and his colleagues Philippe Grangier, Gérard Robert, and Jean Dalibard were able to take advantage of the progress that had occurred in laser technology since the experiments of Fry and Thompson, so there was an improvement of a factor of about 10 in excitation rate, leading to greater accuracy in results.

Also, whereas the previous experiments had used single-channel polarizers, some of Aspect's experiments used two-channel polarizers, which made it possible to detect either mode of polarization. The two-channel polarizers consisted of polarizing cubes with dielectric layers that transmitted one polarization and reflected the other. This change also made analysis much speedier than it had been with single-channel polarizers.

Naturally, as for all the prior experiments, there was an auxiliary assumption; here the most natural one was to consider the same photons being detected in the experiment whatever the orientation of the polarizers—a reasonable enough assumption, although, of course, one that could be debated.

Three experiments were reported. In the first [62], which was reported in 1981, single-channel analysers were used but the detectors were far enough apart that light could not move from one measurement to the other, although, of course, information about the direction of each polarizer had plenty of time

Figure 5.13 The Bells and Reinhold Bertlmann, on a rainy day about 1985. © Renate Bertlmann.

to reach the other one, as the directions of the polarizers themselves were static. In the second experiment, double-channel analysers were used [63].

In both cases, local causality was convincingly ruled out; in the second experiment, for which local realism said that the appropriate parameter, S, should be between -2 and $+2$, it was found to be equal to 2.697 ± 0.015, clearly violating Bell's inequality by a substantial amount. In contrast, the quantum prediction was 2.70 ± 0.05 and so was definitely upheld. The experiments actually tested the predictions for a wide range of cases, always with the same result.

The third experiment [64] was the crucial one in which the settings of the polarizers were changed while the photons were in flight, and the distance apart was again large enough that information on these settings could not be exchanged between the two wings of the experiment at a speed less than or equal to that of light.

In this experiment, the authors had to revert to using single-channel analysers. In fact, each polarizer was replaced by an optical switching device followed by two polarizers in the two different orientations required for that experimental run in that wing. The optical switch was able to switch the incident light in a periodic fashion from being incident on one polarizer to the other extremely quickly.

To test whether the switch would be quick enough for the desired experiment, we may compare the period of the switching (about 10 ns, where 1 ns = 1 nanosecond, which is 10^{-9} s) and the lifetime of the intermediate level (5 ns) with the time it would take for information to move from one wing of the experiment to the other. The subsystems for the two wings of the experiment were about 12 m apart; so, dividing that distance by the speed of light, we see that the time it would take for information to pass between them would be around 40 ns. Therefore, quite clearly, information on the polarizer angle in one wing of the experiment could not reach the atomic event in the other in time to influence it.

The switching of the light was achieved by acousto-optical interaction with an ultrasonic standing wave in water, the standing wave resulting from interference between two counter-propagating acoustic waves produced by electroacoustical transducers. Light that is incident on the water is transmitted when the amplitude of the standing wave is zero, and it is deflected when that value is at its maximum.

Clearly Aspect's third experiment was an immense achievement, certainly worthy of the highest honours in science. However he himself would agree that the experiment was not ideal; a number of problems emerged while it was being performed. Before the experiment, tests of the switching with a laser beam had worked perfectly but, in the actual experiment, because of the divergence of the light beams, it was found that switching was not complete; some photons did not travel to the correct polarizer.

Also, because single-channel polarizers were used in this experiment, as with experiments of the previous decade , experimental runs had to be carried out four times: once with both polarizers present, once with each in turn absent, and once with both absent. Because the divergence of the beams had been reduced to improve the switching, the collection rate was only a few particles per second, a rate that was reduced by a factor of around 40 in comparison to what had been achieved in the group's previous experiments, and which was not much greater than the accidental rate of around 1 per second. Consequently, experimental runs had to be very much longer than in the previous experiments—around 12,000 s longer—and so ran into problems with experimental drift.

Thus, it was not at all surprising that the results in this experiment were less accurate than those obtained without switching. In this case, the relevant parameter, S' had to be between −1 and 0 for local causality to be upheld. The result was $S' = 0.101 \pm 0.020$ so, although the accuracy of the experiment was lower than in the previous experiments, the inequality was still violated convincingly. The quantum prediction was $S' = 0.113 \pm 0.005$; so, although the central values were a little apart, there was still good general agreement.

Overall, the most obvious problem with the experiment was that the switching was not random, which would have been ideal, but periodic. However, as Aspect pointed out, the switches in the two wings of the experiment were

Figure 5.14   Anton Zeilinger. © Renate Bertlmann; Central Library for Physics, Vienna.

driven by different generators at different frequencies, so drift should occur independently in the different wings of the experiment.

It may be mentioned here that, four years afterwards, Zeilinger [65] (Figure 5.14) was to point out what may be regarded as an unfortunate coincidence in the details of Aspect's third experiment. As has been said, the polarizer accessed by the switch changed every 10 ns, while the time for a signal travelling at the speed of light to get from one wing of the experiment to the other was 40 ns. In that time, the polarizer to which the information was travelling would have changed from, say, A to B, to A again, to B again, and then back to A. It could thus be argued that the switching was irrelevant!

This argument may or may not be taken too seriously—after all, there is no stipulation that any information pass at the speed of light; it could travel more slowly. The argument did show, though, that, for all the immense experimental ability of Aspect and his team, as technology improved, it would be important to repeat the experiment with random rather than periodic switching. Such an experiment was to be carried out by Zeilinger and co-workers 16 years after Aspect's work.

In fact, it perhaps should be said that the problem Zeilinger pointed out was not so much related, or not *just* related, to the locality loophole as to a different loophole, the *memory loophole* [66]—the hidden variables related to the particle in one wing of the experiment being able to *remember* measurement choices and outcomes in that wing or the other one. Memory of settings is rendered irrelevant by random switching, while the problem of memory of results may be dealt with by using a statistical analysis in which independence of different results in a sequence of measurements is not assumed.

Turning back to 1982, despite the fact that arguments could be made that neither the detector nor the locality loopholes had been conclusively closed—indeed, at the time of writing this book, the struggle for a completely loophole-free test is still proceeding—there is little doubt that Bell recognized that the verdict had been given.

He could not believe that marginally imperfect experiments would follow quantum theory, but the results of perfect ones would be entirely different and agree with local causality. He had to accept that the physical world violated his inequality, and local causality had not been upheld. Even in these most crucial experiments, quantum theory had triumphed. He would, of course, continue to think about these matters for the rest of his life, but he would have to work with, or perhaps round, the quantum theoretical results in his subsequent ideas, rather than challenging them.

Aspect's results drew steadily increasing interest to the topic and to Bell's ideas. Google Scholar shows the number of citations each year for *On the Einstein Podolsky Rosen Paradox* reaching the 20s in the 1970s, and then 50 for the first time in 1983, 100 in 1994, 200 in 2000, and 500 in 2008. The total by 2015 is around 9,500. The conclusions of Aspect's work, mainly expressed in the simple terms that 'local realism was ruled out', were mentioned in popular journals, newspapers, and private conversations.

Naturally, this situation was extremely aggravating to those we may call the *realists*—Selleri, whom we met in two previous chapters, Emilio Santos, Tom Marshall, and, a little later, Caroline Thompson, who, for dearly held philosophical reasons, were dedicated to the traditional virtues of locality and realism. In the year after Aspect's work, the first three published a paper titled 'Local Realism Has Not Been Refuted by Atomic Cascade Experiments' [67], which made use of the loopholes in the experiments that had been performed to date and claimed to produce a model which obeyed locality and realism and which fitted Aspect's data as well as quantum theory did.

Selleri also wrote *Quantum Paradoxes and Physical Reality* [68], which was mentioned in Chapter 3 and which is a sustained and powerful argument for his traditional viewpoint across the whole of science; in it, he said that profound distortions had been introduced into the contemporary views of modern physics as a result of 'the ideological and philosophical ideas of the Copenhagen and Göttingen schools' [69]. Bell was, of course, a great believer in realism, but it is doubtful that he felt that this forthright approach of the 'realists' had any chance of being successful.

Let us now turn to the papers authored by Bell alone in the 1980s. It is interesting and fortunate that many of these, whether deliberately or not, could be seen to have the nature of final statements of different types.

The first paper to be discussed [70] had the double merit of being the paper which Bell suggested [71] would be the best point of entry to his work, even for quantum experts, although he stressed that they should not omit the slightly more technical material at the end, and of including, without doubt, the best joke in his published work.

It will be appreciated that Bell was always on the lookout for examples where things or people were spatially separated but some of their properties were intimately connected—analogies, of course, to EPR—and his aim was always to make Bohrian analysis look a little silly and to reinforce the EPR position. In Chapter 3, we met the example of the twins, and around 1980, Bell came across another example that he thought would both be amusing and help his argument.

He had noticed that Bertlmann [10] had decided long before that it was 'crazy' to wear matching socks—socks should be worn 'correctly', as he put it, with different colours on each foot. This did not escape Bell's notice, and he took advantage of the idea when he visited Vienna in autumn 1980 as a Schrödinger guest professor to give three talks; he gave one on his work on particle physics with Bertlmann, and two on quantum theory, the first dealing with his general views on the Copenhagen interpretation and its competitors, and the second discussing non-locality.

It must have been in the second that he remarked that the EPR system behaved exactly as Bertlmann's socks did. A little laughing and joking ensued, but Bertlmann thought no more about it. It seems, though, that, unknown to Bertlmann, Bell had already used the joke in a more public arena than the lecture hall; however, as Bertlmann said, Bell was not in the habit of announcing his jokes.

Bell had been asked to give a talk at the Hugo Foundation in June 1980, with publication to follow in 1981, and his paper was titled 'Bertlmann's Socks and the Nature of Reality'. He started by saying that 'the philosopher in the street', that is, one who has not suffered from instruction in quantum mechanics, is totally unimpressed by EPR correlations; it continued with 'The case of Bertlmann's socks is often cited.' We are informed that Dr Bertlmann likes to wear socks of different colours. While what colour sock he will have on a particular foot on a given day is totally unpredictable, what you *do* know is that, if you see one sock and it is pink, the second sock will *not* be pink. Although, as Bell added, 'there is no accounting for tastes', there was surely no mystery that the sock colours are correlated.

For the benefit of the audience, Bell provided a beautiful sketch (Figure 5.15), which fortunately the journal retained, of 'M. Bertlmann' emerging into a

Les chaussettes
de M. Bertlmann
et la nature
de la réalité

Fondation Hugot
juin 17 1980

pink →

not
pink →

Figure 5.15 The famous socks. J. S. Bell. Bertlmann's socks and the nature of reality, *Journal de Physique* Colloque C2, pp. 41–61 (1981).

room, with the sock which was visible labelled as 'pink', and an arrow help-fully labelled 'not pink' pointing to the sock which was not yet visible.

Bell then explained the EPR (Bohm) experiment and said that the general belief among physicists—he mentioned Pascual Jordan in particular—was that it was the *observation* that *produced* an experimental result. Thus, obser-vation of 'spin-up' in one wing of the experiment actually *produced* 'spin-up' in that wing and hence must also produce 'spin-down' in the other. As Bell said, it is if we were to deny the reality of the colour of Bertlmann's socks until they were observed; as a child might ask, how does the second sock know what colour the first one has chosen or been presented with in the observation?

'Paradox indeed!', Bell remarked—but only for the quantum theorist, not for the man in the street, or indeed for EPR, who took it for granted that, just as the colours of Bertlmann's socks had been fixed when he put them on and, of course, he arranged that the colours were different, so there must have been an extra parameter—a hidden variable—that determined the result that would be obtained in each wing of the EPR experiment; and, it would be arranged so that different results would be obtained in the two wings.

A point that would have been much more important for Einstein than for the man in the street would have been related to the transfer of information from one sock to another, or from one spin to another that was required under Jordan's approach. While Bertlmann's feet were necessarily close enough together to make this unproblematic in the time required (if still puzzling conceptually), for the EPR case it could involve information travelling across the universe, and it was this that Einstein called 'spooky action at a distance' [72].

Bell continued with the sock motif as he moved on to explain his own theorem. For washing pairs of socks at particular temperatures, he asserted that the sum of (the probability of one of the socks surviving 1,000 washes at 0°C but the other sock not surviving 1,000 washes at 45°C) and (the probability of one of the socks surviving 1,000 washes at 45°C but the other sock not surviving 1,000 washes at 90°C) must not be less than (the probability of one sock surviving 1,000 washes at 0°C but the other sock not surviving 1,000 washes at 90°C).

This result, Bell continued, is trivial, and it is just as trivial when it is translated to the EPR situation, with Celsius degrees replaced by angular measure, pairs of socks replaced by EPR pairs, and 'surviving 1,000 washes' becoming 'registering with the polarizer at a particular angle'. Yet, according to quantum theory, this result is not respected by quantum mechanics. The parallel between socks and EPR pairs, which works so well when the directions travelled in the two wings of the experiment are in the same line, fails when they are at certain angles.

Bell, by the way, acknowledged d'Espagnat's paper, which was mentioned at the very end of Chapter 4, for the argument about the EPR pairs (not the socks). He also commented that, on the failure of the 'trivial' argument and the success of quantum theory, it would be wrong to say that 'Bohr wins again', as the argument just given was not known to him and the other opponents of EPR. Nevertheless, Bell admitted that it would no longer be so easy for Einstein to say of local causality: 'I still cannot find any fact anywhere which would make it appear likely that that requirement will have to be abandoned.'

As Bell continued, his presentation up to this point had aimed at simplicity, but from this point on he would aim at generality. First, he listed some of the previous example's aspects which could be dispensed with: the perfection of alignment of the two measurements, determinism, the mention of spins, particle, fields or any other picture of what goes on at the microscopic level, and any talk of a 'quantum mechanical system'. Rather, the argument would depend solely on predictions about correlations in visible outputs of conceivable experimental set-ups.

His system consisted of a long box of unspecified equipment with three inputs and three outputs, the outputs registering 'yes' or 'no', and the central input being a 'go' signal that would commence an experimental run. Immediately after 'go', the central output would register 'yes', indicating a successful process, and very soon afterwards the other outputs would register 'yes' or 'no', indicating, for example, whether a measurement of the direction of spin in that wing, by an appropriate device, had given a particular reading—say, 'up', rather than 'down'.

However, just before these signals, information would be inserted at the two outer inputs, dictating the orientations of the magnets in these devices. The box would be sufficiently long that information about the orientation at one end could not, even travelling at the speed of light, reach the other end before the 'yes' or 'no' had been registered at that end.

Many repetitions of the experiment would build up statistics, which, as Bell continued, could be tested by the CHSH inequality. However, as he showed, there are results which, although allowed by quantum theory, would fail the CHSH test; consequently, 'the quantum correlations are locally inexplicable'. He stressed that the argument assumed *only* that the outputs and the inputs were 'well localized'.

He then suggested four possible positions that could be taken on these matters.

First, the experiments might actually support local realism against quantum theory; even though this paper was published before the results of Aspect's work were known, Bell admitted that the situation was not encouraging.

Second, the input parameters, treated above as separate, might not actually be independent. We may have arranged for them to be generated by 'apparently random radioactive devices, housed in separate boxes and thickly shielded, or by Swiss national lottery machines, or by elaborate computer programmes, or by apparently free willed experimental physicists', but we cannot be *sure* that they are not influenced by the same factors that cause the outputs. This is the freedom of choice or free will loophole mentioned in Chapter 4. But, as Bell said, this result would be more mind-boggling than that of causal chains travelling faster than light. Apparently separate parts of the world would be 'deeply and conspiratorially entangled and our apparent free will would be entangled with them'.

The third possibility is that causal influences might indeed go faster than light. As mentioned in Chapter 4, Bell thought that a return to a pre-Einstein relativity might be the neatest solution, although the fact that the ether was unobservable would be disturbing, as would indeed be the very question of messages being sent faster than light.

The fourth solution would be to admit, against all Bell's instincts, that Bohr had been right all along and that there is no reality below some 'macroscopic' level. Then, Bell said, fundamental theory would remain vague until concepts like 'macroscopic' could be made sharper than they were at that time.

Bertlmann, of course, was in Vienna and knew nothing about the paper until a colleague, who kept his eye on the preprints arriving in the department, rushed up to him to tell him that Bell had written a paper about his socks [10]. Bertlmann was, of course, shocked, delighted, honoured... He immediately phoned Bell at CERN, and asked: 'What have you done?' Bell was totally unapologetic and just said: 'I am going to make your socks as famous as Schrödinger's cat!'

Up to this time, like most people working on particle physics at CERN or elsewhere, Bertlmann had known nothing of Bell's work in quantum theory, but he thought that now he must become an expert on it, although he was always to tell people that he was *not* an expert on quantum mechanics but rather a victim of it! So he studied and then made it his business to give talks in this area of physics and send out his own preprints. Since then, much of his published work has been in the area of the foundations of quantum theory, although often the papers have also been connected to particle physics [73, 74].

In the next paper to be mentioned [75], which was written in honour of de Broglie, Bell made his final statement about the de Broglie–Bohm theory, the pilot wave, with his famous sentence: 'In 1952 I saw the impossible done.' This paper was discussed in Chapter 2 when Bohm's work was introduced; this chapter will just reiterate Bell's disappointment that Born and von Neumann chose to ignore the theory, and his intense anger that the physicists of power, Pauli, Rosenfeld, and Heisenberg, resorted to smug name-calling rather than admit that Bohm's reinvention of de Broglie's work showed the falsehood of their creed that hidden variables could not exist.

In the paper, Bell presented a fairly standard account of the de Broglie–Bohm argument, and also showed where the various impossibility 'proofs' failed; two we have met before, those of von Neumann and Gleason–Jauch, and what he called 'the most recently published' by Res Jost [76]. The latter 'proof' concerned unstable 'identical' particles. Jost had suggested that, if the decay times of these particles were determined in advance by hidden variables, which would differ from particle to particle, the particles could not really be regarded as identical, even though being identical would be essential for them to obey, as they would, quantum statistical mechanics.

Bell pointed out that this question had been dealt with by Bohm in his original papers. While beta decay cannot be discussed in this non-relativistic version of the theory, alpha and gamma decay and fission are explained by regarding unstable nuclei as composites of stable protons and neutrons, and the wave function then respects the rule of symmetry or antisymmetry. Consequently, all the statistical predictions would be retained.

In this paper, Bell cited two interesting papers [77, 78] written a few years earlier by three members of Bohm's group at Birkbeck: Hiley, who had worked closely with Bohm for many years, Chris Dewdney, and Chris Philippidis. With progress in computers, they were able to compute Bohm theory trajectories for two extremely interesting cases: two-slit interference, and tunnelling.

It was a dramatic shift from writing equations on paper, however meaningful they might be, to seeing graphs of the particles emerging from the slits and travelling along a range of trajectories, deflected even in free space by the quantum potential, to the maxima of the interference pattern. In the tunnelling case, it is fascinating to see particles approaching the barrier, some being reflected by the quantum potential without reaching the barrier, some entering the barrier but still being deflected, and others tunnelling through.

The next of Bell's papers [79] to be discussed briefly consists of his opening remarks at a conference titled 'New Avenues of Quantum Theory and General Relativity', which was held at Amalfi in May 1984; the paper is called 'Speakable and Unspeakable in Quantum Mechanics'. In this paper, he started by talking about Arthur Koestler's 1959 book *The Sleepwalkers* [80], which discusses the Copernican revolution and claims that the architects of the revolution were driven by religiously motivated prejudices and obsessions, that they made mistakes that luckily cancelled at the important places, and that they failed to understand what was important in their own results. All this, Koestler said, was essential for the eventual success of the endeavour—the architects of the Copernican revolution had sleepwalked their way into the scientific future.

Bell then asked whether any elements of Koestler's picture applied to the development of quantum theory and relativity. There are no obsessions, he suggested; today's theorists take up and put down hypotheses with light hearts, and they are never in fear of religious authorities—and he insisted, presumably with a twinkle in his eye, that they never make mistakes, and they are always and immediately able to distinguish between what is important and what less so.

However, he did see some likeness between Koestler's use of the term 'sleepwalking' and the way in which theoretical physicists had made immense progress despite the fundamental obscurity in their understanding of the central theory of quantum mechanics. Were they sleepwalking, and, if so, since progress had been so rapid, would it be wise to wake them up?

Rather than doing so, he argued against a myth—that quantum theory had somehow undone the Copernican revolution—that Copernicus and his followers showed that man was not at the centre of the universe but that, with its emphasis on observers, quantum theory had reversed the process. A few popular presentations, he said, have implied that the very existence of the universe depends on us being here to observe it [81]!

He continued that he did not agree that the public should be told that a central role for consciousness is integrated into modern physics, or that 'information' is the real stuff of physics, or that the technical features of contemporary physics had been anticipated by the saints of ancient religions via introspection.

He pointed out that, once experimenters have decided on the experiment and set up the equipment, their job is done, and the apparatus functions well without them, and he then explained why the physicists who had first studied quantum theory had found it impossible to describe the 'quantum world', and so had decided that all that could be done was to analyse the results of measurement—and, indeed, that this analysis was all that was required.

It would be as if, he suggested in an amusing but also highly useful explanation, our friends had no words to describe the strange places where they go on holiday. All we could learn from them is whether they had come back browner or fatter than before. If they had, and other friends wished on their own holidays to become browner or fatter than they had been before they went on holiday, we may advise them to opt for the same location. In a similar way, he said, our apparatus visits the (unspeakable) microscopic world, and we see what happens to it (speakable) when it is there. But this leaves the problem of how to divide the world into 'speakable' apparatus and 'unspeakable' quantum system.

Bell then said that the 'founding fathers' of quantum theory were wrong. There was a way out, which was, of course, the de Broglie–Bohm interpretation, which he then described, although he admitted that this theory itself had a problem: non-locality. Although non-locality was not relevant for signalling, Bell still found it a gross violation of relativistic causality. In addition, Bell said that he found it of major concern that our two 'fundamental pillars of contemporary theory', relativity and quantum theory, were incompatible.

The title of this paper was subsequently used by Bell for a collection of his papers [82]; this collection was put together and published by Simon Capelin of Cambridge University Press in 1987. Doubtless, Bell was pleased at this acclamation and, much more recently, Capelin [83] has said that publishing this collection of Bell's papers was one of things he was most proud of in his career. Two more points from Bell's preface to the book are worth mentioning. The first is that he amusingly dismissed the 'impossibility proofs' as 'unspeakable'; the second, touchingly, and as mentioned in Chapter 2, is that he renewed very especially his warm thanks to Mary Bell: 'When I look through these papers again I see her everywhere.'

The next paper to be discussed here, published in *Physics Reports*, was originally Bell's own contribution to the Amalfi conference [84]; a version with an expanded introduction and additional references was then published in a book honouring Bohm [85]. In the paper, Bell wrote that its original title had been

'Quantum Field Theory Without Observers, or Observables, or Measurements, or Systems, or Apparatus, or Wave Function Collapse, or Anything Like That'—a title, which, he said, might suggest that the issue discussed in the paper was a philosophical one. On the contrary, according to Bell, it was quantum theory's conventional formulations, and in particular those of quantum field theory, which revelled in philosophical terminology, and which were unprofessionally vague and ambitious. Professional theoretical physicists, he suggested, should do better than that and he said that Bohm had shown a way of doing so.

Bell continued that, even when physicists are practically forced to admit that Bohm's reasoning is cogent, a final protest is that it is all non-relativistic—despite the fact that Bohm himself, in one of his original papers, had applied his scheme to the electromagnetic field, and, with Hiley, had subsequently applied it to scalar fields. However, Fermi fields had not been covered; while fermions *might* be composite structures and so no theory would be actually required (and here Bell cited a Skyrme paper of 1961), they *might not*; so, he then constructed a theory for Fermi fields.

As always in his arguments, some 'observables' were promoted to 'beables', although most were considered redundant, and all that was required was to be able to define the positions of things, including instrument pointers, and ink on paper (or the modern equivalent). For this purpose, he said, he chose to work with fermion number density, with the numbers on a dense lattice being the local beables of the system. The vector representing the state of the system would also be a beable, although not a local one.

For development in time, Bell followed the Schrödinger equation. As he said in the paper, although he was forced to allow the fermion number density to evolve probabilistically, he was hoping that this feature of his analysis would disappear if the lattice were replaced by a continuous distribution. The foregoing is an account of what Bell called his new theory 'BQFT', for 'de Broglie–Bohm beable quantum field theory'.

'To what extent', he continued, 'does it agree with OQFT, that is, "ordinary" "orthodox" "observable" quantum field theory?'—the words 'ordinary', 'orthodox', and 'observable' being his, with the additional comment, 'whatever that may mean'. It is clear, he said, that OQFT would be judged by its agreement or otherwise with BQFT, not the other way round! The difficulty in answering this question, he said, is the absence of a sharp formulation of OQFT.

Bell's first attempt at reducing this ambiguity, called OQFT1, proposes a reality in which God prepares the universe and then lets it run. After a period of time, She returns to judge the outcome, and the probability that She observes any particular configuration is identical to the BQFT probability that the outcome *is* that configuration. To that extent, there is complete agreement between BQFT and OQFT1 on the results of 'God's experiment'.

However, unlike BQFT, OQFT1 says nothing about events between 'creation' and the 'last judgement' and, however adequate the account may be 'from the Olympian point of view', it is unsatisfactory for us who live in this period and do at least imagine that we experience actual events.

In OQFT2, the vector representing the state is a sum of terms that are macroscopically different; the vector then collapses to one of these terms, thus providing some account of 'what actually happens'. Would this account agree with the one provided by BQFT? It would, Bell said, provided we accept that OQFT2, despite the inevitably vague words 'collapse' and 'macroscopic', provides a satisfactory solution to the measurement problem of quantum theory (discussed in Chapter 3 in particular).

Bell considered BQFT to be superior to both OQFT1 and OQFT2, but did not regard it as completely satisfactory because, although fermion number density was a sensible choice for a basic local beable, it was not unique; also there were great difficulties with special relativity. Bell then said that he strongly suspected that a sharp formulation of quantum field theory must require, as discussed in Chapter 4, that the rest frame of the universe be experimentally indistinguishable from frames moving with constant speed with respect to it. As he concluded, 'It seems an eccentric way to make a world.'

The next paper to be considered here [86] was presented at a 1985 conference in honour of Wigner, and was titled 'New Techniques and Ideas in Quantum Measurement Theory'. The proceedings were edited by Danny Greenberger, and it is interesting that, only about 20 years after Bell's work, so many people wished to take part in a conference on quantum measurement—there were 84 talks, in all!—and that quantum measurement was thought to be the topic most appropriate for honouring a physicist of such wide achievement.

This paper was titled 'EPR Correlations and EPW Distributions', a play on the fact that Wigner (Eugene Paul Wigner) had many years before proposed a very well-known distribution [87] to represent the quantum theoretical probabilities of the values of positions and momenta in an ensemble (or collection) of particles. In many ways, this representation is extremely interesting, and it has been of use in many applications; however, it can never be completely successful, because the vagaries of quantum theory mean that some of the probabilities must have negative values, which are forbidden for any probability.

In the paper, Bell returned to the original EPR situation (i.e. before Bohm) but replaced the settings of polarizers used, for example, in the Clauser or Aspect experiments, with measurements of the positions of the particles in each wing of the experiment at arbitrary times. These times could be used to generate terms for an inequality analogous to the CHSH inequality and thus be used to distinguish between quantum mechanics and local causality.

Bell then showed that the evolution of the Wigner function was exactly the same as that of the probability distribution of a pair of freely moving classical particles, and that, if it were initially non-negative anywhere, its evolution would remain non-negative. An example of this behaviour is that of the original EPR wave function, which is indeed non-negative everywhere at the beginning of the process and, as Bell showed, must remain so throughout its evolution. Therefore, in this case, there can be no problem of non-locality, as the Wigner distribution provides a local classical model of the correlations.

However, this is not necessarily the case if the Wigner distribution is negative in some regions. Bell concocted such an initial function and studied its evolution in time. He showed that at later times the function still had some negative regions, and he also demonstrated that it exhibited non-locality, although he remarked that he did not know whether the fact that the function taking negative values *entailed* that it had a locality problem.

Bell's ideas from this paper have received a fair amount of comment. For example, Lars Johansen [88] has argued, contrary to Bell, that an initially non-negative Wigner distribution does not necessarily remain non-negative indefinitely, while Ulf Leonhardt and John Vaccaro [89] have used Bell's ideas as the basis of an experiment in quantum optics.

The next paper reviewed here [90] is called 'Six Possible Worlds of Quantum Mechanics', which was published in 1986, but, like many of Bell's papers, had been presented in much the same form at several lectures around the same time. It is probably his most beautiful non-technical introduction to his ideas.

In the first part of the paper, he concentrated on two-slit interference, contrasting the wavelike way in which the electron (the quantum 'system') propagates, with the classical nature of the electron gun ('apparatus') and the screen (more 'apparatus') on which we see the point of scintillation. Yet, we may wish to move some of the gun or the screen from the 'apparatus' into the 'system'. In this way, and as so often, Bell stressed the 'shifty split' between system and apparatus. He then proceeded to outline six 'worlds' with which we may attempt to resolve our problems: three 'unromantic' ones, of which he rather approved, and three 'romantic' ones, which he said might make good copy for journalists, but which should not be taken seriously by professional physicists.

The first unromantic world is based on the purely pragmatic view, which suggests that, as we study the regions of the world remote from everyday experience, for example the very small or the very large, we have no right to expect to retain our usual concepts such as space, time, causality, or even perhaps unambiguity, and we certainly should not demand a clear picture of physics at the atomic level. We should be grateful that we have rules that work for calculating the results of experiments, and in practice we can deal with the shifty split pragmatically moving enough into the quantum system that moving more makes no appreciable further difference. This view, Bell said, was

the working philosophy of all those who used quantum mechanics in a practical way, the only difference being that some—nearly all!—were complacent
about it, while others were concerned—himself and, by 1986, a few others.

The romantic counterpart of the pragmatic world, he says, is that of complementarity and Niels Bohr. Bohr, he feels, rather than being disturbed by the
shifty split, rather revels in it, and delights in the bizarre, and in contradictions
and ambiguities. What Bohr calls 'complementarity', Bell would call 'contra-
dictoriness', and Bell is particularly aghast at such sayings of Bohr as: 'The
opposite of a deep truth is also a deep truth'.

Bell's second unromantic world corresponds to an attempt to bridge the
gap between classical and quantum, not with words but mathematically, by
adding a non-linear term to the Schrödinger equation. It was known that, if
non-linearity was added, there would have to be some element of probability,
known as *stochasticity*, present as well, because non-linearity without stochasticity would allow signals to be sent at speeds greater than that of light [91].
With non-linearity *and* stochasticity, there was perhaps a possibility of creating
a theory in which electrons could act as waves when propagating, but tables
and black marks on photographs could be classical.

As yet there had been no breakthrough in this task, although Bell appreciated Pearle's work, and he was pleased that the task would require mathematical work by theoretical physicists rather than interpretation by philosophers.
As we shall see shortly, Bell was soon to be delighted by further substantial
progress.

The romantic alternative to the second unromantic world puts the division, not between small and big, but between 'matter' and 'mind'. Mind, it
says, is surely different in principle from anything else in the chain from atom
to strictly classical apparatus. Bell reported that eminent physicists such as
Wigner and Wheeler had supported this idea, but that 'unfortunately' it had
not been possible to develop the ideas in a precise way. Bell said that he was, in
fact, convinced that mind had a central role in the ultimate nature of reality,
but he doubted whether contemporary physics was at the stage where the idea
would soon be professionally useful.

Bell's third unromantic world is that of de Broglie and Bohm, and thus, of
course, combines the waviness of electron interference and the definite nature
of 'happenings'. In his opinion, it undoubtedly showed the best craftsmanship
of the various worlds, but he added: 'Is that a virtue in our time?'

Its romantic partner is the many-worlds interpretation, which was discussed at some length in Chapter 4. Bell called it 'an extravagant, and above all
an extravagantly vague, hypothesis' and 'almost silly'; yet he also said that it
might have something distinctive to say about EPR and that it might also make
us more comfortable about the existence of our own world, 'which seems to be
in some ways a highly improbable one'. The latter point refers to the so-called

cosmological anthropic principle, which tells us that, if many of the fundamental constants of physics had values only slightly different from what they have in our universe, life of any form would be impossible [81]. Bell also felt that it might be worthwhile to formulate a precise version of the many-worlds interpretation to check up on these points.

As a more general thought, he hoped that, with technical pragmatic progress, the solution to the quantum measurement problem, invisible 'from the front', might become apparent 'from the back'.

Let us now break off from discussing these papers to describe the important advances which occurred in the foundations of quantum theory during the second half of the 1980s, work to which Bell himself responded with great pleasure.

The first advance was the result of work carried out exactly as Bell had hoped—the addition of a non-linear (but stochastic or probabilistic) term to the Schrödinger term to join up the quantum and the classical regions—and it was carried out in 1986 by three Italian physicists: Giancarlo Ghirardi (Figure 5.16), Alberto Rimini, and Tullio Weber, always known as GRW [92, 93].

Their argument was that, since von Neumann collapse is stated to take place at a measurement but, as we have seen, the term 'measurement' is ill-defined, his approach is fundamentally unsatisfactory. GRW, in contrast, add rigorously to the Schrödinger equation a term that causes collapse, which they

Figure 5.16 Giancarlo Ghirardi. © Renate Bertlmann; Central Library for Physics, Vienna.

describe as a *spontaneous localization process*. At random intervals, the spread of the wave function is reduced to a very small distance, which is not precisely zero, as that would obviously cause difficulties for the Heisenberg principle, but is around 10^{-7} m. The term is a formal addition to the evolution of the wave function, so the scheme is mathematically respectable, unlike the ad hoc collapse of von Neumann.

The crucial point in their scheme is that the rate of localization depends on the size of the composite particle involved. The time between localizations is written as τ/N, where τ is a new parameter estimated by GRW to be about 10^{15} s, or about 10^8 years, and N is the number of particles in the system. So, for a single particle, we would have to wait around 10^8 years for collapse, just as we require; however, for a macroscopic system, with N perhaps equal to 10^{20}, collapse would take place in about 10^{-5} s, again as one would wish. As Bell [94], who loved the scheme, said: 'Schrödinger's cat is not both dead and alive for more than a split second.' Ghirardi [93] has spelled out the enormous support and help Bell gave to GRW once their paper had been published.

The second exciting event of the late 1980s in quantum foundations was the announcement of 'Bell's theorem without inequalities'. By this time, Horne had teamed up with Zeilinger and Greenberger (Figure 5.17). At first, the three

Figure 5.17 Danny Greenberger (centre); Mary Bell is on the left. © Renate Bertlmann; Central Library for Physics, Vienna.

Figure 5.18 Helmut Rauch. © Renate Bertlmann; Central Library for Physics, Vienna.

of them had carried out important work on the neutron interferometer, which had been invented independently around 1974 by Helmut Rauch (Figure 5.18) in Vienna, and Sam Werner at the University of Missouri [95]. Zeilinger worked in Vienna with Rauch, and Horne worked at MIT under Clifford Shull, who, at the age of 79, was to receive the Nobel Prize for Physics in 1994 for his work on neutron diffraction; Horne's actual appointment was at Stonehill College, and Greenberger's at the City College in New York, but there were many trips across the Atlantic for the three to work together.

It was, of course, obvious in principle that neutrons, like photons and electrons, could interfere; however, achieving this result in practice was a major achievement for Rauch and Werner, and neutron interferometry displayed many of the conceptual features of interferometry more directly than had been achieved previously; indeed, it may be said that it has succeeded more than any other device in making the thought experiments of theory into real experiments carried out in the laboratory [69, 95–97]. However, in 1986, the MIT neutron laboratory closed down.

Horne, of course, had been in at the very beginning of work on Bell-type inequalities, and his comrades were also extremely interested in the foundations of quantum theory, so they were enthusiastic about moving to this rather new area of research. In particular, they decided to consider the case of three, rather than two, entangled particles.

After examining very many cases, Greenberger, greatly to his surprise, came up with an example that was not an inequality at all—it was an equality! In contrast to all examples up to that point, for which the clash between local causality and the quantum requirement was statistical, here there was a direct contradiction. Since all three were working on these examples, there was no hesitation in calling the result after all three, and it is always known as the GHZ state.

The GHZ state is a linear combination of the state in which all three spins are (+), and the state where they are all (−). Clearly, there is a high degree of entanglement. As the argument goes, it is possible to construct a certain mathematical expression which we may calculate either via an EPR-type argument or by formal quantum mechanics. The first way we get +1 but the second, we get −1. Certainly there is a direct conflict between quantum theory and local causality—and it does not require any inequalities!

GHZ were a little casual about publication [98] and, although they published a conference paper in 1989, their first formal publication on the topic was not until 1990, when, since they were perhaps slightly edgy about philosophers already showing an interest, the team was buttressed by Shimony [99].

However there was also an intervention by Mermin (Figure 5.19), who had already made major contributions to the physics of solids, and was becoming extremely interested in the foundations of quantum theory. In June 1990, he contributed an account of GHZ to *Physics Today*, a periodical sent each month to all members of the American Physical Society; however, in doing so, and of course giving full credit to GHZ, he came to the conclusion that he himself had a deeper understanding of the scheme than they did. He then published a paper on the equality himself, still crediting GHZ but actually managing to get his paper [100] published four months before that of GHZ and Shimony.

Later on, Mermin recalled [101] sending a copy of his work to Bell and receiving the reply that the argument 'filled Bell with admiration'. Another interaction between the two was to come in 1989, when Mermin told him that, in a book on special relativity [102] he had written long before, he had followed Bell's suggested path of starting with the pre-Einsteinian ideas (described in Chapter 4). Bell was polite but sceptical at the time; however, later on he sent Mermin a letter in which he said that Mermin 'did do it right'.

We now return to Bell's last papers on quantum theory, the first derived from his talk [94] at the Schrödinger Centenary Conference in London in 1987. For the paper, Bell borrowed a title from Schrödinger, 'Are there Quantum Jumps?', 'jumps' for Schrödinger being anything other than the smooth evolution of 'his' waves. If we are to avoid jumps but to steer clear of hidden variables, Bell suggested, we must allow a change to 'his' equation, and this point was Bell's cue for a passionate account of the GRW argument—Bell introduced Ghirardi to the conference practically with a fanfare!

Figure 5.19 David Mermin (left) with Reinhold Bertlmann (right), in Vienna, 2104.
An examination of socks. © Reinhold Bertlmann; Central Library for Physics, Vienna.

Bell then gave a beautifully simple account of the GRW argument, stressing that, in the GRW state, everything, including 'measurement', proceeds according to the fundamental mathematical equations, which are not 'disregarded' from time to time on the basis of 'supplementary, imprecise, verbal prescriptions'. He suggested that detailed studies of models with particular values of the two parameters—the characteristic time, τ, which was mentioned previously in this section, and the size of the region to which the system collapsed—would build up confidence, and that GRW's jumps might be the 'local beables' of the theory. The problem of symmetry for identical particles, he felt, might be left for the moment, to be dealt with when the theory had been developed in the field context.

Bell discussed entanglement, admitting that, by this time, it seemed very difficult to maintain the hope that hidden variables might restore local causality. The GRW argument, he said, at least made the predictions of ordinary quantum theory in EPR situations more precise than they had been before. Mention of EPR brought up the question of special relativity, and while, of course, this could not be discussed in the context of the non-relativistic account given by GRW, Bell was able to show, by quite a lengthy argument, that a 'residue' of special relativity, which he called 'invariance with respect to displacement in *relative* time', did survive the GRW jumps. Bell concluded the paper by saying that the GRW argument was a very nice example of how quantum mechanics required a change, which might be very small ('on some measures'), to become 'rational', and that it also had taken away his fear that any exact formulation must conflict with special relativity.

The next paper to be discussed here [103] is often called 'polemical', being titled 'Against "Measurement" ' and being Bell's final and most savage criticism of what may be called broadly the 'orthodox' approach to quantum theory, in which the results of 'measurement' are all that are considered. The paper was presented at a 1989 meeting which was held at Erice and titled 'Sixty-Two Years of Uncertainty'; Mermin [101] subsequently described Bell's talk as being 'close to the most spell-binding lecture I have ever heard', the only competitors in his view being Feynman's 1965 'Messenger' lectures at Cornell. As well as being published in the proceedings of the meeting itself, the talk was published in *Physics World*, the publication received by all members of the Institute of Physics, and thus it received considerable attention, although, sadly, the publication was followed shortly by Bell's death.

In this paper, while Bell admitted that quantum theory was fine 'FAPP' (which was his acronym for 'FOR ALL PRACTICAL PURPOSES' and which, along with QED FAPP, has since become famous), he called for a theory that would be fully formulated mathematically, with nothing left to the discretion of the theoretical physicist and in which the 'apparatus' would not be separated from the rest of the world into black boxes, as if it were not made of atoms and subject to quantum mechanics.

Bell put forward a list of words that he believed should never be used in any formulation with the slightest pretension to physical precision: *system, apparatus, environment, microscopic, macroscopic, reversible, irreversible, observable, information,* and *measurement.*

The first three words in this list, he said, implied an artificial division of the world, and an intention to neglect, or perhaps take only schematic account of, interactions across the divide. The next four, from 'microscopic' to 'irreversible', cannot be defined precisely. On 'observable', he recalled Einstein's remark that *theory* determines what is observable; observation is a complicated and theory-laden business. On 'information', he merely remarked: 'Information? Whose information? Information about what?'

All this, though, was just warming up for his attack on the word 'measurement'. Bell complained that accounts of quantum theory gave the impression that it was exclusively concerned with the 'results of measurement', which was, he said, extremely disappointing. In the beginning, he said, natural philosophers hit on the idea of experiment to help them understand the world about them. However, he insisted, experiment is a tool. To restrict quantum mechanics to be exclusively about 'piddling laboratory operations' was to betray the whole enterprise.

He then brought in some of his favourite arguments. To carry out a 'measurement', does one just have to be alive or must one have a PhD? What about the shifty split between system and apparatus? Surely the word 'measurement' suggests ascertaining some pre-existing property. Should one not, at the very least, talk of an 'experiment' rather than a 'measurement'?

Bell went on to criticize the rather straightforward application of collapse in the classic books by Dirac [104] and by Landau and Lifshitz (LL) [105], but is perhaps even more critical of the seemingly more sophisticated treatments of measurement in a paper by the expert on statistical mechanics, Nico van Kampen [106], and in the very well-known textbook on quantum mechanics written by his old friend from CERN, Gottfried [107].

Van Kampen used the idea of collapse but, unusually, he believed that, rather than this idea being in conflict with the Schrödinger equation, it might be obtained in a totally straightforward way from a consideration of system and apparatus together. However, Bell believed that, just like Dirac and LL, he changes 'at a strategically well chosen point' from using a 'quantum' linear combination or superposition of different states to using a 'classical' mixture. Van Kampen has since published several replies to Bell's criticisms, as well as detailed accounts of his own ideas [108–111]; however, it has been suggested [112] that his argument still only produced the superposition of terms involving measured and measuring systems from which collapse should occur, and that it failed to produce a rigorous argument for collapse itself.

Gottfried's account is mathematically more sophisticated. He used an ensemble interpretation, in which a wave function relates not to a single system but to an ensemble of systems. It is a superficially plausible approach, and indeed it was often mentioned favourably by Einstein, although of course his real hopes were for a much more drastic solution to the problems of quantum theory than that provided by Gottfried. Ensemble interpretations have also been advocated by Leslie Ballentine [113, 114], but a view of them that is more nuanced than the one he presented has also been published [115].

The plausibility of ensemble interpretations can be increased by use of the *density matrix*; in this approach, the state of an ensemble of systems is represented by a square array of numbers. The great advantage of this technique is that the density matrix may be used to represent both superpositions and mixtures.

Gottfried [107] claimed that he could manage the step from superposition to mixture by taking account of the interaction between measured and measuring systems. This interaction, he argued, changes the density matrix to a diagonal form in which only the elements in the leading diagonal from the top left to the bottom right are non-zero. Bell, however, was unconvinced, even though he agreed with Gottfried about the elusiveness of the elements of the density matrix that Gottfried equated to zero.

However he sharply disagreed with Gottfried about the idea that even this diagonal density matrix, which Bell calls 'butchered', actually represents the probabilities of different results. 'If one were not actually on the lookout for probabilities,' he said, one would surely interpret the density matrix as still referring to a superposition rather than a mixture. He did not agree that Gottfried had achieved the all-important change from 'and' to 'or'.

Bell and Gottfried discussed their views at a workshop at Amherst in June 1990. They continued to disagree, although Gottfried [116] was to say that the fact that Bell saw fit to devote so much space to decrying his efforts was the highest compliment that he had every been paid.

His reply [116] came in a symposium on quantum theory, held at CERN in memory of John Bell in 1991. In this reply, Gottfried admitted that he was guilty of some of Bell's charges but was still prepared in general to stand for his own views. He said that he considered it a blessing that we may refrain from considering 'the big world outside the laboratory', that he was not squeamish about QED FAPP, and that he refused to be concerned about the 'shifty split'. He still believed that, after a certain threshold has been crossed, the combination of measured system and measuring apparatus has properties independent of, and [even] before, observation by more complex systems such as those 'that draw a CERN Grade 14 salary' (the top grade at CERN). In other words, he still believed that he had demonstrated a collapse from a superposition to a mixture of distinct results.

He did not agree with Bell's suggestion that a crisis in physics was imminent, but he felt that, even if Bell was right, any replacement for quantum theory would be even further from our conceptual prejudices, and that Bell would be left pining for the time when he had 'good old' quantum theory to kick around. However he admitted that he did not understand properly how much of the mathematical formalism of quantum theory it was legitimate to use in his demonstration of consistency between the formalism itself and its interpretation.

Over the next few years, Gottfried became less satisfied with his initial response to Bell, and he produced what he considered an improved version [117], which was also included in a new edition of his book [118], which had Tung-Mow Yan as a co-author. An interesting question raised in this work was the question of whether, if given the Schrödinger equation, Maxwell, from the previous century, could have derived the statistical interpretation of quantum theory, in other words, the fundamental role of probability and the Born rule, without having an occasional word of clarification from a physicist of today.

In his answer, Gottfried said that, in the classical limit, quantum theory relates only to ensembles, and he argued that, if Maxwell had been given information about the behaviour of spin-½ particles, he could have come very close to the Heisenberg principle and would have been able to deduce the statistical interpretation of quantum theory for discrete variables such as spin, although Gottfried admitted that Maxwell could not have achieved this interpretation for continuous variables such as position or momentum. Gottfried's later argument has also been questioned [119].

Other replies to Bell's paper came from Squires [120], who supported his former colleague, and in a rather surprising article from Peierls [121]. Peierls had long made it clear that he was a strong advocate of an orthodox approach to the interpretation of quantum theory, to the extent that he disapproved of the words 'Copenhagen interpretation', because they gave the impression that it was just one interpretation perhaps among many. For Peierls [122], the Copenhagen interpretation was not really an *interpretation* of quantum theory—it was part of the theory. Thus, it was no surprise that his response to Bell's article was titled 'In Defence of "Measurement"' [121].

It *was* very surprising, though, that Peierls's actual words seemed to bear little relationship to the very well-known ideas of Bohr or Heisenberg. Rather, he based his approach on the word 'knowledge', which we may regard, in the present context, as having the same meaning as one of those outlawed by Bell—'information'.

It is common for those who learn initially about the conceptual difficulties of quantum mechanics to decide that there is no real problem—it is just that the formalism deals with information. Before a measurement, they suggest, the state of a spin may be a linear combination of (+) and (−), which means that the observer does not know which it is. When a measurement is made, it becomes (+) *or* (−), which means that the observer now knows which it is. At the measurement, the state collapses to the result obtained in the measurement; consequently, a second measurement will give the same result.

The novice may well think in terms of all quantities *having* precise values, but the observer *knows* only some of them; however, this idea, of course, only leads to the usual problems with hidden variables. Certainly the *knowledge interpretation* or *epistemic interpretation* is not a simple solution to the difficulties, as the beginner may feel; however, as Peierls argues, it may be a start.

Bell's questions, though, are challenges to the proponent of this interpretation. 'Information about what?' is intended to make us ask, if information is about something, why not concentrate on what it is about, rather than on the information itself? To put things another way, with a knowledge interpretation, there may be confusion between what we cannot know, and what we could know but have just not bothered to find out. Bell's other question, 'Whose information?', makes it clear that our mention of 'the observer' above is extremely naïve.

Peierls began his reply to Bell by agreeing with him that a precise formulation of quantum theory was required, and that no textbook of the day explained things properly, but he disagreed with Bell in saying that he did *not* think that it is all very difficult!

If our knowledge of the system is complete, he continued, we represent this knowledge with a wave function; if it is not complete, we use a density matrix. An uncontrolled disturbance may reduce our knowledge, while measurement may increase it, but if we start off with the maximum allowed knowledge, and then gain some additional knowledge, we must compensate by losing some of our original knowledge. For example, if we know the value of the z-component of spin and then measure the x-component, we will end up knowing the value of the x-component but having lost all knowledge of the z-component.

Peierls pointed out that, when our knowledge changes, the density matrix must change, but that this process is *not* physical, and so the change does *not* obey the Schrödinger equation. In a measurement, the density matrix will have changed when the actual experiment is completed, but it will only reach its final form when we actually *know* the result.

An awkward question that is often put to proponents of knowledge interpretations is, how does quantum theory apply in the early universe, when there were no observers? Peierls's reply is that, when we draw conclusions about the early universe today, *we* are acting as observers.

On Bell's question 'Information about what?', Peierls's answer has already been given: knowledge (information) about the system we are trying to describe. This answer is a little vague, but it does make it clear that there *is* a system. This may seem obvious but, in Chapter 6, we shall reach a different suggestion.

On 'Whose information?', Peierls's answer is subtle and clear. Many people may have some information about the state of a system, but their individual information may vary, so their density matrices may be different. However, the matrices must be consistent. For example, it must be impossible that one person may know the value of the *x*-component of spin, and the other that of the *z*-component.

Peierls was able to come up with a mathematical requirement for this situation to be the case. If the density matrices for the two observers are A and B, then we must demand that they *commute*, or in other words, that $AB = BA$, a relationship that is obeyed by matrices only in special circumstances.

At the time that this paper was written, information (or knowledge) was very much a side issue in quantum theory. However, as we shall see in Chapter 6, it was in a few years to become exceptionally important. Incidentally, Bell himself would not have been surprised by Peierls's stance in this paper because, through the 1980s, the two had been having a friendly but frank correspondence [123] on these matters.

Having discussed these responses to Bell's paper, let us return to his own conclusions. He thought that there were two possible approaches that could be used to go beyond the approach constrained by 'measurement', both of which eliminate the shifty split. The de Broglie–Bohm approach retains the linear wave equation and so must add complementary variables, while the GRW approach uses only the wave-function and so must modify the linear Schrödinger equation. Bell's big question then was, which of these two precise pictures, if either, could be developed in a way obeying special relativity?

The last paper to be discussed here is Bell's final rigorous statement [124] on quantum mechanics and local causality; this paper discusses non-locality, concentrating very much on what can or cannot travel faster than light. It begins with Hendrik Casimir's argument concerned the 'coincidence' that the unrelated events of his egg being boiled and his alarm ringing occur at the same time—he said that he, 'the great chef', had imposed a structure on his kitchen. Bell called his own paper 'La Nouvelle Cuisine' and dedicated it respectfully to 'the great chef'.

In this paper, Bell started the argument by mentioning some things that *do* travel faster than light. British sovereignty is one; when the Queen dies in London ('May it long be delayed,' Bell insists), Prince Charles instantaneously becomes king, even if he is in Australia. And, while electric and magnetic fields propagate at speed c, scalar and vector potentials, which Bell described as 'mathematical conveniences and to an extent arbitrary' (being fixed by one convention or other), may travel at any speed up to an infinite one.

So we need to know what is convention and what is not, or in Bell's terminology, what are the *local beables*; however he admitted that, if space-time is quantized, or if one considers *string theory*, or even with the vagueness of quantum theory, there may be no local beables in the most serious theories. However, if we assume that the results of observation are real and localized, we may be able to make do with *local observables*, although still hoping for a reformulation of quantum mechanics in which the local beables are 'explicit and mathematical rather than implicit and vague'.

With the concept of particle no longer sharp, the statement 'particles cannot go faster than light' cannot be sharp either, but we may perhaps replace it with 'no signals travel faster than light'. Without this principle, cause may follow effect in some frames, and Bell presented an amusing example in which he might commit murder with a '*tachyon* gun' (a tachyon being a hypothetical particle that travels faster than light) but, in the frame of the courts of justice (which Bell included in his diagram), he would be seen merely as catching an antitachyon in the barrel of his gun to prevent harm to passers-by and deserves a medal!

And fortunately, as Bell spelled out, ordinary local quantum field theory does seem to have the required causal structure. There are two types of 'external interventions'—measurements, and the imposition of external fields—and neither of them, at least in a simple approach, allow superluminal signalling. 'Who could ask for anything more?', Bell asked, but immediately did so! He presented his usual concerns about measurement—where does it take place, and who is privileged to carry it out?—but he also recognized that 'external' fields must, in an accurate treatment, be incorporated into the quantum system—perhaps truly 'external' fields may be found only at the interface between the brain and the mind.

A different approach would be to study the principle of *local causality*. A first approach to this might be to say that we expect the probabilities of an event occurring in region A at a particular time to depend only on events at region B and at times prior to the time of the event in region A, such that light can travel from B to A in the time that has elapsed. (We may express this idea by saying that B must be in the *backward light cone* of A.)

However, the ringing of Casimir's alarm does tell us instantaneously that the egg is boiled, despite its not being in the backward light cone of the egg. To avoid this difficulty in the argument, we must say that events in A must be *completely* determined by events in its own backward light cone. If we already *know* that the egg is boiled, the ringing of the alarm does not affect the situation.

Bell then showed, using the EPR approach and his usual argument, that ordinary quantum mechanics is *not* locally causal. He further stated the CHSH result that, from a Bell-inequality type of analysis, quantum mechanics cannot be embedded in a locally causal theory.

Nevertheless he showed that we 'cannot signal faster than light', by the following argument. Each experimenter in a Bell test may control the polarizer setting in that wing of the apparatus, but cannot control the output of the counter, and it is easy to show that the two experimenters cannot communicate with each other by means of the experimental apparatus. Bell then asked whether we should fall back on 'no signalling faster than light' as the expression of the fundamental causal structure of contemporary theoretical physics.

Such a statement would concur with Shimony's idea [48] that, despite the clash between quantum theory and local causality, one could speak of a 'peaceful coexistence' between quantum theory and special relativity. However, Bell dismissed this idea for two reasons. First correlations have not been explained; but secondly, and more importantly, the concentration on 'signalling' rests on concepts that are 'desperately vague or vaguely applicable'. To stress that 'we' cannot signal immediately suggests the question, highlighted by Bell, 'Who do we think *we* are'?

Who are the 'we' who can make 'measurements', manipulate 'external fields', or 'signal' at all, even if not faster than light? Do 'we' include chemists or only physicists, plants or only animals, pocket calculators or only mainframe computers? It seemed unlikely to Bell that we would get a sharp answer to this question.

Perhaps, he said, causality might be analogous to thermodynamics, which does not appear in the fundamental laws of physics, only making an appearance for large, complicated systems. Could causal structure similarly emerge only in a 'thermodynamic' approximation when the notions of 'measurement' and 'external field' become legitimate approximations? Bell was not convinced, and stated that he preferred to work with sharp internal concepts, with the way perhaps being shown by GRW.

He ended the paper by returning to the idea of cooking [125], quoting Einstein's judgement on the new 'cookery' of quantum mechanics, as translated by Casimir: 'In my opinion it contains all the same a certain unpalatability.'

In this paper, Bell provided an interesting Appendix in which he discussed the historical question of when it first became understood that the velocity of light was a limit, while his last paper in QHA (mentioned in 'Bell and quantum theory in the 1970s' in Chapter 4) is a short account of FitzGerald, who was very much involved with matters connected with the velocity of light.

Honours and endings

As has been said, Bell's renown grew after the results of Aspect's experiments were published. His major awards came in the following years [126], which, sadly, were the last few years of his life (Figures 5.20 and 5.21).

In 1987 he became an Honorary Foreign Member of the American Academy of Arts and Sciences and, in the following year, he was awarded the Dirac Medal for Theoretical Physics of the Institute of Physics in London. In 1988 he received the very unusual distinction of being awarded two honorary degrees, both DSc's, from Irish universities: Queen's University Belfast, which was his alma mater, and Trinity College, Dublin (Figures 5.22–5.24). Then, in 1989, he was awarded the Dannie Heinemann Prize for Mathematical Physics of the American Physical Society, and the Hughes Medal of the Royal Society.

It may be of interest that Clauser, Aspect, and Zeilinger were to share the Wolf Prize of the Wolf Foundation in 2010 for their work on the tests of Bell's theorem; this prize is often looked on as a forerunner to the Nobel Prize. Also, Jackiw won the Heinemann Prize in 1995, and he and Steven Adler shared the Dirac Medal of the Abdus Salam International Centre for Theoretical Physics in 1998.

The question is often asked whether Bell could or should have been awarded the Nobel Prize for Physics. A quarter of a century after his death, it would scarcely be doubted. For example, his former collaborator, Tony Hey [127] has written that: 'There are no posthumous Nobel Prizes, but if there were, it is clear that John Bell would top the shortlist for many physicists'. As we shall see in Chapter 6, foundational work in quantum theory has proceeded apace—it is strongly rumoured that, once a loophole-free Bell test has been conducted, a trip to Stockholm will be very likely for those who carried it out, let alone the one who invented it, had he been alive, and the development of quantum information theory has attracted perhaps even more attention than quantum theory itself.

Before Bell's death, though, a Nobel Prize was probably unlikely. It is well known how difficult the Nobel authorities find it to come to grips with epoch-making contributions to theoretical physics; Einstein, Heisenberg, and Schrödinger all had to wait several years to receive the prize, long after their contributions were fully understood and appreciated by the great body of physicists. Up to the time of Bell's death, his work was probably more esoteric and less understood and appreciated than theirs; so, although he was not only nominated but shortlisted in 1989 (and perhaps in previous years) [128], few would have expected him to gain the accolade at that time.

Bell himself was somewhat ambivalent about the Nobel Prize [10]. On the one hand, he felt that those awarding the prize should keep close to the original requirements of Nobel and the statutes; they should give the prize for

Figure 5.20 John Bell cooking. © Renate Bertlmann.

Figure 5.21 John Bell with a *Schneekugel* (snowing ball) made by Renate Bertlmann; in the Bells' flat, 1989. © Renate Bertlmann.

Figure 5.22 John Bell as an honorary graduate at Trinity College, Dublin, 1988. Courtesy Trinity College Dublin.

work done in the year the prize was given, not over past years, and it should be given for work which benefited the welfare or prosperity of mankind. Particle physics, he felt, would be right out of it—Mr Nobel had much more practical things in mind than that!

(It may be said that this was his hair-shirt side. Rather similarly [129], he claimed that those who enjoyed their work, such as particle physicists, should be paid less than those whose work was hard grind. At one stage, he was put on a committee to discuss the salaries of employees at CERN. Very quickly, his fellow physicists had him removed from the committee, as it was rumoured that he was suggesting that their salaries should be steadily shrunk until only those who really valued and enjoyed their work would remain.)

To return to the Nobel Prize, he did, of course, recognize that fundamental work in physics was what was nearly always rewarded. In addition, while he was far from being conceited or self-serving, he was clear-sighted and honest enough to know that he had made a major contribution to a central area of the subject.

Figure 5.23 John Bell as an honorary graduate at Queen's University Belfast, 1988. Courtesy Queen's University Belfast.

Einstein and Bohr were justifiably regarded as the two greatest theoretical physicists of the century, but their discussions on the fundamental aspects of quantum theory had come to an impasse. It was not so much that they could not agree, as that there seemed no way forward by the methods of physics for agreement to be conceivable—only perhaps by armchair philosophy. Of course, it is true that the great majority of the physics community would have recognized Bohr as the winner of the debate, but very few would have been able to give any coherent justification for this belief.

Bell was able to take the question forward, to make a theoretical advance, and, perhaps most surprisingly, to show how experimental physics could

Figure 5.24 The garden party following the graduation at Queen's University Belfast in 1988; (from left) Jackie Bell, Mary Bell, Annie Bell, Ruby McConkey, and Sam McConkey. Courtesy Ruby McConkey.

contribute to a solution. Yet, he must have known that he was still extremely unlikely to receive the reward he must have known he deserved.

Bertlmann [10] remembers a sunny September day in 1987, when he and Bell had been working, and had then sat outside to have a drink. Bell ordered 'deux infusions verveines', as always, and they proceeded to discuss Bell's inequalities and the Nobel Prize. It was fairly well accepted that, if Clauser's experiment and the subsequent ones had violated quantum mechanics, Bell would have been awarded the prize, and Bell added—perhaps 'wryly' would be the best word to describe his tone—'And I would be famous.'

Bertlmann insisted that there were two reasons why Bell should be awarded the prize. The first was that it would provide encouragement for theoreticians and experimentalists to study the physics of entanglement. Bertlmann would particularly stress Aspect's experiment, which was a new and very important test for quantum mechanics. The fact that quantum mechanics *passed* the test should in no way have denied Bell the Nobel Prize.

The second reason was the whole connection with non-locality—the fact that an experimental test could be performed to check local causality. Bell had explained that, when he invented his test, he was really practically convinced that the experimental results would follow quantum theory, and he was concerned more with the implication of the work, which was non-locality—what it meant and how it might be used—than with the experimental results themselves. But Bertlmann said that, at that point in the conversation, the sun began to go down; he remembers red sunlight, Bell's red hair, and Bell saying rather sadly, 'Who cares about non-locality?'

Bell could be scathing about certain quite well-known physicists whose achievements in publicity were as good as—sometimes perhaps better than!—their abilities in physics. As we have seen throughout the book, particularly in such work as the CPT paper and the important quantum mechanics papers, he was some combination of unable and unwilling to push himself in this way.

But of course this did not mean that he did not hope for the appreciation that he deserved—no more than he deserved! And perhaps, despite his great successes and the enormous interest in his quantum work, at least in certain quarters, although not in every case the ones he would have chosen, among the leaders in physics policy and assessment he felt just a little undervalued. In Chapter 2, we met the comment by Jackiw and Shimony [48] about the reception of Bell's CPT work, and despite his position and obvious authority at CERN, he may have felt that he put in an enormous effort to support the institution and its employees—the 'Oracle of CERN'—and received just a little less appreciation than he deserved.

A considerable sadness for Bell in this decade was the sudden death of Jun John Sakurai. As we saw in 'Pasupathy, Bertlmann, and Rajaraman', Sakurai was born in Tokyo but came to the United States in 1949 at the age of 16. After being awarded his PhD by Cornell, he was based at the University of Chicago from 1959 to 1970 and then moved to UCLA. He also made a number of visits to CERN.

He was undoubtedly one of the leading particle physicists of his time, making an important contribution to the theory of the weak interactions, and in 1960 publishing the first important theory of strong interactions based on non-Abelian gauge invariance. He was working in the field of duality at the same time as Bell and Bertlmann, and we saw in 'Pasupathy, Bertlmann, and Rajaraman' how Bell avoided drawing more attention than necessary to the fact that a paper of which Sakurai was joint author contained some computational errors. Sakurai appreciated this gracious gesture, and the two became close friends.

Sakurai also became friends with Bertlmann, and invited him to the United States for research visits. Indeed, Noriko Sakurai, Jun John's wife, felt that her husband's karma was closely linked to that of Bertlmann, and Bertlmann has said that he feels that Bell's and Sakurai's were linked as well.

Early in his career, Sakurai had published two excellent textbooks [130, 131] and, in the early 1980s, while he was working at CERN on a new book, *Modern*

Quantum Mechanics, he told Bell that he was including the Bell inequalities—the first book to do so. Naturally, Bell was extremely pleased, particularly because Sakurai was so well respected by the whole physics community.

Then one evening at the beginning of November in 1972, the Bertlmanns and Sakurai were all at CERN, and the Bells and Sakurai had been invited to the Bertlmanns for dinner. Sakurai did not arrive, which seemed very strange as he was a very meticulous person and would certainly have apologized if he had been unable to make it. The others thought that perhaps he had forgotten, and Bell insisted that the secretaries at CERN be contacted to see if they had any information. However, none was forthcoming, and those present went ahead and had their dinner without Sakurai.

On Monday, Sakurai did not show up at CERN, and Bertlmann told an administrator, who went to visit Sakurai's address but could not get in. In the end, Bertlmann was given the address; he went round himself and found Sakurai —who was only 49!—dead. Bell already knew of Bertlmann's visit, and Bertlmann at once phoned him. Both men were extremely shocked, and Bertlmann had certainly never seen Bell in such a state before. Bell actually commented to Bertlmann that he hoped Bertlmann would not find him in such a state someday, and that statement also shocked Bertlmann.

When Bell and Bertlmann returned to CERN, they realized that Sakurai had been writing his latest book by hand on sheets of paper, which were now lying loose on his desk. Bell checked with Noriko, and then looked through the sheets of paper to see whether the manuscript was sufficiently complete that it could be edited into publishable form. Although only three chapters were complete, it was clear, in Bell's words, that 'the bulk of the creative work had been done'; in addition, there was a large collection of exercises. Bell made sure the papers were copied and sent to Noriko, and he also sent her a report to say that the book should definitely be completed.

Bell himself wrote a foreword for the book [132], in which he praised highly the determination of Noriko Sakurai in achieving the final publication of the book. He also praised the great efforts of Sakurai's associate, San Fu Tuan, who had acted as editor. He did not, of course mention himself, but it is quite clear that he also made a large contribution to the book finally being published.

Bertlmann was in awe not only of Bell's physics and personality but also of his wit—wit which always illuminated beautifully the serious topic under discussion. During the 1980s, Bertlmann prepared a talk titled 'Bell's Theorem and the Nature of Reality' and presented it at several places in Europe, including Berne. When the Berne list of seminars was sent to CERN, Bell saw it and phoned Bertlmann to say how pleased he was. He suggested the talk should be called 'Bertlmann's Socks and ...' or, if Bertlmann was modest, just 'Socks and ...' rather than 'Bell's Theorem and ...'; however, Bertlmann was sure that his original title was the right one!

Bertlmann also wrote a paper on Bell's theorem specifically for particle physicists, so few of whom knew anything about it, and he sent a copy to just about every individual particle physicist. Bell was pleased about that too and commented that Bertlmann had 'Bell's authorised version of Bell's Theorem'.

Then, Bertlmann realized that Bell's sixtieth birthday was coming up in 1988 and, since Bell was such a very quiet person, nobody knew about it, so there would be no celebration. Bertlmann had a piece of theoretical quantum mechanics hidden away and he used it to write an article that he dedicated to 'J. S. Bell on his 60th birthday' [133]. He included a general account of Bell's work and, at the end, he 'got his own back' for the socks paper by including as a 'Conclusion' a delightful cartoon of Einstein and his saying 'spooky action at a distance', with the spook itself holding in widespread hands a pair of socks; as a pièce de résistance, the spook was summoned up from a bottle of Bell's whisky (Figure 5.25).

Bell enjoyed the paper very much, despite the fact that he rarely drank—and certainly he never drank Bell's whisky! There were many phone calls between the two at the time; in one of them, Bertlmann has recalled, Bell remarked that he particularly liked the conclusion. At first, Bertlmann thought Bell must have meant the concluding remarks on non-locality, but then he realized that Bell actually meant the cartoon!

Bertlmann sent copies of the paper, stressing the fact that it was for Bell's sixtieth birthday, to those who he thought might be interested, including Shimony, whom he had not at that time met. Shimony in turn contacted Cushing, who was both Professor of Physics and Professor of Philosophy at the University of Notre Dame, and a great supporter of both Bell and Bohm. Cushing arranged that the conference on quantum theory at Notre Dame, already being organized by him together with Ernan McMullin, would be dedicated to Bell on the occasion of his sixtieth birthday; nearly all the planned talks were already in the relevant area of research. The papers from the conference were subsequently published in 1989 [134], with the addition of the words 'dedicated to John Bell with enthusiastic concurrence of all those whose essays appear here'.

Meanwhile, Bertlmann knew that the journal *Foundations of Physics* had on several occasions published special editions dedicated to various prominent physicists, so he contacted Alwyn van der Merwe, who was co-editor of the journal, to see if he would be interested in publishing an edition honouring Bell. Van der Merwe was extremely positive and agreed that the two of them should both solicit contributors. The only drawback was that the journal was filled up until 1990, but that allowed time for the writing of the articles. In the end, five issues of the journal were dedicated to Bell, with the first being published on 1 October 1990.

Figure 5.25 Reinhold Bertlmann's 'Conclusion'. © Reinhold Bertlmann.

It was in May 1990 that the Bertlmanns and the Bells met in Paris, where Bertlmann was Visiting Professor at the University of Marseilles. Bertlmann had just finished the article [135] that would be the introduction to these issues; in that article, he described Bell's character and contributions to physics, as well as his own interaction with Bell. However he kept the organization of the special issues a secret, to be revealed only when Bell would be presented with the first issue. Later on he was sad that Bell never knew of this honour.

In the summer of that year, there was a meeting on the nature of quantum mechanics at Amherst (Figure 5.26), organized by George Greenstein and

Figure 5.26 Group picture at Amherst College, August 1990, showing Tony Leggett (first row, far left), John Bell (first row, far right), Philip Pearle (second row, far left), Jon Jarrett (second row, third from the left), David Mermin (second row, second from the right), Victor Weisskopf (second row, far right), Michael Horne (third row, far left), Kurt Gottfried (third row, far right), Jeremy Bernstein (fourth row, far left), Dan Greenberger (fourth row, far right), Abner Shimony (fifth row, third from the left), George Greenstein (fifth row, third from the right), and Anton Zeilinger (fifth row, second from the right). Courtesy Emilio Segré Visual Archives.

Arthur Zajonc [136]. There were only about a dozen people invited, including John and Mary Bell, Gottfried, Weisskopf, Pearle, Tony Leggett, Greenberger, Horne, Zeilinger, and Mermin. There were no prepared talks, no schedule, and no proceedings—just, in Mermin's words [101], 'wonderful conversations'. There were 'lots of entertaining discussions and arguments', and Mermin subsequently said: 'It was far and away the finest conference I have ever been to.'

It should not be thought that all was sweetness and light! Greenstein and Zajonc report that Bell and Gottfried, the latter, of course, believing that Bell's concerns about quantum theory were, at the least, rather exaggerated, at one point 'squared off'. The debate, they say, had reached 'a passionate intensity'. Later on, Greenstein grew perturbed as Bell and Gottfried stood apart from the throng with their heads together—'Were they at it again?' He sidled over unobtrusively to eavesdrop—only to find them comparing their cameras!

For all the intellectual pleasure and indeed fun of the week, one point could not be denied—at least in retrospect, the pictures taken of Bell show him looking tired. Bertlmann has also said that, at their meeting at Paris, he did not think that Bell was in good health.

Again in retrospect, it seems that his health had been less than perfect for much of his life. Mary Bell [137] has said that he suffered from migraine attacks, sometimes lasting a week, all his life. Although they nearly disappeared for a number of years, she says that towards the end of his life he had many short attacks. It may be remembered that Bertlmann's first visit to the Bell's apartment was because John Bell had been not well enough to go to CERN.

When we note the (happily) long lives of his parents, who sadly did what parents should not have to do—experience a child dying—and the pleasant fact that his three siblings are (again happily) flourishing a quarter of a century after his own death, then, as well as, of course, feeling extremely sad about his death, we cannot help wondering why he should have suffered in this way. As has been said before, not only did he work hard but he faced perhaps a continuing stressful division between the work he was paid for and certainly enjoyed doing at CERN, and where his heart took him—to the study of the basic ideas of quantum theory.

In both areas of work, he may have felt under a certain amount of pressure he could certainly have done without. On the particle side, there were those who were sure that he was quite happy to look through their manuscripts and sort out their misconceptions. On the quantum side, there were innumerable people who wanted to understand his ideas, but, far more often, wanted just to expound their own ideas, which *may* have been of interest; however, perhaps it was much more likely that they did not even appreciate the strength of the quantum theory that they were criticizing. Although as mentioned before, he claimed the freedom to refrain from comments on ideas that did not interest him, there must have been many occasions when he did not take advantage of this freedom.

He may have prided himself on his calm appearance and, when one heard him lecture, it was difficult not to take to take for granted that the composed exterior was a perfect reflection of the state of mind inside. However, it is common that a desire to maintain a calm external appearance can mask and even accentuate stress within, and it seems likely enough that Bell may have been an example of this. Certainly he took every opportunity for calm away from any limelight, particularly with Mary—walking, skiing, and just living. However, it has been suggested that he was unwilling to go further and actually start a programme where he might have learned to fight stress and relax.

Like most Irish men, and perhaps like most men, Bell was unenthusiastic about seeking medical help unless it was absolutely necessary. It seems he did in the end obtain medical attention at a hospital in Geneva, but no particular warnings or instructions were given.

Nevertheless, on Monday, 1 October 1990 (ironically, the very day when the first issue of *Foundations of Physics*, dedicated to him, appeared), Bell had a stroke. He was taken to hospital and, although everything was done to keep him alive, in the end, tragically, it was not possible.

It goes without saying that Mary was devastated. John Ellis, a colleague in the Theoretical Physics Division at CERN, took a lot of responsibility sending telegrams and messages, telling everybody that the funeral would be on Thursday, 4 October. It was unthinkable to contact John's parents directly, so instead John's brother, Robert was contacted, and he and his wife Nancy went round to break the sad news to John's parents.

Bertlmann was hugely distressed that the telex sent to Vienna had been mislaid in the experimental section of the physics department. As he did not receive it until the day of the funeral, he could not attend.

Bell's death was, of course, a great tragedy for Mary and for all his family, but also an event of enormous sadness for all who knew him and had such great respect for him. Moreover, it was a great blow for all in the physics community, for whom he had been such an important figurehead.

On the day of the funeral, a message sent from Peierls was read, which included the following words:

Ever since John Bell worked with me on his PhD thesis, I have known his outstanding ability, originality and clarity of thinking. We all know how he fulfilled this early promise by many contributions to several branches of physics. He raised important problems about the interpretation of quantum mechanics. His articles on this subject were always provocative, and even those who did not agree with his point of view found them thought-provoking and instructive. We shall sadly miss his incisive contributions to this discussion as to other areas of physics. His many friends will miss his warmth, his cheerful spirit and his dry sense of humour. His loss leaves a sad gap in the physics community'.

Llewellyn Smith added:

I shall speak myself from a purely personal point of view. The time since John's death is too short and the shock too great to be able to step back and speak in a broad way of his life, personality and work. Whenever I returned to CERN, I always tried to see him for the pleasure of his company, his gentle humour, his delightful—indeed beautiful—use of the English language, his advice, and of course his views on physics. Many of us who were Fellows at CERN in the 60s regard John as one of our teachers and we strive to emulate his achievements and his style as a physicist. We would join him and Mary whenever possible for lunch which would often develop into a symposium conducted unobtrusively by John in Socratic style on politics, quantum mechanics, vegetarianism, or some topical or classical problem in physics.

John was a profound and original scientist with a deep interest in the fundamentals of both classical and quantum mechanics. Total honesty and decency characterized all John's acts. He was unfailingly kind, considerate and helpful to his younger colleagues. His marriage to Mary was obviously exceptionally close and happy. All our thoughts and sympathy are with her and with John's family. Physics, and the world of scholarship generally, has lost a major figure and many of us have lost a close friend.

6

The Work Continues

Taking Bell's work forward

Since Bell's death, the amount of interest and work, both theoretical and experimental, on fundamental issues in quantum theory has continued to increase steadily [1, 2].

One very important task, of course, has been to continue work to remove the loopholes on experiments testing Bell's inequalities. An important technical advance in support of this aim has been the invention of the technique of parametric down-conversion, which produces pairs of entangled photons very much more copiously than the cascade method used in the previous experiments. In parametric down-conversion, high-frequency light from a laser is shone on a particular type of crystal, of which a well-known example is beta barium oxide, or BBO. In this process, the incoming photon breaks into two photons, conserving energy of course.

Although the technique was used as early as 1970, it was not until 1998 that the so-called type-II scheme was invented by Paul Kwiat and other members of Zeilinger's group [3], which was working at the time in Innsbruck, before Zeilinger's move to Vienna in 1999. In this scheme, the photons are in a genuinely entangled state, and the brightness of this source enables coincidence detection rates as high as $1,500$ s^{-1} and allows Bell's inequalities to be violated by as many as 100 standard deviations. This method of creating entangled pairs would also be important in quantum information theory.

A major advance in the removal of the loopholes came again from Zeilinger's group when Gregor Weihs (Figure 6.1) and co-workers [4, 5] were able to clear the problems remaining from Aspect's work in closing the locality loophole. This result was achieved through arranging that the measurement stations in the two wings of the experiment were sufficiently far apart that the registration of data in each wing was made completely independent.

To ensure that, and defining a 'measurement' as beginning when the decision on measurement direction is taken in that wing of the experiment and ending with the registration of the photon, the 'measurement' must be carried out fast enough that, during the time that it takes, no information may travel to the other wing of the experiment. In the Weihs experiment, the two wings were separated by 400 m, so the time taken for light to travel between wings

Figure 6.1 Gregor Weihs. © Renate Bertlmann; Central Library for Physics, Vienna.

was 1.3×10^{-6} s; thus, the time for the 'measurement' had to be less than that. This requirement was achieved through the direction of the measurement in each wing being determined by a physical random number generator which was sampled every 10^{-7} s.

There were two other requirements, the first being that the photons should reach the two wings at the same time with the least possible leeway; in the Weihs experiment, any difference in time was less than 5×10^{-9} s. The second was that the measurements also had to be synchronized, in this case, by a pulse of precision 3×10^{-9} s; however, apart from that, all procedures in each wing were totally independent.

In these experiments, the locality loophole was convincingly closed and, in addition, Bell's inequalities were violated by around 30 standard deviations. Also, the fact that switching of the polarizer directions was random rather than periodic meant that the part of the memory loophole referring to the memory of previous directions was closed, although, of course, the part related to memory of previous results was not.

The detector loophole was closed a few years later by Mary Rowe (Figure 6.2) and co-workers [6] in David Wineland's group (Figure 6.3). Over the previous decade or so, the technology of the ion trap had developed to the extent that it was possible to trap a single ion for a period of weeks. In Rowe's experiment, pairs of beryllium ions were confined in an ion trap, and two laser beams

Figure 6.2 Mary Rowe. Courtesy Emilio Segré Visual Archives.

Figure 6.3 David Wineland. Courtesy Emilio Segré Visual Archives.

were used to stimulate a transition leading to entanglement of the two ions. Measurements were then performed, using light from a laser beam acting as a detector, and statistics were built up of the occupancies of different states of superposition of the two ions; using these statistics, it was calculated that Bell's inequality had been violated, by around eight standard deviations.

More importantly, though, the crucial difference between this experiment and previous ones is that the particles were not destroyed in the measurement but recycled for the next run. Thus, no experiment 'failed', since a result could be obtained for each pair of ions. Thus no fair-sampling assumption was required, and the detector loophole had certainly been closed.

Rowe's ions were extremely close together, so there was no possibility of adapting this experiment to add closure of the locality loophole. As Jan-Åke Larsson [7] has pointed out, closing the locality loophole requires a system that is easy to transport without destroying entanglement; the obvious choice is photons, but the detectors in straightforward photon correlation experiments, which were the only ones carried out until fairly recently, are notably inefficient.

However, work was being done on checking that entanglement and violation of Bell's inequalities did indeed survive over long distances, principally by Wolfgang Tittel and co-workers [8] in Nicolas Gisin's group [9] at Geneva. Gisin (Figure 6.4) has had important experience in Swiss telecommunications, so he has been able to combine industrial and academic

Figure 6.4 Nicolas Gisin. © Renate Bertlmann; Central Library for Physics, Vienna.

expertise in his many experimental and theoretical contributions to the foundations of quantum theory and quantum information theory. In these particular experiments, the inequalities were found to be maintained over a distance of 10.9 km.

In the last few years, supreme efforts have been made to take steps towards provision of the technology to carry out tests without any loopholes. For example, Thomas Scheidl and others in Zeilinger's group [10] have carried out a Bell test between two Canary Islands; according to them, this test simultaneously closes both the locality and the freedom of choice loopholes. For the latter loophole, they have stressed that light does not have enough time to travel between the emission of the particles and the setting of the polarizers, so these settings cannot be influenced by the imagined hidden variables of the system. (The authors agree with Larsson [7] that *no* experiment could rule out the supreme lack of freedom of choice, known as *superdeterminism*, according to which *all* events in the universe have a common cause.)

Another experiment carried out by the same group [11] was the first to use photons to close the detector loophole. The experiment used high-efficiency superconductor detectors to increase the actual detection of photons practically to 100%, although there were losses in other parts of the system.

A more recent idea to close the freedom of choice loophole (which the authors call the 'setting independence' loophole), has come from Kaiser and co-workers [12]. Their idea is to use cosmic photons from quasars or from patches of the cosmic microwave background.

Several groups are now competing to obtain the first totally loophole-free test—the article 'Violation of Local Realism with Freedom of Choice' [10] lists some of the people and groups involved. When it comes, it will be a fitting tribute to Bell that so much time and effort of so many highly talented people has been used to test his original idea.

In Chapter 5, we met the GHZ argument which was first presented in 1989; however, it took until 1999 for experiments to show that, in this case, as for Bell's inequality, quantum theory is found to be correct, and local causality is violated. The experiment was carried out by Dirk (Dik) Bouwmeester and colleagues [13] in Zeilinger's group at Innsbruck. In their experiment, two pairs of entangled photons were transformed into three entangled photons in the GHZ state and a fourth photons known as the *trigger photon*. For this experiment, it was essential that both the trigger photon and the three photons in the GHZ state were actually detected. An interesting point is that, while, as we shall see, in many cases experiments designed for early tests of Bell's theorem led on to important experiments in quantum information theory, this experiment was an example of the reverse, as it had been based on previous experiments by the same group on quantum teleportation and entanglement swapping, which will be discussed later in this chapter.

One thing that would certainly have pleased Bell is that, rather than the Copenhagen interpretation having total supremacy, as was the case when he entered physics, several other interpretations have been put forward and are now discussed freely. It would especially have pleased him that many of them are based securely on physics, rather than on less tangible or more philosophical speculation. Of course, it goes without saying that Bell would probably have disapproved most thoroughly of many of these new interpretations, but he would still have appreciated the fact that opposing Copenhagen did not immediately rule one out of serious consideration as a physicist.

Over the past few years, David Deutsch [14] (Figure 6.5) has been the most prominent advocate of many-worlds interpretations, while also, as we shall see shortly, being the founder of quantum computation, and David Wallace [15] and Simon Saunders [16] have also been extremely active in putting forward a range of different arguments and positions over many worlds. As mentioned earlier, many astrophysicists, in particular, have given their support to one or another member of this range of interpretations.

Other interpretations which have been mentioned in previous chapters—the Bohm interpretation, the ensemble interpretation, the GRW interpretation, and the knowledge interpretation—have retained or, in some cases, gained popularity. By a knowledge interpretation, we mean Peierls's version of

Figure 6.5 David Deutsch. © Corbis.

the Copenhagen interpretation (as we saw in Chapter 4), and others have put forward similar ideas.

A newer interpretation that has received a lot of attention is that of *consistent histories* or *decoherent histories*, with different aspects being developed in particular by Robert Griffiths [17] and Roland Omnès [18]. Broadly, the intention is to combine different wave functions and their probabilities in a consistent manner without relying on measurements. A history is a series of quantum events, and it may be assigned a probability, provided certain consistency conditions are obeyed; for example, one cannot deal with cases where interference between different histories may occur. Murray Gell-Mann and James Hartle [19] have gone further than Griffiths and Omnès along the same general lines, by attempting to produce a complete cosmology for the universe, explaining how the observer, which they call an IGUS, or *information gathering and utilizing system*, has evolved.

Another class of interpretation, stochastic theories, have been developed, particularly by Edward Nelson [20]. Those producing these theories have attempted to derive a direct analogy between quantum mechanics and classical statistical mechanics; however, their idea was essentially shown to be in violation of locality by Bell's theorem, and has been renounced by Nelson himself, although the same type of analysis has been revived by Guido Bacciagaluppi [21], who is, of course, well aware of the problems the ideas face. He regards such work as shedding some light on quantum theory, but not as providing an actual interpretation.

Other interesting interpretations include John Cramer's *transactional interpretation* [22], which describes quantum theory in terms of waves going backwards as well as forwards in time; Roger Penrose's idea [23] that variation in gravitational fields may cause a collapse of wave function; and Shigeru Machida and Mikio Namiki's Hilbert space approach [24], where collapse is a result of statistical fluctuations in the measuring apparatus.

Other interesting developments include the further investigation of Zeh's work on environmental decoherence (mentioned in Chapter 3); this work is being done by Zeh himself, Erich Joos, and Zurek [25, 26]. Environmental decoherence should not be regarded as an explanation of wave function collapse, as is sometimes thought; however, the use of this concept does help to prevent quantum drift of macroscopic objects from classicality.

Another area of interesting work has been macroscopic quantum theory. Tony Leggett (Figure 6.6), a Nobel Prize winner for his work on liquid helium, has discussed the ideas of *macroscopic realism* and *non-invasive measurability* at the conceptual level [27] and also, with Anupam Garg [28], has produced an inequality, analogous to Bell's inequality, to confirm the presence of these requirements in experiment. Just as for the Bell case, quantum theory does not respect this inequality.

Figure 6.6 Tony Leggett. Courtesy Tony Leggett.

On the experimental side, Zeilinger and members of his group [29] have been able to demonstrate interference with the 'buckyball', C_{60}, which clearly therefore cannot be regarded as macroscopic. John Bell would have loved to know whether, as the mass of a particle increases further, there is a limit (the shifty split) where there is a switch to a macroscopic nature.

While this book has picked out some of the most interesting features of the last quarter-century, the most striking general point is that, quite contrary to when Bell entered the field and it was felt that Bohr had explained everything, the foundations of quantum theory are now an extremely popular area of study. In *Physical Review Letters*, which is the most prestigious journal for short communications and which is published each week, there would usually be several articles in this area out of maybe 70 in total in each issue, while in *Physical Review A*, perhaps the leading journal for articles in general areas of physics, there are often ten or more articles in any monthly issue. These articles are, of course, in addition to very many articles in other journals, while the journal *Foundations of Physics* publishes many articles on the foundations of quantum theory, and a journal has recently been founded solely devoted to this area of study—the *International Journal of Quantum Foundations*.

The birth of quantum information, and Bell's contribution

As we saw in Chapter 5, Bell did not appreciate the idea that information played any role in the conceptual difficulties of quantum theory. Yet, a quarter of a century after his death, the subject of *quantum information theory* has been established as being of great interest, both practically and theoretically, and is a major area of current research in physics. Indeed, several very well-known physicists [30–33] have suggested that quantum physics is in principle a theory of information, and indeed that information is the fundamental quantity to be considered when one wishes to explain the universe and develop scientific theories.

As to Bell's question, 'Information about what?', Mermin's reply [30] is that this is 'a fundamentally metaphysical question that ought not to distract tough-minded physicists'. He says that there is no way that one can decide whether 'information is about something objective, or is merely information about other information'.

To discuss quantum information, we need to contrast it with classical information theory. This was developed by Claude Shannon [34] (Figure 6.7), who was working at Bell Telephone Laboratories in the 1930s; his work demonstrated, for example, how to process a number of telephone calls efficiently.

The classical theory of computation was developed principally by Alan Turing [35] (Figure 6.8), best known, of course for his work at Bletchley Park. Before the war, he had developed the idea of the *universal Turing machine*, which essentially worked with a sequence of symbols on a paper tape, moving step by step according to the appropriate symbol on the tape. Simple as the machine was, it was the prototype of a *universal computer*, any example of which may carry out the same range of tasks—some, of course, much faster than others.

Of course, these theories were not called *classical* at the time. It was not so much quantum theory that was not taken into account as *physics* itself. While it was recognized that information and computation had to be manifested physically, by means of marks on paper, for example, or electronic signals, it was

Figure 6.7 Claude Shannon; ©IEEE; from N.J.A. Sloane and A.D. Wyner (eds), Claude Elwood Shannon: Collected Papers (IEEE Press, 1993).

taken for granted that intrinsically information was mathematical or abstract in nature. In contrast, it was Rolf Landauer [36] of IBM who insisted, from the 1960s onwards, that 'Information is physical', and that 'Information is inevitably tied to a physical representation.'

In 1985 Deutsch [37] took the next step of pointing out that the same idea applied to computation: 'The theory of computation has traditionally been studied almost entirely in the abstract, as a topic in pure mathematics. That is to miss the point of it. Computers are physical objects, and computations are physical processes. What computers can or cannot compute is determined by the laws of physics alone, and not by pure mathematics'.

And, of course, once we have decided that computation is physics, we must use the correct physics. Just as classical mechanics is always only an approximation, although often, of course, an extremely good one, to the correct theory of quantum mechanics, so classical computation will always be an approximation to the correct theory of computation; in 1985, the latter was the about-to-emerge theory of *quantum computation*. In this way, Deutsch introduced the age of the *universal quantum computer*.

Figure 6.8 Alan Turing; from Andrew Hodges, *Alan Turing: The Enigma* (Burnett Books, 1983).

Before we continue with Deutsch's important work, let us discuss briefly the work of Feynman in the first part of the 1980s (Figure 6.9). As mentioned in Chapter 4, he wrote several important and influential papers discussing how quantum theory might be applied to computation, and he also stimulated the work of several others, including Rolf Landauer and Charles Bennett.

Here we concentrate on a paper derived from a 1981 conference [38] which was published the following year; in this paper, Feynman discussed whether a classical computer with local connections could simulate exactly either a classical system or a quantum system, without the amount of work increasing exponentially as the size of the system increased. Such a simulation was quite possible in the classical case, but for the quantum case, the amount of work did indeed increase exponentially.

He next discussed whether a classical computer could simulate a quantum computer *probabilistically*, and again the answer was negative, the reasoning underlying the answer being extremely interesting. Feynman started

Figure 6.9 Richard Feynman. Courtesy California Institute of Technology.

with an EPR type of system and worked in detail through all the probabilities involved. Everything worked out fine—except that some of the probabilities involved had to be negative. Feynman had effectively worked through a proof of Bell's theorem, although without mentioning Bell's name! In his editorial comments, Hey [39] remarked that 'Only Feynman could discuss "hidden variables", the Einstein-Podolsky-Rosen paradox and produce a proof of Bell's Theorem without mentioning John Bell.'

Hey assumed that Feynman had heard or read of Bell's theorem and had perhaps forgotten his source. Feynman did have form for this kind of behaviour—in particular, with Gell-Mann he 'borrowed' the V-A structure of the weak interaction from Robert Marshak and George Sudarshan [40]. However, it is also quite possible that he re-invented the theorem himself—he obviously had the ability!

It may be stressed then that, for Feynman, it is Bell's theorem that makes the necessity to go beyond classical computers inevitable. However, even more interestingly, he remarked that he often had fun trying to squeeze the difficulty of quantum mechanics into an increasingly small place—to isolate the essential difficulty so as to give the possibility of analysing it in detail. He felt that he had located it in the contrast between two numbers—one required by quantum theory, the other by classical theory—a direct result of a Bell type of analysis. Thus, the significance of Bell's work, according to Feynman, could scarcely be over-exaggerated. Indeed, he thought that Bell's work was the core element of quantum theory!

We now return to Deutsch, who not only rewrote Turing's arguments, which had been at the centre of classical computation, so as to produce the new laws of quantum computation, but began its study. The main difference between classical and quantum computation is that, while the basic unit for the classical case is the *bit*, which can take the value 0 or 1, for the quantum case it is the *qubit*, which is essentially a linear combination of both. For any practical computation, we need a *register*, which contains a number of qubits, while a *gate* carries out operations on qubits. Gates are assembled into *networks* to carry out quantum computations.

At first sight, it might appear that quantum computation would immediately and obviously have enormous benefits over classical computation because, while the latter may produce results for a starting bit of *either* 0 or 1, the quantum case may seem to be able to perform *both* simultaneously, since it handles the linear combination; in fact, for n qubits, the advantage would seem to be a factor of 2^n. However, we must remember that, although this work may in a sense be carried out by the quantum computer, we must finish the computation, in standard quantum fashion, by taking a measurement, which means that one result from the 2^n is selected for us. Quantum computation does have great possible applications [41, 42] but they have to be much more subtly achieved than suggested by these first thoughts.

The first technique of quantum computation was invented by Deutsch [37] himself—the so-called Deutsch algorithm. It was hugely important conceptually because it did go beyond classical computation, but it did not really seem to be of much, if any, use. It studied the rather hypothetical situation where our input is 0 or 1 and, for each input, we know that our output, which is obtained by a long computation, will be either 0 or 1. The question to be answered is, are the two possible outputs the same? Classically, we would have to perform two computations, but the Deutsch algorithm is able to reach the answer with just one. However, it actually works in only 50% of cases, a result which again suggests that the gains from quantum computation may not be large, perhaps not even positive.

In fact, in 1992, with Richard Jozsa [43], Deutsch was able to show how a problem that was more complicated, although not more useful, than the one described above might be solved. In this case, there are n initial qubits, and the number of classical runs required would be 2^n, while the quantum computer still needs only one run. Then, in 1998, Artur Ekert and colleagues [44] at Oxford showed how the original Deutsch algorithm could be made to succeed every run. That was only after, though, the exciting events of the next paragraph. Until 1994 it really seemed that the idea of quantum computation was extremely interesting but of very little practical importance.

Then in that year, Peter Shor [45] made a stupendous advance. It should be explained that finding the prime factors of a large number is believed to be effectively impossible on a classical computer, because the time required increases exponentially with the length of the number. This fact is important because much of our electronic security relies on the RSA algorithm, named after Ronald Rivest, Adi Shamir, and Len Adleman, and this algorithm is built around that impossibility.

Shor invented an algorithm which uses quantum rather than classical computation that can factorize large numbers in a time that increases only polynomially, not exponentially, as the number increases. The startling thing is that the time to factorize a number considered impregnable when using a classical computer may take only a few minutes using Shor's algorithm.

An equally startling quantum algorithm was discovered by Lov Grover [46] in 1996—Grover's algorithm. Searching through a list of N items would take around $N/2$ steps classically, but Grover showed how to reduce the number, using a quantum computer, to about $\sqrt{N}$.

It goes without saying that these algorithms totally transformed the view of the science community towards quantum computation, and for the last 20 years, an enormous amount of work has been carried out on both theoretical and experimental approaches. On the experimental side, the difficulties have been immense, and a start has been made on systems of a very few qubits and gates. Many different physical systems have been tried, one of the main criteria being that the *decoherence time*, the time for which a superposition of different states may be maintained, must be as large as possible and must be larger than the gate operation time. Overall, steady progress has been made.

Another component of quantum information theory is *quantum cryptography* or *quantum key distribution*. We just saw that *public-key cryptography*, in the form of the RSA algorithm, may be in difficulties because of quantum computation; however, *private-key cryptography*, or the *one-time pad*, remains safe. In this method, the sender and the receiver of the message share a 'pad' of keys; using each key in turn, the sender (traditionally referred to as 'Alice') encrypts the message, and the receiver (traditionally referred to as 'Bob') decrypts it. However, the difficulty is the distribution of sufficient keys—as the name 'one-time pad' suggests, each key must only be used once.

In the so-called BB84 protocol for quantum cryptography, invented by Charles Bennett (Figure 6.10) and Giles Brassard [47] in 1984, Alice sends Bob a stream of photons, each of which is polarized along the horizontal direction, the vertical direction, or one of the directions bisecting them, which we may call $+45°$ or $-45°$. Bob uses randomly one of two polarizers, one of which

Figure 6.10 Charles Bennett. © Renate Bertlmann; Central Library for Physics, Vienna.

enables him to distinguish between horizontal and vertical polarizations, the other between the +45° and −45° polarizations. Subsequently, Bob sends Alice a list of the polarizers he used by public channel, and they disregard the mismatches, with the matches giving them the required key.

However, they also exchange a portion of the list of their matches to discover whether Eve, the eavesdropper, has been at work. If she has, there will be occasions (50% of cases) when she has used the wrong polarizer; in 50% *of these*, or 25% overall, she will leave the photon with the wrong polarization. This fact will be clear to Alice and Bob in their check, and this set of results will have to be disregarded.

BB84 was rather unusual among quantum information techniques in that it uses the most basic ideas of quantum theory—so it uses neither entanglement nor any of Bell's ideas. So, it is interesting that an alternative protocol was suggested by Ekert (Figure 6.11) [48] in 1991; in this protocol, Alice and Bob share an EPR pair, and they each make a measurement along one axis chosen randomly; the cases where they choose the same axes provide the key, and the other results are used for a test of Bell's inequality. If there is *no* violation, that is a sign that eavesdropping has taken place and effectively created a hidden variable.

It should be mentioned that the no-cloning theorem, mentioned in Chapter 4, is central to the whole scheme. If Eve, on receipt of a signal, could clone it, naturally she could keep one copy intact and investigate the others.

Figure 6.11 Artur Ekert. © Renate Bertlmann; Central Library for Physics, Vienna.

Antonio Acín, Nicolas Gisin, and Lluis Masanes [49] argued in 2008 that, while any protocol for quantum key distribution relies on sequences of measurements producing correlations, if no Bell's inequality is violated the results could be produced by measurements acting in a mathematical space of higher dimension, and thus would imply a loss of secrecy, as they show is the case for BB84. To provide complete security, a Bell's inequality must be violated, as in Ekert's method. Again, this argument shows how absolutely fundamental Bell's work is in the entire fabric of quantum theory.

While Bennett and Brassard themselves carried out rather primitive, but conceptually highly significant experiments, the effort to apply the technique practically has been made chiefly by Gisin's group [50] using optical fibre placed across Lake Geneva, in experiments that followed on from their previous work testing Bell's inequalities.

The other well-known technique of quantum information theory is *quantum teleportation* [51], in which particle 1 is initially in the possession of Alice, but its state is sent to Bob. The protocol uses an EPR pair of particles, particles 2 and 3; particle 2 goes to Alice, and particle 3 goes to Bob. By a measurement on particles 1 and 2, Alice is able to leave Bob's particle 3 *either* in the original state of particle 1, *or* in one of three simply related states. A message from Alice to Bob over a public channel, based on the result of her own measurement, then tells him how to create the original state of particle 1 (or, in other words, which of the four possibilities has occurred).

The first experiments demonstrating the technique were again performed by Zeilinger and co-workers [52], although several other groups were not far behind with their own variations. As already mentioned, these experiments were also the basis of the same group's demonstration of GHZ. As before, Gisin [53] was the first to demonstrate the technique over distances of several kilometres of fibre.

Lastly, there is the technique of *entanglement swapping* or *teleporting entanglement*, the name of which indicates precisely what is achieved. The technique was invented by Marek Żukowski, Zeilinger, Horne, and Ekert [54], and the first experiments were again carried out by Zeilinger's group [55].

An interesting question is how much credit Bell deserves for the emergence of quantum information theory [56]. Many might suggest that the work, theoretical and experimental, connected with Bell's theorem led fairly directly onto quantum information theory, and so he may have considerable responsibility for the existence of the whole area of research. This idea has led to the names of certain books [1, 2, 57] and to the prominence of those already working on Bell's inequalities, including Zeilinger and Gisin, in the new field of quantum information.

Against that, it might be argued that the field of quantum computation started with Deutsch, who was using ideas that were rather remote from those of Bell, and that, of the large numbers now working in the field, although they may use EPR and Bell states, few have much idea of their fundamental significance. Indeed, Mermin [58] has pointed out how restricted a slice of quantum theory those involved in quantum computation need to be familiar with.

Another argument may concern the basic issues in quantum information theory, an obviously crucial one being the source of the speed-up in, for example, the Shor algorithm. One important opinion is that of Deutsch [14], who, of course, was a fervent supporter of the many-worlds interpretation. He felt that it was obvious that calculations were being carried out in each of these worlds, an idea called *parallelism*, and this would easily explain the increase in speed.

However, this opinion was very much disputed by Steane [59], who argued that a quantum computer required only one universe, and that it would be misleading to say that a quantum computer performs more operations than the number allowed in a single universe. His opinion is that the source of the speed-up is entanglement. Since entanglement is indeed present and plays a central part in many of the schemes we have studied, this assumption does indeed seem natural.

To the extent that it is true, this suggestion does reflect the influence of Bell. It may not be true to say that entanglement was 'invented' by EPR—as the wave function for the helium molecule has the same general form, but they and Schrödinger were certainly the only conspicuous users of the property until Bell revived it in the 1960s, since when its use has grown so enormously.

However various arguments suggest that the role of entanglement is, at the very least, less than that argued by Steane, as it is present in many important cases but not in all [60–62]. An alternative suggestion is that contextuality is the 'magic' ingredient that causes speed-up [63]. This idea also, of course, would reflect considerable credit onto Bell, who drew attention to the idea in his answer to Kochen and Specker during his all-important 1964 paper.

Other suggestions have been that interference or superposition is crucial; however, Philip Ball [64] has shown that none of the suggested factors seem to be either necessary or sufficient for the speed-up in quantum computation. It may be best to say that the power of quantum computation is a result of a fusion of all these aspects of quantum theory, with different elements of quantum computation relying on different aspects of quantum theory. With Bell's wide and deep contribution to the conceptual understanding of quantum theory, it is clear that he has made a hugely important contribution to the structure of quantum computation and quantum information theory.

Note added in proof: Loophole-free tests of Bell's Theorem

As forecast in this chapter, several reports of loophole-free tests of Bell's Theorem have been made towards the end of 2015. They are:

Marissa Giustina and 21 others in the Vienna group of Anton Zeilinger, Significant-loophole-free test of Bell's theorem, Physical Review Letters 115, 250401 (2015).

Bas Henson and 18 others led by Ronald Hanson in Delft. Loophole-free Bell inequality, Nature 526, 682–6 (2015).

Lynden K. Skalm of National Institute of Standards and Technology (NIST), Colorado and 33 others, Strong loophole-free test of local realism, Physical Review Letters 115, 250402 (2015).

An account of the experiments and their place in quantum theory has been given by Alain Aspect in Viewpoint: Closing the door on Einstein and Bohr's quantum debate, https://physics.aps.org/articles/ v8/123.

Note added in proof: Richard Feynman and Bell's Theorem

Further discussion of the views of Richard Feynman on Bell's work is included in a note to be published in the American Journal of Physics in mid-2016: Richard Feynman and Bell's Theorem by Andrew Whitaker.

(I would like to thank David Jackson, Editor of the journal for speeding the refereeing of this note so that this information may be included here.)

7

Work of the Highest Calibre, and a Fine Life

John Bell's work on accelerators, nuclear physics, and elementary particle physics alone would have made it clear that he was an excellent physicist—it may be remembered that, for his Royal Society Fellowship, quantum theory played at most a minimal part, and Llewellyn Smith [1] felt that Bell himself thought that his non-quantum work was undervalued—but any still higher rating of his achievement must definitely rest on his work on quantum theory.

One may start by saying that he breathed life into the crucially important, but in fact for three decades moribund, debate between the two most important theoretical physicists of the century, Niels Bohr and Albert Einstein, and demonstrated that the result was not just theoretically absorbing but experimentally challenging.

In fact physicists, while of course applying the formalism of quantum theory magnificently to a wide range of areas of physics, had in a sense for those decades ignored major aspects of the theory itself, broadly at the demand of Bohr, Heisenberg, and Pauli. As we saw in Chapter 6, there are so many significant elements of quantum theory now being investigated thanks to Bell that he should surely be regarded as just as important a figure for the theory as its founders Heisenberg, Schrödinger, and Dirac. Also in Chapter 6, we saw that his principal result, the Bell's inequalities, are not *just* an interesting and perhaps perplexing phenomenon, but the central element in many of the crucial components and applications of quantum theory, for example, Feynman's discussion of the need for quantum computation, and Gisin and his colleagues' ability to distribute quantum keys.

Schrödinger is famous for his highly perceptive remark, made as early as 1935, that entanglement was 'not *one* but *the* characteristic feature of quantum mechanics, the one that enforces the entire departure from classical thought'. Yet, in a sense, Einstein, Podolsky, and Rosen showed with the EPR paper that, for the case they considered, entanglement as such was not necessarily a surprise at all, providing, of course, that one was prepared to go beyond the Copenhagen interpretation and accept the existence of hidden variables.

It was Bell who showed that EPR's riposte to entanglement just does not work in general, and Bell's inequalities are left, for good or ill, for conceptual concern or provident application, as the central element of the theory. It is plausible to replace Schrödinger's remark with 'The characteristic feature ... is Bell's inequalities.'

Where might Bell be placed in the hierarchy of great physicists? It is generally taken for granted that Newton, Maxwell, and Einstein would occupy the leading places, and perhaps Bohr would be positioned somewhere between those three and the next group. (It might be fair to suggest that, as late as the 1960s, Bohr could have been regarded as a fourth in the top group; however, some of the arguments of Bell and others have perhaps diminished Bohr's reputation to an extent, although this is not to forget his many fine achievements.)

Then, it could be proposed that Bell might be in the next grouping, which might include such names as Galileo, the founders of thermodynamics and of quantum theory, Michael Faraday, Rutherford, Fermi, Lawrence, Feynman, Gell-Mann, and perhaps a few others to be added according to taste. It is high company but a responsible study of Bell's achievements suggests that such an assignment is quite fair.

We now turn to Bell's character, noting Shimony's opinion [2] that 'Bell's moral character is primarily responsible for his discovery of Bell's Theorem.' Bell, of course, was by no means opposed to the 'pragmatic philosophy' of quantum theory when working, but 'out of working hours' he was concerned with 'the intrinsic ambiguity in principle of the theory'.

And, when in this mode, he was completely unprepared to accept arguments he felt to be mainly wishful thinking, whether from Dr Sloane at Queen's, or later on from Bohr and Heisenberg, or indeed from many others. He recognized the significance of Bohm's hidden variable model, was disgusted with those who dismissed and disregarded it, and realized that he must be able to use it as a guide to defeating the theorems of von Neumann and Gleason (or Kochen and Specker).

He was unafraid of challenging the 'big beasts' of Copenhagen and, 50 years on, it is easy to forget that this did take real courage. In previous chapters, it was noted that he did so rather obliquely—and this was certainly no more than good sense!—but over the years it is fair to say that his strategy has had considerable success. The Copenhagen interpretation has by no means been 'defeated', and indeed perhaps it should not be, but it is by no means beyond challenge, and that, for Bell, was probably the most important result.

Rather than merely being put off by the non-locality of the Bohm model, Bell was able to utilize that aspect of the model, together with the hidden variables or the realism he hoped to recover from the demise of the 'shifty split' that he thought and hoped would come, and the entanglement from the, at that time, almost totally neglected EPR paper, to produce his famous inequality. At the time, almost every aspect of his ideas was a challenge to authority.

In more general ways than this, Bell also demonstrated honesty, scrupulousness, and—indeed—simple kindness. A glance through his collected papers shows his frequent apologies for missing a previous paper which had anticipated some of his own ideas, occurrences that even the 'victim' would scarcely have thought more than the cut-and-thrust of academic publishing. He was scrupulous to colleagues and, particularly once established, extremely helpful towards possible competitors—just for example we saw in Chapter 4 his friendly hint to Shimony that would have allowed him credit for the stochastic generalization of Bell's theorem.

In personal and family matters, he was always kind and thoughtful. He was always, of course, generous to his parents, particularly perhaps to his mother, to whom he would regularly send air tickets to Geneva so that she could pay a visit (as his father was not a traveller!) and he was able to buy for them the Belfast house that they had previously been renting.

As another example, when his sister Ruby's daughters, Isabel and Dorothy, were teenagers, they were invited over to Geneva for a few weeks—almost as a rite of passage. They realized that Mary and John would be working, and were expecting and quite happy to spend the time amusing themselves. They were delighted when they found that they were going to be taken to Sienna, Florence, and Pisa for a week [3]. Doubtless, these kindnesses of the Bells were a few among very many, the great majority of which have gone unremarked.

Bell was, of course, not a saint! Particularly when he was young, he had a temper. Even apart from the 'Dr Sloane incident', Mary [4] has pointed out that, when Bell went to work for Walkinshaw at Harwell, 'Bill [Walkinshaw] did not mind John's Celtic temperament'. Later in life, he could be angered by cruelty to animals, or what he saw as the immorality concerning the founding of Israel. However, despite his temper, there is no doubt that he matched his excellence in physics with his fine qualities as a human being.

The Nobel Prize winner Jack Steinberger [5] has beautifully summed up much of what has been said here about Bell:

> One of the greatest privileges of my rewarding life in physics has been the contact with John Bell. Trying to learn the behaviour of neutral kaons in the light of CP violation, I had the pleasure of benefiting from John's penetrating understanding and insight, and of his readiness to share this. Bell was among the most brilliant physicists I have known; in addition he had the very important human qualities of being accessible, unpretentious and kind to his less gifted colleagues.

It is interesting to consider whether Bell could be regarded as being similar to any other important physicists. A previous suggestion [6] has been that he could be compared to Faraday and, by coincidence or otherwise, it has emerged [7] that, in his visit to Robert Bell's shop, Dr Sloane made exactly the same suggestion.

It might be said that the study of electromagnetism was progressing in a satisfactory manner before Faraday's work, which showed that the well-worn ideas could be reinterpreted in a completely new way by use of the field concept, thus bringing immense added depth to the understanding of the subject. Similarly, quantum theory had been brilliantly successful in its range of applications before Bell emerged, but he undoubtedly took it in new and highly interesting and productive directions. It is possible even to suggest that, if Faraday and Bell had not lived, physics might still be lacking the major insights that their work had provided.

Figure 7.1 Perhaps the best-known picture of John Bell at CERN. © CERN.

Faraday was a man of great integrity and dignity, and a great communicator of science, willing to give his time and effort for others. Exactly the same could be said of Bell.

Faraday's character was, of course, based on an immensely strong religious commitment. While the same could not be said of Bell, it might be said that his personal integrity and his willingness to devote himself to science and those interested in science were precisely analogous for the twentieth century to those characteristics that would automatically have flowed from religious belief in the nineteenth.

John Bell's honesty in his science, and his willingness to work steadily for its progress and to devote his time and efforts to help others to do the same, stand as guiding lights to those coming after him (Figure 7.1).

References

QHA John Bell, *Quantum Mechanics, High Energy Physics and Accelerators: Selected Papers of John S. Bell* (Mary Bell, Kurt Gottfried, and Martinus Veltman, editors) (World Scientific, Singapore, 1995).

FQM John Bell, *John S. Bell on the Foundations of Quantum Mechanics* (Mary Bell, Kurt Gottfried, and Martinus Veltman, editors) (World Scientific, Singapore, 2001) (excerpt from QHA above).

SUQM J. S. Bell, *Speakable and Unspeakable in Quantum Mechanics* (Cambridge University Press, Cambridge, 1st edition, 1987; 2nd edition, with two added papers and foreword by Alain Aspect, 2004).

QUS Reinhold Bertlmann and Anton Zeilinger, editors, *Quantum [Un]Speakables: From Bell to Quantum Information* (Springer-Verlag, Berlin/Heidelberg, 2002).

Chapter 1 A Tough Start but a Good One

[1] Robert Lynch, *The Northern IRA and the Early Years of Partition 1920–1922* (Irish Academic Press, Dublin, 2006).

[2] J. C. Beckett, *The Making of Modern Ireland 1603–1923* (Faber, London, 1966).

[3] Brian Inglis, *The Story of Ireland* (Faber, London, 3rd edition, 1970).

[4] Jonathon Bardon, *A History of Ireland in 250 Episodes* (Gill and Macmillan, Dublin, 2008).

[5] Much of the information in this section and the following ones is obtained from interviews by the author and Phil Burke with Annie Bell on 20 February 1998 and 26 March 1998; interviews by the author with David Bell (cousin of JSB) on 20 October 2013, with Ruby (Bell) McConkey and Dorothy (McConkey) Whiteside on 4 December 2014, and with Robert Bell on 23 February 2015; from a memoir of David Andrew Bell, dated 31 October 2005; and from Jeremy Bernstein, *Quantum Profiles* (Princeton University Press, Princeton, NJ, 1991) and the tapes Bernstein made when interviewing Bell in connection with the writing of that book; these tapes have been published under the title *John Bell and the Identical Twins* and are online at http://cds.cern.ch/record/1092112?ln=en.

[6] The birth was not registered until 13 August and, strangely, the original birth certificate gave the date of birth as 20 July. This error was officially corrected by the Assistant Superintendent of Registration on production of a statutory declaration made by Bell's mother on 26 April 1944, presumably in connection with his entry to Queen's University.

[7] A personal anecdote may help this point. Towards the end of her life, and after the death of John, I became very friendly with Annie. I remember the first time I took my wife to meet her; it must be mentioned that Annie was old and somewhat deaf. I felt that they had a very pleasant but very ordinary conversation, mainly about Woolworth's, where they had both worked at different times. On leaving the house, my wife's first words were: 'So that's where he got his brains from!'

[8]   See the article on Billy Bell (politician) in *Wikipedia*; http://en.wikipedia.org/wiki/ Billy_Bell_(politician).

[9]   Derek Mahon, *Spring in Belfast*; http://anglisztika.ektf.hu/new/content/letoltesek/ angnyir/segedanyagok/an612/Mahon.pdf.

[10] Brian Barton, *Northern Ireland in the Second World War* (Ulster Historical Foundation, Belfast, 1995).

[11] Crosbie Smith and Norton Wise, *Energy and Empire: A Biographical Study of Lord Kelvin* (Cambridge University Press, Cambridge, 1989).

[12] The Royal Belfast Academical Institution: History; http://www.rbai.org.uk/index. php?option=com_content&view=article&id=3&Itemid=108.

[13] Belfast Technical College: Steam engines, cuckoos, and monkeys! http://www. bbc.co.uk/legacies/heritage/northern_ireland/tech/.

[14] Bell told Bernstein that Dorothy was 'the first lady civil engineer in Ireland, supervising building operations'. Dorothy says that this was something of an exaggeration.

[15] Geoffrey Thomas, *Cyril Joad* (Birkbeck University Publications, London, 1992); see also Cyril Joad at http://manwithoutqualities.com/2009/03/23/cyril-joad/.

[16] Albert Einstein, *Relativity: The Special and General Theory* (Routledge, London, 1916, and many more editions up to 1954).

[17] Robert Millikan, Duane Roller, and Earnest Watson, *Mechanics, Molecular Physics, Heat and Sound* (Finn, Boston, MA, 1937).

[18] J. J. Thomson, *Elements of the Mathematical Theory of Electricity and Magnetism* (Cambridge University Press, Cambridge, 1897, and several later editions).

[19] Letter from Reggie Scott to Andrew Whitaker, dated 2 February 1998.

[20] T. W. Moody and J. C. Beckett, *Queen's Belfast: The Making of a University* (Faber, London, 2 volumes, 1959).

[21] A. C. B. Obituary Notices of Fellows Deceased: Thomas Andrews, *Proceedings of the Royal Society* **107**, xi–xv (1886).

[22] Andrew Motion, *Philip Larkin: A Writer's Life* (Faber, London, 1993).

[23] Brian Walker and Alf McCreary, *Degrees of Excellence: The Story of Queen's, Belfast, 1845– 1995* (Institute of Irish Studies, Queen's University, Belfast, 1995).

[24] Information obtained from the Queen's University Belfast Calendar for various years.

[25] Leon Mestel and Bernard E. J. Pagel. Sir William Hunter McCrea, *Biographical Memoirs of Fellows of the Royal Society* **53**, 224–36 (2007).

[26] David Bates, Robert Boyd, and D. G. Davies. Sir Harrie Stewart Wilson Massey, *Biographical Memoirs of Fellows of the Royal Society* **30**, 444–511 (1984).

[27] C. A. Coulson. Samuel Francis Boys, *Biographical Memoirs of Fellows of the Royal Society* **19**, 94–115 (1973).

[28] Alexander Dalgarno. Sir David Robert Bates, *Biographical Memoirs of Fellows of the Royal Society* **43**, 48–71 (1997).

[29] A substantial memoir of the life and work of James Hamilton, prepared by his son Andrew Hamilton, is at http://www.jameshamiltonphysicist.com/.

[30] Richard Arthur Buckingham, *Who Was Who 1991–1995* (A and C Black, London, 1996).

[31] I would like to thank Professor Robert Cormac, then Pro Vice Chancellor at Queen's University Belfast, for permission to consult the student archives of the university.

[32] L. W. Kerr, *Recollections of John Stewart Bell*, document prepared by Dr Kerr for Andrew Whitaker, 23 January 1998.

[33] A. J. Ayer, *Language, Truth and Logic* (Victor Gollancz, London, 1936).

[34] Bertrand Russell, *History of Western Philosophy* (Routledge, London, 1946).

[35] The papers from the CERN symposium were published as John Ellis and Daniele Amati, editors, *Quantum Reflections* (Cambridge University Press, Cambridge, 2000).

[36] The best source of information about Professor Emeléus is his obituary written by Dr J. R. M. (Robin) Coulter, who worked with Emeléus for many years. It was published in *The Independent* on 27 June 1989. I would like to thank Dr Coulter for providing me with a copy of this obituary. A photo of a painting of Professsor Emeléus is online at http://www.bbc.co.uk/arts/yourpaintings/paintings/professor-karl-george-emeleus-19011989-169052.

[37] I would like to thank Professor Brian Gilbody for letting me make use of a statement on the career of Dr Sloane; this statement was originally written for Professor Gilbody by Professor Emeléus on 4 December 1976. An account of his life, presented at his funeral following his death on 5 December 1982, was also used.

[38] K. G. Emeléus, *The Conduction of Electricity through Gases* (Methuen, London, 1929).

[39] Norman N. Greenwood. Harry Julius Emeleus, *Biographical Memoirs of Fellows of the Royal Society* **42**, 124–50 (1996).

[40] Sinclair McKay, *The Secret Life of Bletchley Park* (Aurum, London, 2010).

[41] John Herivel, *Herivelismus and the German Military Enigma* (Baldwin, Kidderminster, 2008).

[42] John Herivel, *Joseph Fourier: The Man and the Physicist* (Clarendon, Oxford, 1975).

[43] F. J. Weinberg. Alfred Rene Jean Paul Ubbelohde, *Biographical Memoirs of Fellows of the Royal Society* **35**, 382–402 (1990).

[44] Max Born, *Atomic Physics* (Blackie, London, 1935).

[45] Leonard Schiff, *Quantum Mechanics* (McGraw-Hill, New York, NY, 1949).

[46] P. A. M. Dirac, *The Principles of Quantum Mechanics* (Clarendon, Oxford, 1930).

[47] Andrew Whitaker. John Bell in Belfast: Early years and education, in QUS, pp. 7–20.

[48] Anja Skaar Jacobsen, *Léon Rosenfeld: Physics, Philosophy and Politics in the Twentieth Century* (World Scientific, Singapore, 2012).

[49] Euan Squires, *Mystery of the Quantum World* (Institute of Physics, Bristol, 1986).

[50] Euan Squires, *To Acknowledge the Wonder* (Institute of Physics, Bristol, 1985).

[51] Euan Squires, *Conscious Mind in the Physical World* (Institute of Physics, Bristol, 1990).

[52] Léon Rosenfeld. Misunderstandings about the foundations of quantum theory, in S. Körner, editor, *Observation and Interpretation in the Philosophy of Physics* (Dover, New York, NY, 1957), pp. 41–5.

[53] Rudolf Peierls, in P. C. W. Davies and J. R. Brown, editors, *The Ghost in the Atom* (Cambridge University Press, Cambridge, 1986), pp. 70–82.

[54] A. Pais, *Niels Bohr's Times in Physics, Philosophy and Polity* (Clarendon, Oxford, 1991).

[55] John von Neumann, *Mathematische Grundlagen der Quantenmechanik* (Springer, Berlin, 1932); translated as *Mathematical Foundations of Quantum Theory* (Princeton University Press, Princeton, NY, 1955).

[56] Readers who have been brought up on the statement that 'Bell was the man who proved Einstein wrong' may like to wait for Chapter 3 for clarification.

[57] Dipankar Home and Andrew Whitaker, *Einstein's Struggles with Quantum Theory: A Reappraisal* (Springer, New York, NY, 2011).

[58] Andrew Whitaker. Kelvin: The legacy, in Raymond Flood, Mark McCartney, and Andrew Whitaker, editors, *Kelvin: Life, Labours and Legacy* (Oxford University Press, Oxford, 2008), pp. 278–306, 343–5.

[59] Andrew Whitaker. Maxwell's famous (or infamous) demon, in Raymond Flood, Mark McCartney, and Andrew Whitaker, editors, *James Clerk Maxwell: Perspectives on His Life and Work* (Oxford University Press, Oxford, 2014), pp. 163–86, 325–8.

[60] C. Cercignani, *Ludwig Boltzmann: The Man Who Trusted Atoms* (Oxford University Press, Oxford, 1998).

[61] Helge Kragh. The origin of radioactivity: From solvable problem to unsolved non-problem, *Archive for History of Exact Sciences* **50**, 331–58 (1987).

[62] Paul Forman. Weimar culture, causality and quantum theory: Adaptation by German physicists and mathematicians to a hostile environment, *Historical Studies in the Physical Sciences* **3**, 1–115 (1971).

[63] H. A. Bethe and G. Hildebrandt. Paul Peter Ewald, *Biographical Memoirs of Fellows of the Royal Society* **34**, 134–76 (1988).

[64] J. J. Dropkin and B. Post. Peter Paul Ewald, 23 January 1888–22 August 1985, *Acta Crystallographica* **A42**, 1–5 (1986).

[65] Paul Peter Ewald. Max von Laue, *Biographical Memoirs of Fellows of the Royal Society* **6**, 134–56 (1960).

[66] J. M. Ziman, *Principles of the Theory of Solids* (Cambridge University Press, Cambridge, 1965).

[67] Jeremy Bernstein, *Hans Bethe, Prophet of Energy* (Basic Books, New York, NY, 1979).

[68] A substantial account of Moyal's life and work is online at http://press.anu.edu.au//maverick/mobile_devices/pr01.html.

[69] Walter Moore, *Schrödinger: Life and Thought* (Cambridge University Press, Cambridge, 1989).

[70] http://openplaques.org/plaques/7044.

[71] Max Born, *Natural Philosophy of Cause and Chance* (Clarendon, Oxford, 1949).

Chapter 2 The 1950s: Progress on All Fronts

[1] Guy Hartcup, *The Effect of Science on the Second World War* (Palgrave, London, 2000).

[2] Daniel J. Kevles, *The Physicists: The History of a Scientific Community in Modern America* (Knopf, New York, NY, 1978).

[3] P. J. Hore, *Nuclear Magnetic Resonance* (Oxford University Press, Oxford, 1995).

[4] Armin Hermann, John Krige, Ulrike Mersits, and Dominique Pestre, *History of CERN* (North-Holland, Amsterdam, Volume 1, 1967).

[5] Abraham Pais, *Inward Bound: Of Matter and Forces in the Physical World* (Clarendon, Oxford, 1986).

[6] Laurie M. Brown, Max Dresden, and Lillian Hoddeson, editors, *Pions to Quarks: Particle Physics in the 1950s* (Cambridge University Press, Cambridge, 1989).

[7] Peter Galison and Bruce Hevly, editors, *Big Science: The Growth of Large-Scale Research* (Stanford University Press, Stanford, CA, 1992).

[8] Brian Cathcart, *Test of Greatness: Britain's Struggles for the Atom Bomb* (John Murray, London, 1994).

[9] Margaret Gowing, *Independence and Deterrence: Britain and Atomic Energy 1945–52* (Palgrave Macmillan, London, 2 volumes, 1974).

[10] Richard Rhodes, *The Making of the Atomic Bomb* (Simon and Schuster, New York, NY, 1986).

[11] Margaret Gowing, *Britain and Atomic Energy* (Palgrave, London, 1964).

[12] Sabine Lee. Rudolf Ernst Peierls, *Biographical Memoirs of Fellows of the Royal Society* **53**, 266–84 (2007).

[13] Rudolf Peierls, *Bird of Passage: Recollections of a Physicist* (Princeton University Press, Princeton, NJ, 1988).

[14] Nancy Arms, *A Prophet on Two Countries: The Life of F. E. Simon* (Pergamon, Oxford, 1960).

[15] Kenneth D. McRae, *F. E. Simon and the Race for Atomic Weapons in World War II* (Oxford University Press, Oxford, 2014).

[16] Mark L. E. Oliphant and Lord Penney, John Douglas Cockcroft, *Biographical Memoirs of Fellows of the Royal Society* **14**, 139–88 (1968).

[17] Abraham Pais, *Niels Bohr's Life in Physics, Philosophy and Polity* (Oxford University Press, Oxford, 1991).

[18] *Harwell: The British Atomic Establishment 1946–1951* (Her Majesty's Stationary Office, London,1952); see also Echo 1946–1996 Special Anniversary Edition (published by the UKAEA); http://www.research-sites.com/UserFiles/File/Archive/ Newsletters%20and%20Brochures/ECHO%20Harwell%201946-1996%20Special%20 Anniversary%20Edition.pdf.

[19] Harry Jones. Herbert Wakefield Banks Skinner, *Biographical Memoirs of Fellows of the Royal Society* **6**, 258–68 (1960).

[20] Guy Hartcup and Thomas Edward Allibone, *Cockcroft and the Atom* (Adam Hilger, Bristol, 1984).

[21] Jeremy Bernstein, *Quantum Profiles* (Princeton University Press, Princeton, NJ, 1991).

[22] Bill Walkinshaw to Phil Burke, 14 July 1998.

[23] L. W. Kerr, *Recollections of John Stewart Bell*, document prepared by Dr Kerr for Andrew Whitaker, 23 January 1998.

[24] Milorad Mladjenović, *The Defining Years in Nuclear Physics 1932–1960s* (Institute of Physics Publishing, Bristol, 1998).

[25] Brian Cathcart, *The Fly and the Cathedral* (Penguin, Harmondsworth, 2005).

[26] Edmund Wilson, *An Introduction to Particle Accelerators* (Oxford University Press, Oxford, 2001).

[27] M. Stanley Livingston, *Particle Accelerators: A Brief History* (Harvard University Press, Cambridge, MA, 1969).

[28] Jennet Conant, *Tuxedo Park* (Simon and Schuster, New York, NY, 2002).

[29] H. A. Bethe and M. E. Rose. The maximum energy obtainable from the cyclotron, *Physical Review* **52**, 1254–5 (1937).

[30] Michael C. Crowley-Milling, *John Bertram Adams: Engineer Extraordinary* (Gordon and Breach, Yverdon, 1993).

[31] Andrew Whitaker, Maxwell and the rings of Saturn, in Raymond Flood, Mark McCartney, and Andrew Whitaker, editors, *James Clerk Maxwell: Perspectives on his Life and Work* (Oxford University Press, Oxford, 2014), pp. 115–38, 319–22.

[32] Edward G. Bowen, *Radar Days* (Adam Hilger, Bristol, 1987).

[33] Reginald V. Jones, *Most Secret War: British Scientific Intelligence 1939–1945* (Hamish Hamilton, London, 1978).

[34] John Riley Holt. Thomas Gerald Pickavance, *Biographical Memoirs of Fellows of the Royal Society* **39**, 304–23 (1994).

[35] J. W. Burren and J. D. Lawson. William Walkinshaw 1916–2001, *CERN Courier* 26 February 2002; http://cerncourier.com/cws/article/cern/28592/2.

[36] Mary Bell. John Bell and accelerator physics, *Europhysics News* **22**, 72 (1991).

[37] Jeremy Bernstein's account of discussions with John Bell in Ref. 21 is based on a series of tapes produced in interviews towards the end of Bell's life. They are available on the CERN website: http://cds.cern.ch/record/1092112?ln=en.

[38] QHA, pp. 8–16.

[39] M. Stanley Livingstone, *High-Energy Accelerators* (Interscience, New York, NY, 1954).

[40] John S. Foster, T. Kenneth Fowler, and Frederick E. Mills. Nicholas C. Christofilos, *Physics Today* **26**, 109–15 (1973).

[41] Robert P. Crease. Crackpots and their convictions, *Physics World* **14** (5), 14 (2001).

[42] J. S. Bell. A new focussing principle applied to the proton linear accelerator, *Nature* **171**, 167–8 (1953).

[43] J. S. Bell. Basic algebra of the strong focussing system, A. E. R. E. report T/R 1114 (1953).

[44] Philip G. Burke and Ian C. Percival. John Stewart Bell, 28 July 1928–1 October 1980, *Biographical Memoirs of Fellows of the Royal Society* **45**, 2–17 (1999).

[45] Klaus Wille, *The Physics of Particle Accelerators: An Introduction* (Oxford University Press, Oxford, 2000).

[46] J. S. Bell. Vertical focussing in the microtron, *Proceedings of the Physical Society B* **66**, 802–4 (1953).

[47] J. S. Bell. Linear accelerator phase oscillations, A. E. R. E. report T/M 114 (1954). QHA, pp. 21–30.

[48] Mary Bell. Some reminiscences, in QUS, pp. 3–5.

[49] R. B. R. Shersby-Harvey, L. B. Mullett, W. Walkinshaw, J. S. Bell, and B. G. Loach. A theoretical and experimental investigation of anisotropic-dielectric-loaded linear electron accelerators, *Proceedings of the Institution of Electrical Engineers B* **104**, 273–90 (1957).

[50] Alec Merrison and Anthony Tucker. Lord Flowers' obituary, *The Guardian*, 29 June 2010; http://www.theguardian.com/science/2010/jun/29/lord-flowers-brian-flowers-obituary.

[51] Basil J. Hiley. David Joseph Bohm, *Biographical Memoirs of Fellows of the Royal Society* **43**, 107–31 (1997).

[52] F. David Peat, *Infinite Potential: The Life and Times of David Bohm.* (Addison-Wesley, Reading, MA, 1997).

[53] David Bohm, *Quantum Theory* (Prentice-Hall, Englewood Cliffs, NJ, 1951).

[54] Niels Bohr, *Atomic Theory and the Description of Matter* (Cambridge University Press, London, 1934).

[55] Andrew Whitaker, *Einstein, Bohr and the Quantum Dilemma: From Quantum Theory to Quantum Information* (Cambridge University Press, Cambridge, 2006).

[56] Dipankar Home and Andrew Whitaker, *Einstein's Struggles with Quantum Theory: A Reappraisal* (Springer, New York, NY, 2007).

[57] J. S. Bell. Einstein-Rosen-Podolsky experiments, *Proceedings of the Symposium on Frontier Problems in High Energy Physics*, Pisa, June 1976, pp. 33–45. QHA, pp. 768–77; SUQM, pp. 81–92.

[58] Bell refers to Albert Einstein. Reply to criticisms, in P. A. Schilpp, editor, *Albert Einstein Philosopher-Scientist* (Tudor, New York, NY, 1949), pp. 665–88,

[59] Max Jammer, *The Philosophy of Quantum Mechanics* (Wiley, New York, NY, 1974).

[60] Arthur Fine, *The Shaky Game* (University of Chicago Press, Chicago, 1986).

[61] Niels Bohr. Discussions with Einstein, in P. A. Schilpp, editor, *Albert Einstein Philosopher-Scientist* (Tudor, New York, NY, 1949), pp. 199–241.

[62] Don Howard. Nicht sein kann was nicht sein darf, or the prehistory of EPR: Einstein's early worries about the quantum mechanics of composite systems, in Arthur Miller, editor, *Sixty-Two Years of Uncertainty* (Plenum, New York, NY, 1990), pp. 61–106.

[63] Erwin Schrödinger. Die gegenwartige Situation in der Quantenmechanik [The present situation in quantum mechanics], *Naturwissenschaften* **23**, 807–12, 823–8, 844–9 (1935). English translation by J.D. Trimmer in Ref. 64, pp. 152–67.

[64] J. A. Wheeler and W. H. Zurek, editors, *Quantum Theory and Measurement* (Princeton University Press, Princeton, NJ, 1983).

[65] Leon Rosenfeld. Niels Bohr in the thirties: Consolidation and extension of the concept of complementarity, in S. Rozental, editor, *Niels Bohr: His Life and Work Seen by his Friends and Colleagues* (North-Holland, Amsterdam, 1967), pp. 14–36.

[66] Niels Bohr. Can quantum-mechanical description of physical reality be considered complete?, *Physical Review* **48**, 696–702 (1935).

[67] Howard Wiseman. From Einstein's theorem to Bell's theorem: A history of quantum nonlocality, *Contemporary Physics* **47**, 79–88 (2006).

[68] Mara Beller, *Quantum Dialogues: The Making of a Revolution* (University of Chicago Press, Chicago, IL, 1999).

[69] M. A. B. Whitaker. The EPR paper and Bohr's response, *Foundations of Physics* **34**, 1305–40 (2004).

[70] Jens Hebor, *The Standard Conception as Genuine Quantum Realism* (University Press of Southern Denmark, Campusvej, 2005).

[71] Albert Einstein. Autobiographical notes, in P. A. Schilpp, editor, *Albert Einstein Philosopher-Scientist* (Tudor, New York, NY, 1949), pp. 1–95.

[72] John Bell. On the Einstein Podolsky Rosen paradox, *Physics* 1, 195–200 (1964); QHA, pp. 701–6; FQM, ppp. 7–12; SUQM, pp. 14–21.

[73] John Bell. Bertlmann's socks and the nature of reality, *Journal de Physique* **42**, 41–62 (1981). QHA, pp. 793–819; FQM, pp. 126–147; SUQM, pp. 139–58.

[74] David Bohm. Hidden variables and the implicate order, in Basil Hiley and F. David Peat, editors, *Quantum Implications: Essays in Honour of David Bohm* (Routledge, London, 1987), pp. 33–45.

[75] Jeremy Bernstein, *Quantum Leaps* (Harvard University Press, Harvard, MA, 2009).

[76] A. Landé, *Quantum Mechanics* (Pitman, London, 1951).

[77] David Pines. David Bohm 1917–92, *Physics World* **6**, 67 (1993).

[78] David Bohm. A suggested interpretation of the quantum theory in terms of 'hidden variables', *Physical Review* **85**, 166–79, 180–93 (1952).

[79] John Bell. On the impossible pilot wave, *Foundations of Physics* **12**, 989–99 (1982). QHA, pp. 842–52; FQM, pp. 148–58; SUQM, pp. 159–68.

[80] Peter Holland, *The Quantum Theory of Motion* (Cambridge University Press, Cambridge, 1993).

[81] Guido Bacciagaluppi and Anthony Valentini, *Quantum Theory at the Crossroads: Reconsidering the 1927 Solvay Conference* (Cambridge University Press, Cambridge, 2009).

[82] James T. Cushing, *Quantum Mechanics: Historical Contingency and the Copenhagen Hegemony* (University of Chicago Press, Chicago, IL, 1994).

[83] Max Born, *The Born–Einstein Letters* (Macmillan, London, 2005).

[84] Werner Heisenberg, *Physics and Philosophy* (Allen and Unwin, London, 1959).

[85] Wolfgang Pauli, Remarques sur le problème des paramètres caches dans la mécanique quantique et sur la théorie de l'onde pilote, in André George (editor), *Louis de Broglie, physicien et penseur* (Albin-Michel, Paris, 1953), pp. 33–42.

[86] Franz Mandl KES (King Edward VII School, Sheffield) 1936–52; http://oldedwardians.org.uk/obits/mandl.html.

[87] Franz Mandl to Phil Burke, 2nd July 1998.

[88] John Bell. John Bell, in Paul Davies and Julian Brown, editors, *The Ghost in the Atom* (Cambridge University Press, Cambridge, 1986), pp. 45–57.

[89] Manchester Physics Series; http://eu.wiley.com/WileyCDA/Section/id-390541.html.

[90] Interviews, conducted by Phil Burke and the author, with Annie Bell, on 20 February 1998 and 26 March 1998.

[91] Interview of Robert Bell by Andrew Whitaker, on 23 February 2015.

[92] John Perring to Phil Burke, 21 July 1998.

[93] History of Theoretical Physics in Birmingham; http://www.theory.bham.ac.uk/history/.

[94] David Sherrington, Paul Golbart, and Nigel Goldenfeld, editors, *Stealing the Gold: Celebrating the Pioneering Physics of Sam Edwards* (Oxford University Press, Oxford, 2004).

[95] T. W. B. Kibble. Paul Taunton Matthews 19 November 1919–26 February 1987, *Biographical Memoirs of Fellow of the Royal Society* **34**, 554–80 (1988).

[96] D. Fishlock and L. E. Roberts. Walter Charles Marshall, Lord Marshall of Goring, *Biographical Memoirs of Fellow of the Royal Society* **44**, 298–312 (1998).

[97] Jeremy Bernstein, *Nuclear Iran* (Harvard University Press, Cambridge, MA, 2014).

[98] Rudolf Peierls, in Paul Davies and Julian Brown, editors, *The Ghost in the Atom* (Cambridge University Press, Cambridge, 1986), pp. 70–82.

[99] Rudolph Peierls. Bell's early work, *Europhysics News* **4**, 69–70 (1991).

[100] Roger Bowey and Mariana Sanchez, *Introductory Statistical Mechanics* (Oxford University Press, Oxford, 1999).

[101] J. S. Bell. Time reversal in field theory, *Proceedings of the Royal Society A*, **231**, 479–95 (1955). QHA, pp. 129–45.

[102] Gerhard Lüders. On the equivalence of invariance under the reversal and under particle–antiparticle conjugation for relativistic field theories, *Kongelige Danske Videnskabernes Selskab, Matematisk-Fysiske Meddelelser* **28**, 1–17 (1954).

[103] Roman Jackiw and Abner Shimony. The depth and breadth of John Bell's physics, *Physics in Perspective* **4**, 78–116 (2002).

[104] Wolfgang Pauli. Exclusion principle, Lorentz group and reflection of space-time and charge, in W. Pauli, editor, *Niels Bohr and the Development of Physics* (Pergamon, London, 1955), pp. 30–51.

[105] Tony Skyrme, *Selected Papers with Commentary* (Gerard E. Brown, editor) (World Scientific, Singapore, 1994).
This book includes the following reprinted paper: R. H. Dalitz. An outline of the life and work of Tony Hilton Royle Skyrme (1922–1987), *International Journal of Modern Physics A* **3**, 2719–44 (1988).

[106] Tony Evans. Phil Elliott, *The Guardian*, 21 January 2009; http://www.theguardian.com/world/2009/jan/21/obituary-phil-elliot?guni=Article:in%20body%20link.

[107] J. K. Perring and T. H. R. Skyrme. The alpha-particle and shell-models of the nucleus, *Proceedings of the Physical Society* **69**, 600–9 (1956).

[108] Charles Clement. Tony Lane, *The Guardian*, 10 March 2011; http://www.theguardian.com/science/2011/mar/10/tony-lane-obituary.

[109] David Thouless. John Hubbard 1931–1980; https://www.royalholloway.ac.uk/physics/documents/pdf/research/cmpc2013talks/cmpc2013thouless.pdf.

[110] Charles Clement to Phil Burke, 2 July 1998.

[111] T. H. R. Skyrme. Quantum field theory, *Proceedings of the Royal Society A* **231**, 321–35 (1955).

[112] J. S. Bell. A variational method in field theory, *Proceedings of the Royal Society A* **242**, 122–8 (1957). QHA, pp. 155–61.

[113] J. S. Bell and T. H. R. Skyrme. The anomalous moments of nucleons, *Proceedings of the Royal Society A* **242**, 129–42 (1957). QHA, pp. 162–75.

[114] Keith Brueckner, Founding Chair of Department of Physics, dies at 90. UC San Diego News Centre: http://ucsdnews.ucsd.edu/pressrelease/keith_brueckner_founding_chair_of_department_of_physics_dies_at_90.

[115] J. S. Bell and T. H. R. Skyrme. The nuclear spin-orbit coupling, *Philosophical Magazine* **1**, 1055–68 (1956).

[116] J. S. Bell, R. J. Eden, and T. H. R. Skyrme. Magnetic moments of nuclei and the nuclear many-body problem, *Nuclear Physics* **2**, 586–92 (1956/57). QHA, 146–52.

[117] J. S. Bell. Nuclear magnetic moments and the many body problem, *Nuclear Physics* **4**, 295–312 (1957).

[118] J. S. Bell. Time reversal in beta-decay, *Proceedings of the Physical Society A* **70**, 552–3 (1957). UHA, pp. 153–4.

[119] J. S. Bell. Many body effects in beta decay, *Nuclear Physics* **5**, 167–72 (1957).

[120] J. S. Bell and R. J. Blin-Stoyle. Mesonic effects in beta-decay, *Nuclear Physics* **6**, 87–99 (1958). UHA, pp. 176–88.

[121] J. P. Elliott. Roger John Blin-Stoyle 24 December 1924–31 January 2007, *Biographical Memoirs of Fellows of the Royal Society* **51**, 41–55 (2008).

[122] J. S. Bell and F. Mandl. The polarization–asymmetry equality, *Proceedings of the Physical Society* **71**, 272–4 (1958). UHA, pp. 189–91.

[123] J. S. Bell and F. Mandl. The polarization–asymmetry equality, *Proceedings of the Physical Society* **71**, 867–8 (1958). UHA, pp. 192–3.

[124] J. S. Bell. Bremsstrahlung from multiple scattering, *Nuclear Physics* **8**, 613–20 (1958). UHA, pp. 194–201.

[125] J. S. Bell. Many body problem with one-body forces, *Proceedings of the Physical Society* **73**, 118 (1959).

[126] J. S. Bell. Particle-hole conjugation in the shell model, *Nuclear Physics* **12**, 117–24 (1959).

[127] J. S. Bell and J. M. Soper. Hard core correlations and nuclear moments, *Nuclear Physics* **13**, 167–76 (1959).

[128] J. S. Bell and E. J. Squires. A formal optical model, *Physical Review Letters* **3**, 96–7 (1959). MHA, pp. 202–3 (1959).

[129] The Bibliography is in Ref. 44.

[130] J. S. Bell and E. J. Squires. The theory of nuclear matter, *Philosophical Magazine Supplement: Advances in Physics* **10**, 211–312 (1961).

[131] Editorial: Skyrmion makeover. *Nature* **465**, 846 (2010).

[132] John Polkinghone, *Rochester Roundabout: The Story of High Energy Physics* (Longman, Harlow, 1989).

[133] John Cockcroft, editor, *The Organization of Research Establishments* (Cambridge University Press, Cambridge, 1965).

Chapter 3 The 1960s: The Decade of Greatest Success

[1] Roman Jackiw and Abner Shimony. The depth and breadth of John Bell's physics, *Physics in Perspective* **4**, 78–116 (2002).

[2] John Iliopoulos. Physics in the CERN Theory Division, in John Krige, editor, *History of CERN, Volume III* (Elsevier, Amsterdam, 1996), pp. 277–326. This article-contains the fullest account of the Theory Division from the beginning of CERN, although there is much relevant information in the first two volumes.

[3] Armin Hermann, John Krige, Ulrike Mersits, and Dominique Pestre, editors, *History of CERN, Volume I: Launching the European Organisation for Nuclear Research* (North-Holland, Amsterdam, 1987).

[4] Carl Rubbia. Edoardo Amaldi 5 September 1908–5 December 1989, *Biographical Memoirs of Fellows of the Royal Society* **37**, 2–31 (1991).

[5] Owen Chamberlain. The discovery of the antiproton, in Laurie M. Brown, Max Dresden, and Lillian Hoddeson, editors, *Pions to Quarks: Particle Physics in the 1950s* (Cambridge University Press, Cambridge, 1989), pp. 273–84.

[6] Oreste Piccioni. On the antiproton discovery, in Laurie M. Brown, Max Dresden, and Lillian Hoddeson, editors, *Pions to Quarks: Particle Physics in the 1950s* (Cambridge University Press, Cambridge, 1989), pp. 285–95.

[7] Simone Turchetti, *The Pontecorvo Affair: A Cold War Defection and Nuclear Physics* (University of Chicago Press, Chicago, IL, 2012).

[8] Pierre V. Auger. Some aspects of French physics in the 1930s, in Laurie M. Brown and Lillian Hoddeson, editors, *The Birth of Particle Physics* (Cambridge University Press, Cambridge, 1983), pp. 173–6.

[9] Donald Perkins, *Particle Astrophysics* (Oxford University Press, Oxford, 2003).

[10] John Polkinghorne, *The Rochester Roundabout: The Story of High Energy Physics* (Longman, Harlow, 1989).

[11] Armin Hermann, John Krige, Ulrike Mersits and Dominique Pestre, editors, *History of CERN, Volume II: Building and Running the Laboratory* (North-Holland, Amsterdam, 1990).

[12] Finn Aaserud, *Redirecting Science: Niels Bohr, Philanthropy and the Rise of Nuclear Physics* (Cambridge University Press, Cambridge, 2003).

[13] Anatole Abragam, *The Principles of Nuclear Magnetism* (Clarendon, Oxford, 1961).

[14] Anatole Abragam and Brebis Bleaney, *Electric Paramagnetic Resonance of Transition Ions* (Oxford University Press, Oxford, 1970).

[15] Michael C. Crowley-Milling, *John Bertram Adams: Engineer Extraordinary* (Gordon and Breach, Yverdon, 1993).

[16] Jeremy Bernstein, *Quantum Profiles* (Princeton University Press, Princeton, NJ, 1991).

[17] Mary Bell. Some reminiscences, in QUS, pp. 3–5.

[18] Martinus Veltman. Notes on Section 2: High energy physics, in QHA, pp. 3–5.

[19] Martinus Veltman. The path to renormalizability, in Lillian Hoddeson, Laurie Brown, Michael Riordan, and Max Dresden, editors, *The Rise of the Standard Model: Particle Physics in the 1960s and 1970s* (Cambridge University Press, Cambridge, 1997), pp. 145–78.

[20] Martinus Veltman, *Facts and Mysteries in Elementary Particle Physics* (World Scientific, Singapore, 2003).

[21] Abraham Pais, *Inward Bound: Of Matter and Forces in the Physical World* (Clarendon, Oxford, 1986).

[22] C. L. Cowan, F. Reines, F. B. Harrison, H. W. Cruse, and A. D. McGuire. Detection of the free neutrino: A confirmation, *Science* **124**, 103–4 (1956).

[23] Bruno Pontecorvo. Electron and muon neutrinos, *Soviet Physics: JETP* **37**, 1751–7 (1959).

[24] Melvin Schwartz. Feasibility of using high-energy neutrinos to study weak interactions, *Physical Review Letters* **4**, 306–7 (1960).

[25] G. Danby, J.-M. Gaillard, K. Goulianos, L. M. Lederman, N. B. Mistry, M. Schwartz, and J. Steinberger. Observation of high-energy neutrino reactions and the existence of two kinds of neutrinos, *Physical Review Letters* **9**, 36–44 (1962).

[26] Lillian Hoddeson, Laurie Brown, Michael Riordan, and Max Dresden, editors, *The Rise of the Standard Model: Particle Physics in the 1960s and 1970s* (Cambridge University Press, Cambridge, 1997).

[27] Dominique Pestre and John Krige. Some thoughts on the early history of CERN, in Peter Galison and Bruce Hevly, editors, *Big Science: The Growth of Large-Scale Research* (Stanford University Press, Stanford, 1992), pp. 78–99.

[28] Melvin Schwartz. The early history of high-energy neutrino physics, in Lillian Hoddeson, Laurie Brown, Michael Riordan, and Max Dresden, editors, *The Rise of the Standard Model: Particle Physics in the 1960s and 1970s* (Cambridge University Press, Cambridge, 1997), pp. 411–27.

[29] M. Veltman. Notes on Section 2: High energy physics, in QHA, pp. 3–5.

[30] J. S. Bell and M. Veltman. Intermediate boson production by neutrinos, *Physics Letters* **5**, 94–6 (1963). QHA, pp. 243–5.

[31] J. S. Bell and M. Veltman. Polarisation of vector bosons produced by neutrinos, *Physics Letters* **5**, 151–2 (1963). QHA, pp. 246–7.

[32] J. S. Bell, J. Løvseth, and M. Veltman. CERN neutrino experiment: Conclusions, in G. Bernardini and G. Puppi, editors, *Proceedings of the 1963 International Conference on Elementary Particles*, pp. 584–90. QHA, pp. 236–42.

[33] Don Perkins. Quarks for real: The 1990 Nobel prize in physics, *Physics World* **3**, 62 (1990). Rather unfortunately, Perkins's criticism appeared on the same page of the journal as an obituary of John Bell by John Ellis (pp. 62–3).

[34] Klaus Winter. Experimental studies of weak interactions, in John Krige, editor, *History of CERN, Volume III* (Elsevier, Amsterdam, 1996), pp. 415–73.

[35] D. Perkins. International Conference, in G. Bernardini and G. Puppi, editors, *Proceedings of the 1963 International Conference on Elementary Particles*, pp. 555–61.

[36] Josef-Maria Jauch and Fritz Rohrlich, *The Theory of Photons and Electrons* (Addison-Wesley, Reading, MA, 1955).

[37] Josef-Maria Jauch and Constantin Piron. On the structure of quantum proposition systems, *Helvetica Physica Acta* **36**, 842–8 (1963).

[38] Andrew M. Gleason. Measures of the closed subspaces of a Hilbert space, *Journal of Mathematics and Mechanics* **6**, 885–93 (1957).

[39] J. J. O'Connor and E. F. Robertson. Andrew Mattei Gleason, *MacTutor History of Mathematics Archive, University of St Andrews*; http://www-history.mcs.st-andrews.ac.uk/Biographies/Gleason.html.

[40] Harvey R. Brown. Bell's other theorem and its connection with nonlocality. Part 1, in A. van der Merwe, F. Selleri, and G. Tarozzi, editors, *Bell's Theorem and the Foundations of Modern Physics* (World Science, Singapore, 1992), pp. 104–16.

[41] S. Kochen and E. P. Specker. The problem of hidden variables in quantum mechanics, *Journal of Mathematics and Mechanics* **17**, 59–87 (1967).

[42] John S. Bell. On the problem of hidden variables in quantum theory, *Reviews of Modern Physics* **38**, 447–52 (1966). QHA, pp. 695–700; FQM, pp. 1–6; SUQM, pp. 1–13.

[43] J. S. Bell. On the Einstein Podolsky Rosen paradox, *Physics* **1**, 195–200 (1964). QHA, pp. 701–6; FQM, pp. 7–12; SUQM, pp. 14–21.

[44] John von Neumann, *Mathematische Grundlagen der Quantenmechanik* (Springer, Berlin, 1932); translated as *Mathematical Foundations of Quantum Theory* (Princeton University Press, Princeton, NJ, 1955).

[45] J. S. Bell. On the impossible pilot wave, *Foundations of Physics* **12**, 989–99 (1982). QHA, pp. 842–52; FQM, pp. 148–58; SUQM, pp. 159–68.

[46] Niels Bohr. Discussions with Einstein, in P. A. Schilpp, editor, *Albert Einstein, Philosopher-Scientist* (Tudor, New York, NY, 1949), pp. 199–241.

[47] Josef-Maria Jauch, *Are Quanta Real: A Galilean Dialogue* (Indiana University Press, Bloomington, IL, 1973).

[48] P. C. W. Davies and J. R. Brown, *The Ghost in the Atom* (Cambridge University Press, Cambridge, 1986). The interview with John Bell is on pp. 45–57.

[49] Kamal Datta to Andrew Whitaker, 12 April 1998.

[50] Albert Einstein. Autobiographical notes, in P. A. Schilpp, editor, *Albert Einstein, Philosopher-Scientist* (Tudor, New York, NY, 1949), pp. 1–95.

[51] Abner Shimony. Metaphysical problems in the foundations of quantum mechanics, *International Philosophical Quarterly* **18**, 3–17 (1978).

[52] Tim Maudlin, *Quantum Non-Locality and Relativity* (Blackwell, Oxford, 1994, with further editions in 2002 and 2011).

[53] Jon Jarrett. On the physical significance of the locality conditions in the Bell arguments, *Noûs* **18**, 569–89 (1984).

[54] John Bell. Bertlmann's socks and the nature of reality, *Journal de Physique* **42**, 41–62 (1981). QHA, pp. 820–41; FQM, 126–47; SUQM, pp. 139–58.

[55] Jeremy Bernstein, *Quantum Profiles* (Princeton University Press, Princeton, NJ, 1991).

[56] Kurt Gottfried and N. David Mermin. John Bell and the moral aspect of quantum mechanics, *Europhysics News* **22**, 67–9 (1991).

[57] Henry P. Stapp. Bell's theorem and world process, *Nuovo Cimento* **19B**, 270–6 (1975).

[58] John F. Clauser. Early history of Bell's theorem, in QUS, pp. 61–98.

[59] David Wick, *The Infamous Boundary: Seven Decades of Controversy in Quantum Physics* (Birkhäuser, Boston, MA, 1995).

[60] John Gribbin. The man who proved Einstein was wrong, *New Scientist* **128**, 43–4 (1990).

[61] Robert Romer. John S. Bell (1928–90), the man who proved Einstein was right, *American Journal of Physics* **59**, 299–300 (1991).

[62] A. Einstein, B. Podolsky, and N. Rosen. Can quantum-mechanical description of physical reality be considered complete? *Physical Review* **47**, 777–80 (1935).

[63] Albert Einstein. Quantum mechanics and reality, *Dialectica* **2**, 320–4 (1948).

[64] Max Born, *The Born–Einstein Letters 1916–1955* (Macmillan, Houndsmills, 2nd edition, 2005).

[65] Dipankar Home and Andrew Whitaker, *Einstein's Struggles with Quantum Theory: A Reappraisal* (Springer, New York, NY, 2007).

[66] Andrew Steane. Quantum computing, *Reports on Progress on Physics* **61**, 117–74 (1998).

[67] Michael A. Nielsen and Isaac L. Chuang, *Quantum Computation and Quantum Information* (Cambridge University Press, Cambridge, 2000).

[68] Interview of Abner Shimony by Joan Lisa Bromberg on 9 September 2002, Niels Bohr Library and Archives, American Institute of Physics, College Park, MD USA; http://www.aip.org/history/ohilist/25643.html.

[69] The interviews, published under the title *John Bell and the Identical Twins*, are online at http://cds.cern.ch/record/1092112?ln=en.

[70] Minnesota Center for Twin and Family Research; https://mctfr.psych.umn.edu/ twinstudy/. See also 'Twins Separated at Birth Reveal Staggering Influence of Genetics'; http://www.livescience.com/47288-twin-study-importance-of-genetics. html.

[71] Euan J. Squires, Lucien Hardy, and Harvey R. Brown. Nonlocality from an analogue of the quantum Zeno effect, *Studies in History and Philosophy of Science* **25**, 425–35 (1994).

[72] J. S. Bell and M. Nauenberg. The moral aspect of quantum mechanics, in A. de Shalit, H. Feshbach, and L. van Hove, editors, *Preludes in Theoretical Physics: In Honor of V. F. Weisskopf* (North-Holland, Amsterdam, 1966). QHA, pp. 707–15; SUQM, pp. 22–8.

[73] Euan J. Squires. Many views of one world, *European Journal of Physics* **8**, 171–3 (1987).

[74] Michael Nauenberg to Andrew Whitaker, December 2014.

[75] J. H. Christenson, J. W. Cronin, V. L. Fitch, and R. Turlay. Evidence for the 2π decay of the K_2^0 meson, *Physical Review Letters* **13**, 138–41 (1964).

[76] B. R. Martin and G. Shaw, *Particle Physics* (Wiley, Chichester, 2nd edition, 1997).

[77] Richard A. Dunlap, *An Introduction to the Physics of Nuclei and Particles* (Brooks/Cole, Belmont, CA, 2004).

[78] Donald Perkins. Gargamelle and the discovery of neutral currents, in Lillian Hoddeson, Laurie Brown, Michael Riordan, and Max Dresden, editors, *The Rise of the Standard Model: Particle Physics in the 1960s and 1970s* (Cambridge University Press, Cambridge, 1997), pp. 428–46.

[79] J. S. Bell and J. K. Perring. 2π decay of the K_2^0 meson, *Physical Review Letters* **13**, 266–7 (1964). QHA, pp. 348–9.

[80] X. de Bouard and 9 others. Two-pion decay of K_2^0 at 10 GeV/c^2, *Physics Letters* **15**, 58–61 (1965).

[81] James Cronin. The discovery of CP violation, in Lillian Hoddeson, Laurie Brown, Michael Riordan, and Max Dresden, editors, *The Rise of the Standard Model: Particle Physics in the 1960s and 1970s* (Cambridge University Press, Cambridge, 1997), pp. 14–36.

[82] L. B. Leipuner, W. Chinowsky, R. Crittenden, R. Adair, B. Musgrave, and F. T. Shively. Anomalous regeneration of K_1^0 mesons from K_2^0 mesons, *Physical Review* **132**, 2285–91 (1963).

[83] J. Bernstein, N. Cabibbo, and T. D. Lee. CP invariance and the 2π decay mode of the K_2^0, *Physics Letters* **12**, 146–8 (1964).

[84] J. S. Bell and J. Steinberger. Weak interactions of kaons, in *3rd European Physical Society Conference on Elementary Particles, Oxford, September 1965* (Rutherford High Energy Laboratory, Chilton, 1966), pp. 147–74. QHA, pp. 280–307.

[85] B. Laurent and M. Roos. On the superposition principle and CP invariance in K^0 decay, *Physics Letters* **13**, 269–70 (1965).

[86] Preface, in QUS, pp. vii–x.

[87] J. S. Bell. Current algebra and gauge invariance, *Nuovo Cimento* **50**, 129–34 (1967). QHA, pp. 322–7.

[88] Abner Shimony, Valentine Telegdi, and Martinus Veltman. John S. Bell, *Physics Today* **44**, 82, 84, 86 (1991).

[89] Andrew Pickering, *Constructing Quarks: A Sociological History of Particle Physics* (Edinburgh University Press, Edinburgh, 1984). This book is not popular with physicists, since it states that quarks have been 'constructed' rather than 'discovered' and that 'there is no obligation upon anyone framing a view of the world to account of what twentieth-century science has to say'. However, the book does present a clear view of the developments in particle physics, and even physicists have been heard to say that, leaving aside the first 19 and the last 13 pages, the book is quite good.

[90] Robert Eisberg and Robert Resnick, *Quantum Physics of Atoms, Molecules, Solids, Nuclei and Particles* (Wiley, New York, NY, 1974, 1985).

[91] Raymond Flood, Mark McCartney, and Andrew Whitaker (editors). *James Clerk Maxwell: Perspectives on his Life and Work* (Oxford University Press, Oxford, 2014).

[92] C. N. Yang and R. Mills. Conservation of isotopic spin and isotopic gauge invariance, *Physical Review* **96**, 191–5 (1954).

[93] S. L. Glashow. Partial symmetries of weak interaction, *Nuclear Physics* **22**, 579–88 (1961).

[94] Ian Sample, *Massive: The Higgs Boson and the Greatest Hunt in Science* (Virgin, London, 2013).

[95] S. Weinberg. A model of leptons, *Physical Review Letters* **19**, 1264–6 (1967).

[96] J. S. Bell and S. M. Berman. On current algebra and CVC in pion beta-decay, *Nuovo Cimento* **47**, 807–10 (1967). QHA, pp. 318–21.

[97] J. S. Bell. Equal-time commutator in a solvable model, *Nuovo Cimento* **47**, 616–25 (1967). QHA, pp. 308–17.

[98] J. S. Bell. Current and density algebra and gauge invariance, in Jack Steinberger (editor), *Rendiconti della Scuola Internazionale di Fisica 'E. Fermi', VarennaI* (1967), pp. 170–7. QHA, pp. 341–8.

[99] Martinus J. G. Veltman. From weak interactions to gravitation (Nobel lecture, 8 December 1999): http://www.nobelprize.org/nobel_prizes/physics/laureates/1999/veltman-lecture.html.

[100] Gerard 't Hooft. Gauge theory and renormalisation. Presented at the International Conference on 'The History of Original Ideas and Basic Discoveries in Particle Physics' at Erice, Italy, 29th July to 4th October 1994; http://arxiv.org/pdf/hep-th/9410038.pdf.

[101] M. Veltman. Perturbation theory of massive Yang-Mills field, *Nuclear Physics B* **7**, 637–50 (1968).

[102] Philip G. Burke and Ian C. Percival. John Stewart Bell, 28 July 1928–1 October 1980, *Biographical Memoirs of Fellows of the Royal Society* **45**, 2–17 (1999).

[103] J. S. Bell. Electromagnetic properties of unstable particles, *Nuovo Cimento* **24**, 452–60 (1962). QHA, pp. 204–12.

[104] J. S. Bell. Rest energy, rest mass, and noncovariant electrodynamics, *Nuovo Cimento* **24**, 554–6 (1962). QHA, pp. 213–5.

[105] J. S. Bell and C. J. Goebel. Double poles and nonexponential decay, *Physical Review* **138**, 1198–1201 (1965). QHA, pp. 268–71.

[106] J. S. Bell. On a conjecture of C. N. Yang, *Physics Letters* **2**, 116 (1962). QHA, p. 228.

[107] J. S. Bell, P. Meyer, and J. Prentki. The Lee theory of intermediate bosons and the K_{t3} decays, *Physics Letters* **2**, 349–50 (1962). QHA, pp. 234–5.

[108] Rudolph Peierls. Bell's early work, *Europhysics News* **4**, 69–70 (1991).

[109] J. S. Bell and S. M. Berman. Pion and strange particle production by 1 GeV neutrinos, *Nuovo Cimento* **25**, 404–15 (1962). QHA, pp. 216–27.

[110] J. S. Bell and J. Løvseth. On muon capture in heavy nuclei, *Nuovo Cimento* **32**, 433–47 (1964). QHA, pp. 248–62.

[111] J. S. Bell. Nuclear optical model for virtual pions, *Physical Review Letters* **13**, 57–9 (1964). QHA, pp. 163–5.

[112] J. S. Bell. Fluctuation compressibility theorem and its application to the pairing model, *Physical Review* **129**, 1896–1900 (1963). QHA, pp. 229–33.

[113] J. S. Bell and H. Ruegg. $\tilde{U}_{12}$, its 143 momenta, its little group $U_6 \times U_6$ and its irregular couplings, *Nuovo Cimento* **39**, 1166–73 (1965). QHA, pp. 272–9.

[114] J. S. Bell. Difficulties of relativistic U(6), in A. Zichichi, editor, *Recent Developments in Particle Symmetries* (Academic Press, New York, NY, 1966).

[115] J. S. Bell. On Singh's lemma for low-energy Compton scattering, *Nuovo Cimento* **52A**, 635–6 (1967). QHA, pp. 339–40.

[116] J. S. Bell and R. Van Royen. On the Low–Burnett–Kroll theorem for soft-photon emission, *Nuovo Cimento* **60A**, 62–8 (1969). QHA, 382–8.

[117] J. S. Bell. Froissart bounds with any spin, *Nuovo Cimento* **61A**, 541–52 (1969). QHA, pp. 389–400.

[118] J. S. Bell and E. de Rafael. Hadronic vacuum polarization and $g_\mu - 2$, *Nuclear Physics* **B11**, 611–20 (1969). QHA, pp. 401–10.

[119] J. S. Bell and R. Jackiw. A PCAC puzzle: $\pi^0 \to \gamma\gamma$ in the σ model, *Nuovo Cimento* **60A**, 47–61 (1969). QHA, pp. 367–81.

[120] Roman Jackiw. John Bell's observations on the chiral anomaly and some of its descendants, in QUS, pp. 377–82.

[121] Reinhold A. Bertlmann, *Anomalies in Quantum Field Theory* (Oxford University Press, Oxford, 1996).

[122] D. G. Sutherland. Current algebra and the decay η→3π, *Physics Letters* **23**, 384–5 (1966).

[123] D. G. Sutherland. Current algebra and some non-strong meson decays, *Nuclear Physics* **B2**, 433–40 (1966).

[124] J. S. Bell and D. G. Sutherland. Current algebra and η→3π, *Nuclear Physics* **B4**, 315–25 (1968). QHA, pp. 328–38.

[125] M. Veltman. Theoretical aspects of high energy neutrino interactions, *Proceedings of the Royal Society A* **301**, 107–12 (1967).

[126] J. Steinberger. On the use of subtraction fields and the lifetimes of some types of meson decay, *Physical Review* **76**, 1180–6 (1949).

[127] Hiroshi Fukuda and Yoneji Miyamoto. On the γ decay of neutral meson, *Progress of Theoretical Physics* **4**, 347–57 (1949).

[128] J. Steinberger. Jack Steinberger: Biographical, in G. Ekspong, editor, *Nobel Lectures in Physics 1981–90* (World Scientific, Singapore, 1993). Online (with updated material) at http://www.nobelprize.org/nobel_prizes/physics/laureates/1988/steinberger-bio.html.

[129] Roman Jackiw. The chiral anomaly, *Europhysics News* **22**, 76–7 (1991).

[130] S. L. Adler. Axial–vector vertex in spinor electrodynamics, *Physical Review* **177**, 2426–38 (1969).

[131] G. 't Hooft. Symmetry breaking through Bell–Jackiw anomalies, *Physical Review Letters* **37**, 8–11 (1976).

[132] C. Bouchiat, J. Iliopoulos, and P. Meyer. An anomaly-free version of Weinberg's model, *Physics Letters* **38**, 519–23 (1972).

[133] Brian Greene, *The Elegant Universe: Superstrings, Hidden Dimensions and the Quest for the Ultimate Theory* (Norton, New York, NY, 2003).

[134] Stephen Hawking, *The Grand Design* (Bantam, New York, NY, 2010).

[135] Peter Woit, *Not Even Wrong: The Failure of String Theory and the Search for Unity in Physical Law* (Cape, London, 2006).

[136] Sam Treiman, Roman Jackiw, and David J. Gross, *Lectures on Current Algebra and its Applications* (Princeton University Press, Princeton, NJ, 1972, reprinted in 2015).

[137] Sam Treiman, Roman Jackiw, Bruno Zumino, and Edward Witten, *Current Algebra and Anomalies* (Princeton University Press, Princeton, NJ, 1986, and World Scientific, Singapore, 1985).

[138] Andrew Whitaker, *The New Quantum Age: From Bell's Theorem to Quantum Computation and Teleportation* (Oxford University Press, Oxford, 2012).

[139] Michael Horne. Preface to Volume One, in Robert S. Cohen, Michael Horne, and John Stachel, editors, *Experimental Metaphysics: Quantum Mechanical Studies for Abner Shimony* (Kluwer, Dordrecht, 1997, 2 volumes), pp. ix–x.

[140] E. P. Wigner. Remarks on the mind–body problem, in L. J. Good, editor, *The Scientist Speculates: An Anthology of Partly Baked Ideas* (Heinemann, London, 1961), pp. 284–301.

[141] A. Shimony. Role of the observer in quantum theory, *American Journal of Physics* **31**, 755–73 (1963).

[142] E. P. Wigner. The problem of measurement, *American Journal of Physics* **31**, 6–15 (1963).

[143] J. F. Clauser. Early history of Bell's theorem, in QUS, pp. 61–98.

[144] C. S. Wu and I. Shaknov. The angular correlation of scattered annihilation radiation, *Physical Review* **77**, 136 (1950).

[145] D. Bohm and Y. Aharanov. Discussion of experimental proof for the paradox of Einstein, Rosen and Podolsky, *Physical Review* **108**, 1070–6 (1957).

[146] C. A. Kocher and E. D. Commins. Polarisation correlation of photons emitted in an atomic cascade, *Physical Review Letters* **18**, 575–7 (1967).

[147] David Wick, *The Infamous Boundary: Seven Decades of Heresy in Quantum Physics* (Springer, New York, NY, 1995).

[148] J. F. Clauser, M. A. Horne, A. Shimony, and R. A. Holt. Experimental test of hidden variable theories, *Physical Review Letters* **23**, 880–4 (1969).

[149] Bernard d'Espagnat. My interaction with John Bell, in QUS, pp. 21–81.

[150] Bernard d'Espagnat. The quantum theory and reality, *Scientific American* **241**, 158–61 (1979).

[151] Bernard d'Espagnat, *Conceptual Foundations of Quantum Theory* (Benjamin, Menlo Park, CA, 1971).

[152] Bernard d'Espagnat, *Veiled Reality: An Analysis of Present-Day Quantum Mechanical Concepts* (Addison-Wesley, Reading, MA, 1995).

[153] Franco Selleri, *Quantum Paradoxes and Quantum Reality* (Kluwer, Dordrecht, 1990).

[154] Oliver Freire. Quantum dissidents: Research on the foundations of quantum theory, *Studies in the History and Philosophy of Modern Physics* **40**, 180–9 (2009).

[155] American Institute of Physics: Oral Physics Project. Franco Selleri; http://www.aip.org/history/ohilist/28003_1.html and http://www.aip.org/history/ohilist/28003_2.html.

[156] Eugene P. Wigner. On hidden variables and quantum mechanical probabilities, *American Journal of Physics* **38**, 1005–8 (1970).

[157] S. Freedman and E. P. Wigner. On Bub's refutation of Bell's locality argument, *Foundations of Physics* **3**, 457–8 (1973).

[158] Hans-Dieter Zeh. On the interpretation of measurement in quantum theory, *Foundations of Physics* **1**, 69–76 (1970).

[159] J. S. Bell. Introduction to the hidden variable question, in Bernard d'Espagnat, editor, *Foundations of Quantum Mechanics, Proceedings of the International School of Physics 'Enrico Fermi', Course IL* (Academic, New York, NY, 1971), pp. 171–81, QHA, pp. 716–26; FQM, pp. 22–32; SUQM, pp. 29–39.

Chapter 4 The 1970s: Interest Increases

[1] Courtesy of the Royal Society.

[2] Thanks to Ruby McConkey.

[3] Donald Perkins. Gargamelle and the discovery of neutral currents, in Lillian Hoddeson, Laurie Brown, Michael Riordan, and Max Dresden, editors, *The Rise of the Standard Model: Particle Physics in the 1960s and 1970s* (Cambridge University Press, Cambridge, 1997), pp. 428–46.

[4] Ian Sample, *Massive: The Higgs Boson and the Greatest Hunt in Science* (Virgin, London, 2010).

[5] Klaus Winter. Experimental studies of weak interactions, in John Krige, editor, *History of CERN, Volume III* (Elsevier, Amsterdam, 2004), pp. 415–73.

[6] Andrew Whitaker, *The New Quantum Age: From Bell's Theorem to Quantum Computation and Teleportation* (Oxford University Press, Oxford, 2012).

[7] Interview of John F. Clauser by Joan Lisa Bromberg on 20 May 2002; http://www.aip.org/history/ohilist/25096.html.

[8] John F. Clauser. Early history of Bell's theorem, in QUS, pp. 61–98.

[9] S. J. Freedman and J. F. Clauser. Experimental test of local hidden-variable theories, *Physical Review Letters*, **28**, 938–41 (1972).

[10] R. A. Holt. Atomic cascade experiments. PhD thesis, Harvard University, Cambridge, MA, 1973.

[11] J. F. Clauser. Experimental investigation of a polarization correlation anomaly, *Physical Review Letters*, **36**, 1223–6 (1976).

[12] E. S. Fry and R. C. Thompson. Experimental test of local hidden-variable theories, *Physical Review A*, **37**, 465–8 (1976).

[13] L. R. Kasday, J. D. Ullman, and C. S. Wu. The Einstein-Podolsky-Rosen argument: Positron annihilation experiment, *Bulletin of the American Physical Society*, **15**, 586 (1970).

[14] M. Lamehi-Rachti and W. Mittig. Quantum mechanics and hidden variables: A test of Bell's inequalities by the measurement of spin correlation in low-energy proton–proton scattering, *Physical Review D*, **14**, 2543–55 (1976).

[15] J. F. Clauser and M. A. Horne. Experimental consequences of objective local theories, *Physical Review D*, **10**, 526–35 (1974).

[16] J. F. Clauser and A. Shimony. Bell's theorem: Experimental test and implications, *Reports on Progress in Physics*, **41**, 1881–1927 (1978).

[17] J. S. Bell. The theory of local beables, presented at the Sixth G. I. F. T. International Seminar on Theoretical Physics at Jaca, June 1975; reproduced in *Epistemological Letters* **9**, 11 (1976) and *Dialectica* **39**, 86–96 (1985). QHA, pp. 754–54; FQM, pp. 50–60; SUQM, pp. 52–62.

[18] J. S. Bell. Einstein-Podolsky-Rosen experiments, in *Proceedings of the Symposium on Frontier Problems in High Energy Physics*, Pisa, June 1976, Annali Della Schola, Normale Superiore di Pisa, pp. 33–45. QHA, pp. 768–77; fQM, pp. 74–83; SUQM, pp. 81–92.

[19] J. S. Bell. Atomic-cascade photons and quantum-mechanical nonlocality, *Comments on Atomic and Molecular Physics* **9**, 121–6 (1980). QHA, pp. 782–7; FQM, pp. 88–93; SUQM, pp. 105–110.

[20] J. S. Bell. Bertlmann's socks and the nature of reality, *Journal de Physique* **42**, 41–62 (1981). QHA, pp. 820–41; FQM, pp. 126–47; SUQM, pp. 139–58.

[21] Alain Aspect. Bell's theorem: The naïve view of an experimentalist, in QUS, pp. 120–53.

[22] A. Aspect. Proposed experiment to test separable hidden-variable theories. *Physics Letters A* **54**, 117–8 (1975).

[23] A. Aspect. Proposed experiment to test the non-separability of quantum mechanics, *Physical Review D* **14**, 1944–51 (1976).

[24] David Kaiser, *How the Hippies Saved Physics: Science, Counterculture and the Quantum Revival* (Norton, New York, NY, 2011).

[25] Andrew Whitaker. ESP and LSD on the CIA's dime, *Physics World* **25**, 42–3 (2012).

[26] Nick Herbert, *Quantum Reality: Beyond the New Physics* (Anchor/Doubleday, Garden City, NJ, 1985).

[27] Nick Herbert, *Faster than Light: Superluminal Loopholes in Physics* (Plume, New York, NY, 1988).

[28] Nick Herbert, *Elemental Mind: Human Consciousness and the New Physics* (Dutton, New York, NY, 1993).

[29] Nick Herbert. FLASH: A superluminal communicator based on a new kind of quantum measurement, *Foundations of Physics* **12**, 1171–9 (1982).

[30] W. K. Wooters and W. H. Zurek. A single quantum cannot be cloned, *Nature* **299**, 802–3 (1982).

[31] D. Dieks. Communication by EPR devices, *Physics Letters A* **92**, 271–2 (1982).

[32] G. C. Ghirardi and T. Weber. Quantum mechanics and faster-than-light communication: Methodological considerations, *Nuovo Cimento B* **78**, 9–20 (1983).

[33] Fritjof Capra, *The Tao of Physics: An Explanation of the Parallels between Modern Physics and Eastern Mysticism* (Shambhala, Boulder, CO, 1975).

[34] Gary Zukav, *The Dancing Wu Li Masters: An Overview of the New Physics* (Morrow, New York, NY, 1979).

[35] Henry Stapp, *Mindful Universe: Quantum Mechanics and the Participating Observer* (Springer, New York, NY, 2011).

[36] H. P. Stapp. Are superluminal connections necessary?, *Nuovo Cimento B* **40**, 191–205 (1977).

[37] Robert Crease and Alfred Goldhaber, *The Quantum Moment: How Planck, Bohr, Einstein, and Heisenberg Taught Us to Love Uncertainty* (Norton, New York, NY, 2014).

[38] Jeremy Bernstein, *Quantum Profiles* (Princeton University Press, Princeton, NJ, 1991) and tapes at http://cds.cern.ch/record/1092112?ln=en.

[39] J. S. Bell. On wave packet reduction in the Coleman–Hepp model, *Helvetica Physica Acta* **48**, 93–8 (1975). QHA, pp. 738–43; SUQM, pp. 45–51.

[40] J. S. Bell. Introduction to the hidden variable question, in Bernard d'Espagnat, editor, *Foundations of Quantum Mechanics, Proceedings of the International School of Physics 'Enrico Fermi', Course IL* (Academic, New York, NY, 1971), pp. 171–81, QHA, pp. 716–26; FQM, pp. 22–32; SUQM, pp. 29–39.

[41] J. S. Bell. Locality in quantum mechanics: Reply to critics, *Epistemological Letters* (November 1975), pp. 2–6. SUQM, pp. 63–6.

[42] J. S. Bell. Free variables and local causality, *Epistemological Letters* **15**, 15 (1977); reprinted in *Dialectica* **39**, 103–6 (1985). QHA, pp. 778–81; FQM, pp. 84–7; SUQM, pp. 100–104.

[43] A. Shimony, M. A. Horne, and J. F. Clauser. Comment of the theory of local beables, *Epistemological Letters* **13**, 1 (1976); *Dialectica* **39**, 96–101 (1985).

[44] A. Shimony. Reply to Bell, *Epistemological Letters* **18**, 1–4 (1978); *Dialectica* **39**, 107–8 (1985).

[45] J. S. Bell. Subject and object, in J. Mehra, editor, *The Physicist's Conception of Nature* (Reidel, Dordrecht, 1993), pp. 687–90. QHA, pp. 734–7; FQM, pp. 40–3; SUQM, pp. 40–44.

[46] J. S. Bell. Quantum mechanics for cosmologists, in C. Isham, R. Penrose, and D. Sciama, editors, *Quantum Gravity 2* (Clarendon, Oxford, 1981), pp. 611–37. QHA, pp. 793–819, FQM, 99–125; SUQM, pp. 117–38.

[47] J. S. Bell. The measurement theory of Everett and de Broglie's pilot wave, in M. Flato, Z. Maric, A. Milojevic, D. Sternheimer, and J. P. Vigier, editors, *Quantum Mechanics, Determinism, Causality and Particles* (Reidel, Dordrecht, 1976), pp. 11–17. QHA, pp. 727–33; FQM, pp. 33–9; SUQM, 93–99.

[48] J. S. Bell. De Broglie-Bohm, delayed-choice double-slit experiment, and density matrix, *International Journal of Quantum Chemistry: Quantum Chemistry Symposium* 14, 155–9 (1980). QHA, pp. 788–92; FQM, pp. 94–8; SUQM, pp. 111–117.

[49] J. S. Bell. How to teach special relativity, *Progress in Scientific Culture* 1, 1–13 (1976). QHA, 755–67; FQM, pp. 61–73; SUQM, 67–80.

[50] H. M. Wiseman. From Einstein's theorem to Bell's theorem: A history of quantum nonlocality, *Contemporary Physics* 47, 79–88 (2006).

[51] Travis Norsen. Against 'realism', *Foundations of Physics* 37, 311–40 (2007).

[52] Gianarlo Ghirardi. Does quantum nonlocality irremediably conflict with special relativity?, *Foundations of Physics* 40, 1379–95 (2010).

[53] M. P. Seevinck and J. Uffink. Not throwing out the baby with the bathwater: Bell's condition of local causality mathematically 'sharp and clean', in Dennis Dieks, Wenceslao J. Gonzalez, Stephan Hartmann, Thomas Uebel, and Marcel Weber, editors, *Explanation, Prediction and Confirmation* (Springer, Berlin, 2011), pp. 425–50.

[54] Nicolas Gisin. Non-realism: Deep thought or soft option?, *Foundations of Physics* 42, 80–85 (2012).

[55] Nicolas Brunner, Otfried Gühne, and Marcus Huber. Fifty years of Bell's theorem, *Journal of Physics A* 47, 420301 (2014).

[56] H. M. Wiseman. The two Bell's theorems of John Bell, *Journal of Physics A* 47, 424001 (2014).

[57] Marek Żukowski and Časlav Brukner. Quantum non-locality: It ain't necessarily so ..., *Journal of Physics A* 47, 424009 (2014).

[58] Tim Maudlin. What Bell did, *Journal of Physics A* 47, 424010 (2014).

[59] Reinhard F. Werner. Comment on 'What Bell did', *Journal of Physics A* 47, 424011 (2014).

[60] Tim Maudlin. Reply to comment on 'What Bell did', *Journal of Physics A* 47, 424012 (2014).

[61] R. F. Werner. What Maudlin replied to, arXiv:1411.2120 (2014).

[62] Travis Norsen. Are there really two different Bell's theorems?, *International Journal of Quantum Foundations* 1, 65–84 (2015).

[63] Howard M. Wiseman and Eleanor G. Rieffel. Reply to Norsen's paper 'Are there really two different Bell's theorems?', *International Journal of Quantum Foundations* 1, 85–99 (2015).

[64] Marek Zukowski. Quantum theory tells us NOT what quantum mechanics IS but what quantum mechanics IS NOT, presented at Quantum [Un]Speakables II: Fifty Years of Bell's Theorem(2014).

[65] Sheldon Goldstein, Travis Norsen, Daniel Tausk, and Nino Xanghi. Bell's Theorem, *Scholarpedia* **6**, 8738 (2011).

[66] Abner Shimony. John S. Bell: Some reminiscences and reflections, in QUS, pp. 51–60.

[67] T. B. Day. Demonstration of quantum mechanics in the large, *Physical Review* **121**, 1204–6 (1961).

[68] G. Lochak. Has Bell's inequality a general meaning for hidden-variable theories? *Foundations of Physics* **6**, 173–84 (1976).

[69] L de la Peña, A. M. Cetto, and T. A. Brody. Hidden-variable theories and Bell's inequality, *Letters to Nuovo Cimento* **5**, 177 (1972).

[70] L. de Broglie. Refutation of Bell's theorem, *Comptes Rendus de l'Academie des Sciences B* **278**, 721–2 (1974).

[71] Max Jammer, *The Philosophy of Quantum Mechanics* (Wiley, New York, NY, 1974).

[72] Philip Pearle. Reduction of the state vector by a nonlinear Schrödinger equation, *Physical Review D* **13**, 857–68 (1976).

[73] N. F. Mott. The wave mechanics of α-ray tracks, *Proceedings of the Royal Society A* **126**, 79–84 (1929).

[74] Werner Heisenberg, *Physical Principles of the Quantum Theory* (University of Chicago Press, Chicago, IL, 1930).

[75] Henry Stapp. The Copenhagen interpretation and the nature of space-time, Lawrence Radiation Laboratory Report UCRL-2094 (1971).

[76] Henry Stapp. Locality and reality, *Foundations of Physics* **10**, 767–95 (1980).

[77] Léon Rosenfeld. The measuring process in quantum mechanics, *Supplement to Progress in Theoretical Physics* **65**, 222–31 (1965).

[78] H. Everett. 'Relative state' formulation of quantum mechanics, *Reviews of Modern Physics* **29**, 454–62 (1957).

[79] E. J. Squires. Many views of one world, *European Journal of Physics* **8**, 171–3 (1987).

[80] B. de Witt and N. Graham, editors, *The Many-Worlds Interpretation of Quantum Mechanics* (Princeton University Press, Princeton, NJ, 1973).

[81] John A. Wheeler. The 'past' and the 'delayed-choice' double-slit experiment, in A. R. Marlow, editor, *Mathematical Foundations of Quantum Mechanics* (Academic, New York, NY, 1978), pp. 9–48.

[82] P. C. W. Davies and J. R. Brown, editors, *The Ghost in the Atom* (Cambridge University Press, Cambridge, 1986). The interview with John Bell is on pp. 45–57.

[83] K. Hepp. Quantum theory or measurement and macroscopic observables, *Helvetica Physica Acta* **45**, 237–48 (1972).

[84] John Bell. George Francis Fitzgerald, *Physics World* **5**, 31–35 (1992). QHA, pp. 829–33; FQM, pp. 235–9.

[85] Franco Selleri. Bell's spaceships and special relativity, in QUS, pp. 413–28.

[86] J. S. Bell. Weak interactions in the nuclear shadow, in H. Ali, editor, *Lectures on Particles and Fields* (Gordon and Breach, London, 1970), pp. 323–40. QHA, pp. 349–66.

[87] J. S. Bell and C. H. Llewelllyn Smith. Near-forward neutrino reactions on nuclear targets, *Nuclear Physics B* **24**, 285–304 (1970). QHA, pp. 411–30.

[88] J. S. Bell and C. H. Llewellyn Smith. Quasistatic neutrino-nucleus interactions. *Nuclear Physics B* **28**, 317–40 (1971). QHA, pp. 444–67.

[89] J. S. Bell. Pseudoscalar meson dominance in neutrino-nucleus reactions, in J. Cummings and H. Osborn, editors, *Hadron Interactions of Electrons and Photons* (Academic Press, London, 1971, pp. 369–94). QHA, pp. 468–94.

[90] J. S. Bell, G. Karl, and C. H. Llewellyn Smith. Isospin bounds for energy partition in $\bar{e}e$ and $\dot{N}N$ annihilation, *Physics Letters* **52**, 363–6 (1974), QHA, pp. 569–72.

[91] J. S. Bell. High energy behaviour of tree diagrams in gauge theories, *Nuclear Physics B* **60**, 427–36. QHA, pp. 494–503.

[92] Martinus Veltman. Notes on Section 2: High energy physics, QHA, pp. 3–5.

[93] H. Jay Melosh IV. Quarks: Currents and constituents, PhD thesis, California Institute of Technology, Pasadena, CA, 1973. Available online at http://thesis.library.caltech.edu/4807/1/Melosh_hj_IV_1973.pdf.

[94] J. S. Bell. The Melosh transformation and the Pryce-Tani-Foldy-Wouthuysen tramsformation, *Acta Physica Austriaca Supplement* **13**, 395–445 (1974). QHA, pp. 516–66.

[95] J. S. Bell and A. J. G. Hey. A theoretical argument for something like the second Melosh transformation, *Physics Letters B* **51**, 365–6. QHA, pp. 567–9.

[96] Tony Hey and Patrick Walters, *The New Quantum Universe* (Cambridge University Press, Cambridge, 2003).

[97] R. P. Feynman, *Feynman Lectures on Computation* (Tony Hey and Robin Allen, editors) (Addison-Wesley, Reading, MA, 1996).

[98] Tony Hey, editor, *Feynman and Computation* (Perseus, Reading, MA, 1999).

[99] J. S. Bell and H. Ruegg, Hydrogen atom on null-plane and Melosh transformation, *Nuclear Physics B* **93**, 12–22 (1975). QHA, pp. 576–86.

[100] J. S. Bell and H. Ruegg, Positronium on null-plane and Melosh transformation, *Nuclear Physics B* **104**, 245–52 (1976). QHA, pp. 587–94.

[101] Charles P. Enz. Ernst Stueckelberg, *Physics Today* **39**, 119–21 (1986).

[102] J. S. Bell and H. Ruegg. Dirac equation with an exact higher symmetry, *Nuclear Physics B* **98**, 151–3 (1975).

[103] G. B. Smith and L. S. Tassie. Excited states of mesons and the quark antiquark interaction, *Annals of Physics* **65**, 352–60 (1971).

[104] J. S. Bell and H. Ruegg. Errata, *Nuclear Physics B* **104**, 546 (1976).

[105] J. S. Bell and A. J. G. Hey. Partons of a one-dimensional box, *Physics Letters B* **74**, 77–80 (1978). QHA, pp. 612–5.

[106] J. S. Bell, A. C. Davies, and J. Rafelski. Partons of a spherical box. *Physics Letters B* **78**, 67–70 (1978). QHA, pp. 616–9.

[107] J. S. Bell and R. K. P. Zia. Final-state interactions and charge asymmetry in $K\ell3^L$ decays, *Nuclear Physics B* **17**, 388–400 (1970). QHA, pp. 431–43.

[108] J. S. Bell. Lorentz contraction and diffractive excitation, *Nuclear Physics B* **66**, 293–304 (1973). QHA, pp. 504–15.

[109] J. S. Bell and G. V. Dass. On neutrino and antineutrino scattering by electrons, *Physics Letters B* **59**, 343–5. QHA, pp. 573–5.

[110] J. S. Bell. Čerenkov and transition radiation from particles with spin, *Nuclear Physics B* **112**, 461–9 (1976). QHA, pp. 595–603.

[111] J. S. Bell and G. Karl. Unstable particles in an electric field, *Nuovo Cimento A* **41**, 487–94 (1977). QHA, pp. 604–11.

[112] Bernard d'Espagnat. The quantum theory and reality, *Scientific American* **241**, 158–81 (1979).

Chapter 5 The 1980s: Final Achievements but Final Tragedies

[1] Klaus Winter. Experimental studies of weak interactions, in John Krige, editor, *History of CERN, Volume III* (Elsevier, Amsterdam, 1996), pp. 459–73.

[2] Peter Watkins, *Story of the W and Z* (Cambridge University Press, Cambridge, 1986).

[3] Michael Crowley-Milling. The development of accelerator art and expertise at CERN: 1960–80. Twenty fruitful years, in John Krige, editor, *History of CERN, Volume III* (Elsevier, Amsterdam, 1996), pp. 477–558.

[4] John Polkinghorne, *Rochester Roundabout: The Story of High-Energy Physics* (Longman, Harlow, 1989).

[5] Arturo Russo. The Interacting Storage Rings: The construction and operation of CERN's second large machine and a survey of its experimental programme, in John Krige, editor, *History of CERN, Volume III* (Elsevier, Amsterdam, 1966), pp. 110–170.

[6] Tim Berners-Lee, *Weaving the Web: The Past, Present and Future of the World Wide Web by its Inventor* (Texere, London, 1999).

[7] Jeremy Bernstein, *Quantum Profiles* (Princeton University Press, Princeton, NJ, 1991).

[8] R. A. Bertlmann and A. Martin. Inequalities on heavy quark–antiquark systems, *Nuclear Physics B* **168**, 111–36 (1980).

[9] Renate Bertlmann at http://www.reactfeminism.org/nr1/artists/bertlmann_en.html.

[10] Reinhold Bertlmann to Andrew Whitaker, 19 September 1988.

[11] Reinhold A. Bertlmann. Magic moments: A collaboration with John Bell, in QUS, pp. 29–47.

[12] GianCarlo Ghirardi. John Stewart Bell and the dynamical reduction programme, in QUS, pp. 287–305.

[13] Jeremy Bernstein, Ref. 7 and the tapes he made when interviewing Bell in connection with the writing of that book, which are published under the title *John Bell and the Identical Twins* and are online at http://cds.cern.ch/record/1092112?ln=en.

[14] Simon Winchester, *The Man Who Loved China: The Fantastic Story of the Eccentric Scientist Who Unlocked the Mysteries of the Middle Kingdom* (Harper Collins, New York, NY, 2008).

[15] Erwin Schrödinger, *My View of the World* (OxBow, Oxford, 1983; first published 1961).

[16] Maharishi European Research University; http://www.meru.ch/index.php?page=kurse-in-seelisberg.

[17] Abner Shimony. John S. Bell: Some reminiscences and reflections, in QUS, pp. 51–60.

[18] John Polkinghorne, *The Rochester Roundabout: The Story of High Energy Physics* (Longman, Harlow, 1989).

[19] Reinhold Bertlmann. Magic moments with John Bell, in Alwyn van der Merwe, Franco Selleri, and Gino Tarozzi, editors, *Bell's Theorem and the Foundations of Modern Physics* (World Scientific, Singapore, 1992), pp. 31–44.

[20] J. J. Sakurai. Duality in $e^+ + e^- \to$ hadrons?, *Physics Letters B* **46**, 207–10 (1973).

[21] J. S. Bell and J. Pasupathy. On duality between resonances and free quark–antiquark pairs, *Physics Letters B* **83**, 389–91 (1979). QHA, pp. 620–2.

[22] K. Ishikawa and J. J. Sakurai. Testing Q^2 duality in charmonium potential models, *Zeitschrift für Physik C* **1**, 117–20 (1979).

[23] J. S. Bell and J. Pasupathy. JWKB connection for radial wave functions and bound-state/free-state duality, *Zeitschrift für Physik C* **2**, 183–5 (1979). QHA, pp. 623–5.

[24] J. S. Bell and R. A. Bertlmann. Testing Q^2 duality with non-relativistic potentials, *Zeitschrift für Physik C* **4**, 11–15 (1980). QHA, pp. 626–30.

[25] M. A. Shifman, V. I. Vainshtein, and V. I. Zakharov. QCD and resonance physics (a) Theoretical foundations; (b) Applications; (c) Rho–omega mixing, *Nuclear Physics B* **147**, (a) 385–447; (b) 448–518; (c) 519–534 (1979).

[26] J. S. Bell and R. A. Bertlmann. Magic moments, *Nuclear Physics B* **177**, 218–36 (1981). QHA, pp. 631–49.

[27] J. S. Bell and R. A. Bertlmann. Shifman-Vainshtein-Zakharov moments and quark–antiquark potentials, *Nuclear Physics B* **187**, 285–92 (1980). QHA, pp. 650–7.

[28] R. A. Bertlmann and J. S. Bell. Gluon condensate potentials, *Nuclear Physics B* **227**, 435–46 (1983). QHA, pp. 662–73.

[29] J. S. Bell and R. A. Bertlmann. SVZ moments for charmonium and potential model, *Physics Letters B* **137**, 107–10 (1983). QHA, pp. 686–9.

[30] R. Rajaraman. Fractional charge, in QUS, pp. 383–99.

[31] R. Jackiw and C. Rebbi. Solitons with fermion number ½, *Physical Review D* **13**, 3398–3409 (1976).

[32] W. P. Su, J. R. Schrieffer, and A. J. Heeger. Solitons in polyacetylene, *Physical Review Letters* **42**, 1698–1701 (1979).

[33] R. Rajaraman and J. S. Bell. On solitons with half integral charge, *Physics Letters B* **116**, 151–4 (1982). QHA, pp. 658–61.

[34] J. S. Bell and R. Rajaraman. On states, on a lattice, with half-integral charge, *Nuclear Physics B* **220**, 1–12 (1983). QHA, pp. 674–85.

[35] Mary Bell. Notes on Section 1: Accelerator physics, in QHA, pp. 2–3.

[36] J. S. Bell and M. Bell. Electron cooling in storage rings, *Particle Accelerators* **11**, 233–8 (1981). QHA, pp. 31–6.

[37] M. Bell and J. S. Bell. Capture of cooling electrons by cool protons, *Particle Accelerators* **12**, 49–52 (1981). QHA, pp. 37–40.

[38] Mary Bell. John Bell and accelerator physics, *Europhysics News* **22**, 72 (1991).

[39] M. Bell and J. S. Bell. Radiation damping and Lagrange invariants. *Particle Accelerators* **13**, 13–23 (1982). QHA, pp. 41–51.

[40] M. Bell and J. S. Bell. Quantum beamstrahlung, *Particle Accelerators* **22**, 301–6 (1988). QHA, pp. 99–104.

[41] M. Bell and J. S. Bell. End effects in quantum beamstrahlung, *Particle Accelerators* **24**, 1–10 (1988). QHA, pp. 105–14.

[42] M. Bell and J. S. Bell. Quantum beamstrahlung in almost uniform fields, *Nuclear Instruments and Methods in Physics Research A* **275**, 258–66 (1989). QHA, 115–23.

[43] Reinhold Schuch. Electron collision experiments with cold ions in storage rings, in John Gillaspie, editor, *Trapping Highly Charged Ions: Fundamentals and Applications* (Nova, Hauppauge, New York, NY, 2000), pp. 167–94.

[44] Lyman Spitzer, *Physics of Fully Ionised Gases* (Interscience, New York, NY, 1956).

[45] M. J. Seaton. Radiative recombination of hydrogenic ions, *Monthly Notes of the Royal Astronomical Society* **119**, 81–9 (1959).

[46] Edmund T. Whittaker, *Analytical Dynamics* (Cambridge University Press, Cambridge, 4th edition, 1952).

[47] Peter Sturrock, *Static and Dynamic Electron Orbits* (Cambridge University Press, Cambridge, 1955).

[48] Roman Jackiw and Abner Shimony. The depth and breadth of John Bell's physics, *Physics in Perspective* **4**, 78–116 (2002).

[49] Jon Leinaas. Hawking radiation, the Unruh effect and the polarization of electrons, *Europhysics News* **22**, 78–80 (1991).

[50] Jon Leinaas. Thermal excitations of accelerated electrons, in QUS, pp. 401–12.

[51] W. G. Unruh. Notes on black-hole evaporation, *Physical Review D* **14**, 870–92.

[52] S. W. Hawking. Black hole explosions?, *Nature* **248**, 30–1 (1974).

[53] J. D. Jackson. On understanding spin-flip synchrotron radiation and the transverse polarization of electrons in storage rings, *Reviews of Modern Physics* **48**, 417–34 (1976).

[54] J. S. Bell and J. M. Leinaas. Electrons as accelerated thermometers, *Nuclear Physics B* **212**, 131–50 (1982). QHA, 52–71.

[55] Richard J. Hughes. Later contributions, *Europhysics News* **22**, 70–72 (1991).

[56] J. S. Bell, Richard J. Hughes, and J. M. Leinaas. The Unruh effect in extended thermometers, *Zeitschrift für Physik C* **28**, 75–80 (1985). QHA, 72–77.

[57] Richard J. Hughes. Uniform acceleration and the quantum theory of vacuum I, *Annals of Physics (New York)* **162**, 1–30 (1985).

[58] Geofrrey L. Sewell. Quantum fields on manifolds: PCT and gravitationally induced thermal states, *Annals of Physics (New York)* **141**, 201–24 (1982).

[59] J. S. Bell and J. M. Leinaas. The Unruh effect and quantum fluctuations of electrons in storage rings, *Nuclear Physics B* **284**, 488–508 (1987). QHA, pp. 78–98.

[60] J. S. Bell. Gravity, in P. Bloch, P. Pavlopoulos, and R. Klapisch, editors, *Fundamental Symmetries* (Plenum, New York, NY, 1987), pp. 1–39.

[61] Alain Aspect. Bell's theorem: The naïve view of an experimentalist, in QUS, pp. 120–53.

[62] A. Aspect, P. Grangier, and G. Roger. Experimental tests of realistic local theories via Bell's theorem, *Physical Review Letters* **47**, 460–3 (1981).

[63] A. Aspect, P. Grangier, and G. Roger. Experimental realisation of Einstein–Podolsky–Rosen–Bohm Gedankenexperiment: A new violation of Bell's inequalities, *Physical Review Letters*, **49**, 91–4 (1982).

[64] A. Aspect, J. Dalibard, and G. Roger. Experimental tests of Bell's inequalities using variable analyzers, *Physical Review Letters* **49**, 1804–7 (1982).

[65] A. Zeilinger. Testing Bell's inequalities with periodic switching, *Physics Letters A* **118**, 1–2 (1986).

[66] Jonathan Barrett, Daniel Collins, Lucien Hardy, Adrian Kent, and Sandu Popescu. Quantum locality, Bell inequalities and the memory loophole, *Physical Review A* **66**, 042111 (2002).

[67] T. W. Marshall, E. Santos, and F. Selleri. Local realism has not been refuted, *Physics Letters A* **98**, 5–9 (1983).

[68] F. Selleri, *Quantum Paradoxes and Physical Reality* (Kluwer, Dordrecht, 1990).

[69] Andrew Whitaker, *The New Quantum Age: From Bell to Quantum Computation and Teleportation* (Oxford University Press, Oxford, 2012).

[70] J. S. Bell. Bertlmann's socks and the nature of reality, *Journal de Physique* **42**, 41–61 (1981). QHA, pp. 820–41; FQM, pp. 126–47; SUQM, pp. 139–58.

[71] J. S. Bell. Preface, in *Speakable and Unspeakable in Quantum Mechanics* (Cambridge University Press, Cambridge, 1987), pp. viii-x.

[72] Max Born, editor, *The Born–Einstein Letters* (Macmillan, London, 1971).

[73] Reinhold Bertlmann and Beatrix Hiesmayr. Bell inequalities for entangled kaons and their unitary time evolution, *Physical Review A* **63**, 06212 (2001).

[74] Reinhold Bertlmann, Heide Narnhofer, and Walter Thirring. Time-ordering independence of measurements in teleportation, *European Physical Journal D* **67**, 62 (2013).

[75] J. S. Bell. On the impossible pilot wave, *Foundations of Physics* **12**, 989–99. QHA, pp. 842–52; FQM, pp. 148–58; SUQM, pp. 159–68.

[76] Res Jost. Comment on 'Einstein on particles, fields and the quantum theory', in Harry Woolf, editor, *Some Strangeness in the Proportion* (Addison-Wesley, Reading, MA, 1980), p. 252.

[77] C. Philipppidis, C. Dewdney, and B. J. Hiley. Quantum interference and the quantum potential, *Nuovo Cimento B* **52**, 15–28 (1979).

[78] C. Dewdney and B. J. Hiley. A quantum potential description of one-dimensional time-dependent scattering from square barriers and square wells, *Foundations of Physics* **12**, 27–48 (1982).

[79] J. S. Bell. Speakable and unspeakable in quantum mechanics, *Physics Reports* **137**, 7–9 (1986). SUQM, pp. 169–72.

[80] Arthur Koestler, *The Sleepwalkers: A History of Man's Changing Vision of the Universe* (Hutchinson, London, 1959).

[81] John Barrow and Frank Tipler, *The Anthropic Cosmological Principle* (Oxford University Press, Oxford, 1988).

[82] J. S. Bell, *Speakable and Unspeakable in Quantum Mechanics* (Cambridge University Press, Cambridge, 1st edition 1987; 2nd edition, with two added papers and foreword by Alain Aspect 2004).

[83] Simon Capelin to Andrew Whitaker, 2005.

[84] J. S. Bell. Quantum field theory without observers, *Physics Reports* **137**, 49–54 (1986).

[85] J. S. Bell. Beables for quantum field theory, in Basil Hiley and F. David Peat, editors, *Quantum Implications* (Routledge and Kegan Paul, London, 1987), pp. 227–34. QHA, pp. 853–60; FQM, pp. 159–66; SUQM, pp. 173–80.

[86] J. S. Bell. EPR correlations and EPW distributions, *Annals of the New York Academy of Sciences* **480**, 263–6. QHA, pp. 861–5; FQM, pp. 167–71; SUQM, pp. 196–200.

[87] E. P. Wigner. On the quantum correction for thermodynamic equilibrium, *Physical Review* **40**, 749–59 (1932).

[88] Lars Johansen. EPR correlations and EPW distributions revisited, *Physics Letters A* **236**, 173–6 (1997).

[89] Ulf Leonhardt and John Vacarro. Bell correlations in phase space: Applications to quantum optics, *Journal of Modern Optics* **42**, 939–43 (1995).

[90] J. S. Bell. Six possible worlds of quantum mechanics, in Sture Allén, editor, *Possible Worlds in Humanities, Arts and Sciences, Proceedings of Nobel Symposium 65* (Walter de Gruyter, Stockholm, 1989), p. 359–73. QHA, pp. 887–901; FQM, pp. 193–207; SUQM, pp. 181–95.

[91] Steven Weinberg, *Dreams of a Final Theory* (Hutchinson, London, 1993).

[92] G.-C. Ghirardi, A. Rimini, and T. Weber. Uniform dynamics for microscopic and macroscopic systems, *Physical Review D* **34**, 470–91 (1986).

[93] GianCarlo Ghirardi. John Stewart Bell and the dynamical reduction program, in QUS, pp. 287–305.

[94] J. S. Bell. Are there quantum jumps?, in Clive Kilmister, editor, *Schrödinger: Centenary Celebration of a Polymath* (Cambridge University Press, Cambridge, 1987), pp. 41–52. QHA, pp. 866–86; FQM, pp. 172–92; SUQM, pp. 201–12.

[95] Helmut Rauch and Samuel A. Warner, *Neutron Interferometry: Lessons in Experimental Quantum Mechanics, Wave–Particle Duality and Entanglement* (Oxford University Press, Oxford, 2nd edition, 2015).

[96] D. M. Greenberger. The neutron interferometer as a device for illustrating the strange properties of quantum systems, *Reviews of Modern Physics* **55** 875–905 (1983).

[97] Helmut Rauch. Towards more complete neutron experiments, in QUS, pp. 351–73.

[98] Daniel Greenberger. The history of the GHZ paper, in QUS, pp. 281–6.

[99] D. M. Greenberger, M. A. Horne, A. Shimony, and A. Zeilinger. Bell's theorem without inequalities, *American Journal of Physics* **58**, 1131–43 (1990).

[100] N. D. Mermin. Quantum mysteries revealed. *American Journal of Physics* **58**, 731–4 (1990).

[101] N. David Mermin. Whose knowledge?, in QUS, pp. 272–280.

[102] N. David Mermin, *Space and Time in Special Relativity* (McGraw Hill, New York, NY, 1968; Waveland Press, Prospect House, IL, 1989).

[103] John Bell. Against 'measurement', in Arthur Miller, editor, *Sixty-Two Years of Uncertainty: Historical, Philosophical and Physical Inquiries into the Foundations of Quantum Mechanics* (Plenum, New York, NY, 1990), pp. 17–31; also *Physics World* **3** (August), 33–40 (1990). QHA, pp. 902–9; FQM, pp. 208–15; SUQM, pp. 213–231.

[104] P. A. M. Dirac, *The Principles of Quantum Mechanics* (Clarendon, Oxford 4th edition, 1981).

[105] L. D. Landau and E. M. Lifshitz, *Quantum Mechanics* (Pergamon, Oxford, 2nd edition, 1977).

[106] N. G. van Kampen. Ten theorems about quantum-mechanical measurements, *Physica A* **153**, 97–113 (1988).

[107] Kurt Gottfried, *Quantum Mechanics* (Benjamin, New York, NY, 1966).

[108] N. G. van Kampen. Quantum criticism, *Physics World* **3**, 20 (1990).

[109] N. G. van Kampen. Mystery of quantum measurement, *Physics World* **4**, 16–17 (1991).

[110] N. G. van Kampen. Macroscopic systems in quantum mechanics, *Physica A* **194**, 542–50 (1993).

[111] N. G. van Kampen, Note on the two-slit experiment, *Journal of Statistical Physics* **77**, 345–50 (1994).

[112] M. A. B. Whitaker. On a model of wave-function collapse, *Physica A* **255**, 455–66 (1998).

[113] L. E. Ballentine. The statistical interpretation of quantum mechanics, *Reviews of Modern Physics* **42**, 358–81 (1970).

[114] Leslie Ballentine, *Quantum Mechanics: A Modern Development* (World Scientific, Singapore, 1998).

[115] D. Home and M. A. B. Whitaker. Ensemble interpretations of quantum mechanics: A modern perspective, *Physics Reports* **210**, 223–317 (1992).

[116] Kurt Gottfried. Does quantum mechanics carry the seeds of its own destruction?, in John Ellis and Daniele Amati, editors, *Quantum Reflections* (Cambridge University Press, Cambridge, 2000), pp. 165–85; also *Physics World* **4**, 34–40 (1991).

[117] Kurt Gottfried. Inferring the statistical interpretation of quantum theory from the classical limit, *Nature* **405**, 533–6 (2000),

[118] Kurt Gottfried and Tung-Mow Yan, *Quantum Mechanics: Fundamentals* (Springer, New York, NY, 2003).

[119] M. A. B. Whitaker. Can the statistical equation of quantum mechanics be inferred from the Schrödinger equation: Bell and Gottfried. *Foundations of Physics* **38**, 436–47 (2008).

[120] E. J. Squires. Quantum challenge, *Physics World* **5**, 18 (1992).

[121] R. Peierls. In defence of 'measurement', *Physics World* **4**, 19–20 (1991).

[122] Rudolf Peierls. Rudolf Peierls, in Paul Davies and Julian Brown, editors, *The Ghost in the Atom* (Cambridge University Press, Cambridge, 1986), pp. 70–82.

[123] Sabine Lee, editor, *Sir Rudolf Peierls: Selected Private and Scientific Correspondence, Volume 2* (World Scientific, Singapore, 2009).

[124] J. S. Bell. La nouvelle cuisine, in A. Sarlemijn and P. Kroes, editors, *Between Science and Technology* (Elsevier, Amsterdam, 1990). QHA, pp. 910–28; FQM, pp. 216–34; SUQM, pp. 232–48.

[125] H. B. G. Casimir. Epistemological considerations, in J. de Boer, E. Dal and O. Ulfbeck, editors, *The Lesson of Quantum Theory: Proceedings of the Niels Bohr Centenary Symposium* (North-Holland, Amsterdam, 1986).

[126] Philip Burke and Ian Percival. John Stewart Bell, *Biographical Memoirs of Fellows of the Royal Society of London* **45**, 1–17 (1999). QHA, p. 2.

[127] Tony Hey. Socks, shifty split and Schrödinger, *Times Higher Education*, 19 September 2003.

[128] Christopher Llewellyn Smith to Andrew Whitaker, 26 July 2015.

[129] Torleif Ericson to Andrew Whitaker, 5 June 2015.

[130] J. J. Sakurai, *Invariance Principles and Elementary Particles* (Princeton University Press, Princeton, NJ, 1964).

[131] J. J. Sakurai, *Advanced Quantum Mechanics* (Addison-Wesley, Reading, MA, 1967).

[132] J. J. Sakurai, *Modern Quantum Mechanics* (Addison-Wesley, Reading, MA, 1985).

[133] R. A. Bertlmann. Bell's theorem and the nature of reality, *Foundations of Physics* **20**, 1191–1212 (1990).

[134] James T. Cushing and Ernan McMullin, *Philosophical Consequences of Quantum Theory: Reflections on Bell's Theorem* (University of Notre-Dame Press, South Bend, IN, 1989).

[135] R. A. Bertlmann. Bell, John Stewart: Physicist and moralizer, *Foundations of Physics* **20**, 1135–8 (1990).

[136] George Greenstein and Arthur G. Zajonc, *The Quantum Challenge: Modern Research on the Foundations of Quantum Mechanics* (Jones and Bartlett, Burlington, MA, 1977).

[137] Mary Bell. Some reminiscences, in QUS, pp. 3–5.

Chapter 6 The Work Continues

[1] Andrew Whitaker, *Einstein, Bohr and the Quantum Dilemma: From Quantum Theory to Quantum Information* (Cambridge University Press, Cambridge, 2006).

[2] Andrew Whitaker, *The New Quantum Age: From Bell's Theorem to Quantum Computation and Teleportation* (Oxford University Press, Oxford, 2012).

[3] P. G. Kwiat, K. Mattle, H. Weinfurter, A. Zeilinger, A. V. Sergienko, and Y. H. Shih. New high-intensity source of polarisation-entangled photon pairs, *Physical Review Letters* **75**, 4337–41 (1995).

[4] G. Weihs, T. Jennewein, C. Simon, H. Weinfurter, and A. Zeilinger. Violation of Bell's inequality under strict Einstein locality conditions, *Physical Review Letters* **81**, 5039–43 (1998).

[5] Gregor Weihs, Bell's theorem for space-like separation, in QUS, pp. 155–62.

[6] M. A. Rowe, D. Klepinski, V. Meyer, C. A. Sackett, W. M. Itano, C. Monroe, and D. J. Wineland. Experimental violation of a Bell's inequality with efficient detection. *Nature*, **398**, 791–4 (2001).

[7] Jan-Åke Larsson. Loopholes in Bell inequality tests of local realism. *Journal of Physics A: Mathematical and Theoretical* **47**, 424003 (2014).

[8] W. Tittel, J. Brendel, H. Zbinden, and N. Gisin. Violations of Bell's inequalities by photons more than 10 km apart, *Physical Review Letters* **81**, 3563–6 (1998).

[9] Nicolas Gisin. Sundays in a quantum engineer's life, in QUS, pp. 199–207.

[10] T. Scheidl, R. and 10 others [including A. Zeilinger]. Violation of local realism with freedom of choice, *Proceedings of the National Academy of Sciences* **107**, 19708–13 (2010).

[11] M. Giustina, and 11 others [including A. Zeilinger]. Bell violation with entangled photons, free of the fair-sampling assumption, *Nature* **497**, 227–30 (2013).

[12] Jason Gallicchio, Andrew S. Friedman, and David I. Kaiser. Testing Bell's inequality with cosmic photons: Closing the setting-independence loophole, *Physical Review Letters* **112**, 110405 (2104).

[13] D. Bouwmeester, J.-W. Pan, M. Daniell, H. Weinfurter, and A. Zeilinger. Observation of three-photon Greenberger–Horne–Zeilinger entanglement, *Physical Review Letters* **82**, 1345–9 (1999).

[14] David Deutsch, *The Fabric of Reality: The Science of Parallel Universes and its Implications* (Allen Lane, Harmondsworth, 1997).

[15] David Wallace, *The Emergent Multiverse: Quantum Theory according to the Everett Interpretation* (University Press, Oxford 2012).

[16] Simon Saunders, Jonathan Barrett, Adrian Kent, and David Wallace, editors, *Many Worlds, Everett, Quantum Theory and Reality* (Oxford University Press, Oxford, 2010).

[17] Robert Griffiths, *Consistent Quantum Theory* (Cambridge University Press, Cambridge, 2003).

[18] Roland Omnès, *Quantum Philosophy: Understanding and Interpreting Contemporary Science* (Princeton University Press, Princeton, NJ, 1999).

[19] M. Gell-Mann and J. Hartle. Classical equations for quantum systems, *Physical Review D* **47**, 3345–82 (1993).

[20] Edward Nelson, *Quantum Fluctuations* (Princeton University Press, Princeton, NJ, 1985).

[21] Guido Bacciagaluppi. Non-equilibrium in stochastic mechanics, *Journal of Physics Conference Series* **361**,012017 (2012).

[22] John Cramer. The transactional interpretation of quantum mechanics, *Reviews of Modern Physics* **58**, 647–88 (1986).

[23] Roger Penrose, *The Road to Reality* (Cape, London, 2004).

[24] M. Namiki and S. Pascazio. Quantum theory of measurement based on the many-Hilbert-space approach, *Physics Reports* **231**, 309–411 (1993).

[25] D. Giulini, E. Joos, C. Kiefer, J. Kupsch, I.-O. Stamatescu, and H.-D. Zeh, *Decoherence and the Appearance of a Classical World in Quantum Theory* (Springer, Berlin, 2003).

[26] W. H. Zurek. Decoherence and the transition from quantum to classical, *Physics Today* **44**, 36–44 (1991).

[27] A. J. Leggett. Schrödinger's cat and her laboratory cousins, *Contemporary Physics* **25**, 583–98 (1984).

[28] A. J. Leggett and A. Garg. Quantum mechanics versus macroscopic realism: Is the flux there when nobody looks?, *Physical Review Letters* **54**, 857–60 (1985).

[29] M. Arndt, K. Hornberger, and A. Zeilinger. Probing the limits of the quantum world, *Physics World* **18**, 35–40 (2005).

[30] N. David Mermin. Whose knowledge?, in QUS, pp. 272–280.

[31] Č. Brukner and A. Zeilinger. Operationally invariant information in quantum measurements, *Physical Review Letters* **83**, 3354–7 (1999).

[32] Vlatko Vedral, *Decoding Reality: The Universe as Quantum Information* (Oxford University Press, Oxford, 2010).

[33] Lee Smolin, *The Trouble with Physics: The Rose of String Theory, the Fall of a Science and What Comes Next* (Penguin, London, 2008).

[34] C. E. Shannon. A mathematical theory of communication, *Bell Systems Technical Journal* **27**, 379–423, 623–56 (1948).

[35] Andrew Hodges, *Alan Turing: The Enigma* (Random House, London, 1992).

[36] Rolf Landauer. Information is physical, *Physics Today* **44**, 23–9 (1991).

[37] David Deutsch. Quantum theory, the Church–Turing principle and the universal quantum computer, *Proceedings of the Royal Society A* **400**, 97–117 (1985).

[38] Richard Feynman, Simulating physics with computers, *International Journal of Theoretical Physics* **21**, 467–88 (1982).

[39] A. J. G. Hey, editor, *Feynman and Computation* (Perseus, Reading, MA, 1999).

[40] S. Glashow. Message for the Sudarshan Symposium, *Sudarshan: Seven Science Quests 2006 ()* **196**, 011003 (2009).

[41] Michael A. Nielsen and Isaac L. Chuang, *Quantum Computation and Quantum Information* (Cambridge University Press, Cambridge, 2000).

[42] Eleanor G. Rieffel and Wolfgang H. Polak, *Quantum Computing: A Gentle Introduction* (MIT Press, Cambridge. MA, 2014).

[43] David Deutsch and Richard Jozsa. Rapid solutions of problems by quantum computation, *Proceedings of the Royal Society A* **439**, 553–8 (1992).

[44] Richard Cleve, Artur Ekert, Chiara Macchiavello, and Michele Mosca. Quantum algorithms revisited, *Proceedings of the Royal Society A* **454**, 339–54 (1998).

[45] P. W. Shor. Polynomial-time algorithms for prime factorisation and discrete logarithms on a quantum computer, in S. Goldwasser, editor, *Proceedings of the 35th Annual Symposium on Foundations of Computer Science* (IEEE, Piscataway, NJ, 1994), pp. 124–49.

[46] L. K. Grover. A fast quantum mechanical algorithm for database search, in *Proceedings of the 28th Annual Symposium on the Theory of Computation* (ACM Press, New York, NY, 1996), pp. 212–19.

[47] C. H. Bennett and G. Brassard. Quantum cryptography: Public-key distribution and coin tossing, in *Proceedings of the 1984 IEEE International Conference on Computers, Systems and Signal Processing* (IEEE, New York, NY, 1984), pp. 175–9.

[48] A. K. Ekert. Quantum cryptography based on Bell's theorem, *Physical Review Letters* **67**, 661–3 (1991).

[49] A. Acin, N. Gisin, and L. Masanes, From Bell's theorem to secure quantum key distribution, *Physical Review Letters* **97**, 120405 (2006).

[50] A. Muller, H. Zbinden, and N. Gisin. Quantum cryptography over 23 km in installed under-lake telecom wire, *Europhysics News* **33**, 335–9 (1986).

[51] C. H. Bennett, G. Brassard, C. Crepeau, R. Jozsa, A. Peres, and W. E. Wootters. Teleporting an unknown quantum state via dual classical and Einstei–Podolsky–Rosen channels, *Physical Review Letters* **70**, 1895–9 (1993).

[52] D. Bouwmeester, J.-W. Pan, K. Mattle, M. Eibl, H. Weinfurter, and A. Zeilinger. Experimental quantum teleportation, *Nature* **390**, 575–9 (1997).

[53] I. Maricikic, H. de Riedmatten, W. Tittel, H. Zbinden, and N. Gisin. Long-distance teleportation of qubits at telecommunication wavelengths, *Nature* **421**, 509–13 (2003).

[54] M. Żukowski, A. Zeilinger, M. A. Horne, and A. K. Ekert. 'Event-ready detectors': Bell experiment via entanglement swapping, *Physical Review Letters* **71**, 4287–90 (1993).

[55] T. Jennewein, G. Weihs, J.-W. Pan, and A. Zeilinger. Experimental nonlocality proof of quantum teleportation and entanglement swapping, *Physical Review Letters* **88**, 017903 (2002).

[56] Andrew Whitaker. John Stewart Bell, quantum information and quantum information theory, presented at Quantum Information II: Fifty Years of Bell's Theorem (2014) (Reinhold Bertlmann and Anton Zeilinger, editors) (to be published).

[57] QUS.

[58] N. David Mermin, *Quantum Computer Science: An Introduction* (Cambridge University Press, Cambridge, 2007).

[59] A. M. Steane, A quantum computer only needs one universe. *Studies in the History and Philosophy of Modern Physics* **34**, 469–78 (2003).

[60] R. Jozsa and N. Linden. On the role of entanglement in quantum computational speed-up, *Proceedings of the Royal Society A* **459**, 2011–32 (2003).

[61] B. P. Lanyon, M. Barbieri, M. P. Almeida, and A. G. White. Experimental quantum computing without entanglement, *Physical Review Letters* **101**, 20051 (2008).

[62] M. Van den Nest. Universal quantum computation with little entanglement, *Physical Review Letters* **110**, 060504 (2013).

[63] M. Howard, J. Wallman, V. Veitch, and J. Emerson. Contextuality supplies the 'magic' for a quantum computation, *Nature* **510**, 351–5 (2014).

[64] Philip Ball, Questioning quantum speed, *Physics World* **27**, 38–41 (2014).

Chapter 7 Work of the Highest Calibre and a Fine Life

[1] Christopher Llewellyn-Smith to Andrew Whitaker, 27 July 2015; also entry in Dictionary of National Biography.

[2] Abner Shimony. John S. Bell: Some reminiscences and reflections, in QUS, pp. 51–60.

[3] Interview with Dorothy (McConkey) Whiteside, on 4 December 2014.

[4] Mary Bell. John Bell and accelerator physics, *Europhysics News* **22**, 72 (1991).

[5] Preface: From Bell to quantum information, in QUS, pp. vii–x.

[6] Andrew Whitaker. John Bell in Belfast: Early years and education, in QUS, pp. 7–20.

[7] Interview of Robert Bell by Andrew Whitaker on 23 February 2015.

Index

Illustrations are represented by numbers in bold